INNOVATIONS IN DESIGN & DECISION SUPPORT SYSTEMS IN ARCHITECTURE AND URBAN PLANNING

Innovations in Design & Decision Support Systems in Architecture and Urban Planning

Edited by

JOS P. VAN LEEUWEN
Eindhoven University of Technology,
The Netherlands

and

HARRY J.P. TIMMERMANS
Eindhoven University of Technology,
The Netherlands

Springer

A C.I.P. Catalogue record for this book is available from the Library of Congress.

ISBN-10 1-4020-5059-3 (HB)
ISBN-13 978-1-4020-5059-6 (HB)
ISBN-10 1-4020-5060-7 (e-book)
ISBN-13 978-1-4020-5060-2 (e-book)

Published by Springer,
P.O. Box 17, 3300 AA Dordrecht, The Netherlands.

www.springer.com

Cover design by Jos van Leeuwen,
Cover image courtesy of Lidia Diappi and Paola Bolchi (see p. 187).

Printed on acid-free paper

Printed in the Netherlands.

TABLE OF CONTENTS

PREFACE

It has been a real pleasure to work with the authors of the papers in this volume; they have contributed with interesting reports on relevant and innovative research projects, which allowed us to compose an inspiring book. Our gratitude goes also to the members of the international scientific committee for their invaluable effort in reviewing and editing these works. Special thanks also go to our colleagues Mandy van de Sande, Marlyn Aretz, and Leo van Veghel, who took great responsibility in organising the practical aspects of the DDSS conference and made it a pleasant and fruitful event.

Eindhoven, July 2006

Jos van Leeuwen and Harry Timmermans
Editors and conference chairs

INTERNATIONAL SCIENTIFIC COMMITTEE

INTRODUCTION

The International Conference on Design & Decision Support Systems in Architecture and Urban Planning is organised bi-annually by the Eindhoven University of Technology. This volume contains a selection of papers presented at the eighth conference that was held at the Kapellerput Conference Centre in the village of Heeze, near Eindhoven, The Netherlands, from 4 to 7 July, 2006.

Traditionally, the DDSS conferences aim to be a platform for both starting and experienced researchers who focus on the development and application of computer support in the areas of urban planning and architectural design. This results in an interesting mix of well-established research projects and first explorations. It also leads to a very valuable cross-over of theories, methods, and technologies for support systems in the two different areas, architecture and urban planning.

This volume contains 31 peer reviewed papers from this year's conference. The volume is organised into two parts: the first part containing three sections in the area of urban planning; the second part containing four sections on design support.

In part one, on urban planning, five of the chapters in section one deal with simulation of land use; the sixth chapter is on visualisation of land use. Section two contains five chapters on multi-agent systems for the computation of movement simulation. The five chapters in the third section are also on multi-agent systems, but relate to processes of urban development. The final chapter in part one discusses the acceptance and application of this type of tools in today's urban planning practice.

Part two, on design support systems, starts with a section containing five chapters on the management and deployment of design knowledge. The second section contains three chapters discussing tools and strategies for urban decision-making. The first three chapters in section three discuss the possibilities to allow designer interaction with systems that automate design; the fourth chapter presents a solution for the exchange of building information models. The final section is on virtual environments, with one chapter focusing on urban environments and another on design environments.

The following paragraphs provide a brief summary of the two parts and the seven sections of this volume.

DEVELOPMENT OF DDSS IN URBAN PLANNING

The literature on design and decision support systems in urban planning is rich of many different tools and modelling approaches that potentially can support decision makers in better understanding the consequences of their decisions or help them in the decision-making process itself. The nature of these tools and systems and their stage of development largely reflect more general trends in science. The first generation of models and systems was often based on statistical input-output relations, weak in terms of behavioural principles, large scale, and implicitly or explicitly based on the assumption of perfect and full information in the sense that uncertainty was rarely addressed. Next, with the increasing popularity of chaos theory, the often complex models were replaced with simple or even simplistic models, based on the belief that simple rules and principles are sufficient to generate complex emerging patterns. Cellular automata models constitute a good example. More recently, realizing that urban developments but also plans are the result of interactive or even strategic choices of actors involved in the process, scholars started to work on multi-agent systems, some still based on simple, primarily data-driven principles, other trying to significantly improve the behavioural foundations of the models. These approaches should not be viewed as successions, but rather are still further developed and applied and still constitute frameworks within which different groups are making progress. This is also reflected in the papers on design and decision support systems in urban planning that were selected for the first part of this volume.

Land Use Simulation and Visualisation

Emberger, Ibesich & Pfaffenbichler describe an integrated land use transportation model system, called MARS. In terms of its key features, it is a conventional approach, put into a new jacket. The authors argue that the system has proven its usefulness. This is also illustrated by another paper on the same model system.

Pfaffenbichler, Monzón, Pardeiro & Vieira describe an application of MARS to Madrid. A series of sustainability indicators is derived from the model, allowing the user to evaluate alternative plans in terms of these indicators. In addition, a cost-benefit analysis is conducted.

As for cellular automata, *Gohnai, Ohgai, Ikaruga, Kato, Hitaka, Murakami & Watanabe* clearly illustrate how useful these models are for simulating dynamical processes such as the spread of fire. This is illustrated by the model itself, but also by experiences gained in the application that is described in this chapter.

Jiao & Boerboom argue that a distinction should be made between transition potential rules and conflict resolution rules. In addition, they provide a useful overview of a few existing methods for deriving these rules from empirical data or for eliciting expert knowledge.

Osaragi & Aoki suggest some valuable improvements for estimating land use transition probabilities, which are central to cellular automata models. In particular, arguing that transition probabilities are required at the level of building lots and assuming raster data are available, they develop an improved method to estimate such transition probabilities and propose a method to estimate time series changes in the land use transition probabilities, based on the concept of land use utility.

Interpretation and correct judgement of the qualities of a landscape on the basis of two-dimensional land use maps is a difficult task, even for experienced professionals. Particularly the implications of changes to land use cannot easily be perceived from maps. *Borsboom-van Beurden, van Lammeren, Hoogwerf & Bouwman* report on their efforts to integrate 3D visualisations of the (changed) landscape with the land use model output. They describe two approaches. One is to insert 3D iconic (exemplary) representations of new land uses into the land use model. The other is to use 3D components of GIS tools to represent existing land use features. The combination of both approaches seems to lead to promising results, but to achieve them technically on a large scale still proofs to be very challenging.

Multi-Agent Models for Movement Simulation

Multi-agent models are potentially relevant to simulate movement. Models of pedestrian movement have recently regained considerable interest, especially due to safety, crowding and feasibility issues. *Bandini, Manzoni & Vizzari* present a general framework and a situated cellular automata model as a special case of a multilayered, multi-agent situated system to address this problem. The advantage of their approach is that both interactions between pedestrians and the interaction between pedestrians and their environment can be easily incorporated in the model. Good examples of 3D visualization to portray simulation dynamics are also presented. The system is illustrated using an underground station as an example.

It could be argued that the behavioural mechanisms underlying pedestrian behaviour in this context are relatively simple in the sense that pedestrians probably wish to leave the station as quickly as possible or go to the train in a convenient way. Behavioural principles in large-scale retail environments are considerably more complex. Pedestrians may be utilitarian or hedonic, some may behave in rational ways, others in satisfying ways, reflecting bounded rationality. There will also be significant heterogeneity in terms of shopping aims, agendas, environmental knowledge, etc. *Zhu &*

Timmermans therefore explore the potential of gene programming to identify the behavioural principles (alternative model specifications) that best describe observed movement patterns. The approach goes beyond genetic programming in that not only parameters are fitted but also the best functional form of the model can in principle be derived. A potential limitation of the suggested approach is that it requires parallel computing, especially for large-large-scale applications.

Chen & Chiu use a less formal approach but show how agent technology can be used to visualize pedestrian movement and guiding users in navigating through streets. Their system, called SCALE, is a platform supporting digital data and a media suitable for advertising and city guide information. Space-tags are virtual objects that can be accessed only within a limited area and for a limited time. The functionality of the system is described and illustrated using a small sample.

Another type of movement is traffic. *Boussier, Estraillier, Sarramia & Augeraud* describe a very interesting hybrid system for traffic simulation. Their multi-agent system is divided into three subsystems: (i) one is concerned urban traffic simulation, with agents representing vehicles, bicycles, pedestrians, etc; (ii) another is with the information system service behaviours, with agents modelling employees and the computing system of the information system itself, and (iii) the third one is dedicated to the decision support objective of the system. Traveller behaviour is modelled using questionnaire data based on Taguchi experimental designs, while belief theory is employed to identifying preference distributions of transport modes. An example involving students illustrates the potential of the system.

Remaining in the context of traffic, *Balmer & Nagel* show how to simulate the interaction between the capacity of an intersection and the shape of a single roundabout using agent technology and evolutionary algorithms. Their contribution is exceptional and extends the typical application area of multi-agent models and traffic simulation. An application to the central roundabout of Zürich shows the potential of this approach.

Multi-Agent Models for Urban Development

Multi-agent models have not only become of interest to simulate movement patterns, such as for example pedestrians, but also to simulate urban change processes. A few such examples are included in this volume. *Diappi & Bolchi* describe an application of a multi-agent system to simulate the gentrification process in Milan, Italy, using Smith's rent gap theory. Different agents represent the actors in this process (homeowners, landlords, tenants and developers). Key in the simulation is the principle that investments are made when the difference between potential rent and capitalised rent is greater than a certain threshold. The rent variables evolve over time. Because expected rents are an average of neighbourhood capitalised rents, a spatially varying process constitutes the result of the simulation. As often the intention in these applications, the model shows evidence of self-organisation in that nonlinear interactions at the local level produce different emerging macro-level configurations.

Zimmermann illustrates the application of the software generation process PROBAnD to a simplified version of Devisch et al's housing market model. In addition to some interesting substantive results, this chapter provides evidence that computing platforms have become available to develop quite complex multi-agent models in a short period of time.

This does not mean that only advanced multi-agent systems are relevant in supporting urban planning in the context of housing. This is adequately shown by *Celikyay,* who implemented in a geographic information system McHarg's ecological evaluation method and Kiemstedt's utilization value analysis to analyse the potential of the natural resources for new residential development. In particular, he reports an application in Bartın city to define potential residential areas and shows that some residential areas have been selected improperly.

Zhou, Kondo, Gordillo & Watanabe address the problem of the location of waste disposal facilities vis-à-vis residential areas. Various probabilistic models, based on distance and testing for sociodemographic differences, are estimated for the city of Chengdu. Differences between the models are small only.

Heurkens also describes a multi-agent tool to support decision-making with respect to urban development related to a possible organisation of the Olympic Games in the Netherlands in 2028. In particular, this paper focuses on a multi-criteria method for choosing the best design.

Based on examples such as the ones described in this volume, one would expect that design and decision support tools (or planning support tools) would have found their way in practice. *Vonk, Geertman & Schot* argue that this has not been the case. They reflect on this lack of application by discussing issues related to the instrumental quality of these systems, user acceptance and extent of diffusion. Results based on interviews show that lack of experience, lack of awareness and lack of instrument quality of PSS are the main bottlenecks blocking user acceptance, and that these effects are enhanced by hampered user acceptance and diffusion. Interesting to note is that this position is at variance with the first paper on urban planning included in this book, which argues and provides evidence of success. Hence, this discussion continues........... and it should.

DEVELOPMENT OF DESIGN SUPPORT SYSTEMS

Whereas in the past decades the tremendous technological innovations in ICT have enabled the development of powerful support tools for architectural and urban design, today there is a common and increasing awareness of the need to remain in control, as designers, of this computational power. It is no longer sufficient for the R&D community in this domain to deliver 'conventional' tools for the generation of design solutions, for the optimisation of design, for the integration or exchange of models, for efficient production of models for architectural and urban designs. While these tools find their way into the practice of designers and the construction industry, reluctance remains to fully accept such tools, rooted in the feeling experienced by many designers of loosing control over the design process. Hence, a focus can be discerned in ongoing research that aims to increase the designer's control of tools. The results expected from this research allow the designer to influence the generation process in generative design and the selection of methods and criteria for the optimisation of design solutions, to have a central, interactive role once again in the design process. Tools will become available that help control the development of corporate knowledge, to control the complex decision-making processes where multiple stakeholders are involved and where the large public claims respect. Tools will provide means to maintain the quality and validity of models and allow designers to take back control over work methods, information flows, and communication routines. The papers selected for the second part of this volume report on the progress made in this line of research.

Managing and Deploying Design Knowledge

Knowledge is probably the most important asset of any design practice. Successful architectural design practices actively manage their corporate knowledge. *Cerulli* introduces a knowledge management tool that offers a bottom-up approach to knowledge acquisition and knowledge sharing. The tool uses a recommender system approach to increase the relevance of the system's knowledge retrieval. It is targeted for use by emergent communities of users across the various hierarchical levels in a design organisation.

Design tasks can be supported by design optimization tools that help the designer to arrive at design solutions in the context of the design problem, the design objectives, and the design constraints. While such tools indeed support the optimization of the design process, the selection and usage of the most suitable tool becomes a problem in itself. *Peng & Gero* present the design and implementation of a so-called 'situated agent' that observes and learns from the user's interaction with a design optimization tool, using this knowledge to improve the behaviour of the tool and thus optimize its future usage. Another issue with optimization tools is to allow for interaction and modification of the problem, the tool itself, and the design process, but in an ongoing optimization process. *Gero & Kannengiesser* address this issue in an approach that is based on parts of the function-behaviour-structure framework.

Machine learning is a powerful mechanism to enhance the capabilities of design & decision support systems. With the assumption that human expert knowledge is based on applying decision criteria in a given domain, systems can be built that can learn to identify these criteria, with sufficient access to domain specific information sources. *Oh, Hwang, Smith & Koile* present the development of such a system as a proof of concept. Their challenge is to identify the relationship between abstract concepts employed by human experts and low level features in machine accessible databases. The project presented describes an experiment that demonstrates how this mapping can be used in teaching machines to identify 'main streets' in urban settings.

Acknowledging the necessity to incorporate environmental issues in design, *El Fiky & Cox* have developed a design support toolbox that offers green architectural design strategies to designers in the early phases of architectural and urban design. Their research combines local, Egyptian, cultural aspects of current and past practices in green architecture with up to date knowledge resources on this topic into a toolbox that helps the designer to decide what design principles to apply by offering feedback on their cultural acceptance and environmental value.

Urban Decision-Making

Providing computational support for complex decision-making in which many stakeholders are involved, such as in urban development questions, is quite a challenge. *Van Loon & Wilms* describe the development of the so-called Urban Decision Room, an approach and pilot system that offers a mathematical optimisation on the basis of constraints. Constraints, in this system, can be defined to describe the design goals as well as the resources or conditions for the urban the development. The system relates resources and goals of all stakeholders involved and can be used to achieve a Pareto optimum in the collective solution space.

Drawing from the experience of a major urban redevelopment project in the city of Auckland, *Hunt* describes how this project developed a strategy and a course of action for achieving participation of the many stakeholders involved, as well as the involvement of the large public through consultations in the process. This project applied the 'open design' model for multi-stakeholder input in the various stages of project development. Hunt draws conclusions on the limits of applicability of this model in large projects with public interest and suggests a number of decision support guidelines for the management of participatory urban development processes.

Facilitating the participation of inhabitants in the decision-making process in redevelopment of neighbourhoods is the objective of the work by *van Leeuwen & van Berlo*. Their approach is to involve inhabitants in a game-like dialogue that helps them to express their desires regarding the liveability of their neighbourhood and, at the same time, to evaluate the consequences of their desires and of the changes they propose. This approach has been implemented in a web-based application that utilises a Bayesian network to predict cause and effect relations between people's desires and experiences and the changes proposed for a neighbourhood.

Design Interactivity and Design Automation

Since the introduction in the early 80's of CAD tools and with their ongoing development, conceptual design in architecture has found ways to address new challenges; one of these is the interactive manipulation of 3D shapes. *Wetzel, Belblidia & Bignon* have investigated the role of morphological operators as has been identified in research and analyse significant examples of rather exceptional architectural works that demonstrate this role in practice. This investigation is used as the point of departure for the definition of an intuitive 3D modelling environment for architectural exploration using morphological operators. A first model for this environment is presented.

One of the problems with the application of genetic algorithms in design is the formulation of the fitness function that will lead the process of genetic evolution to the desired or optimal results. *Cheng* presents an evolutionary procedure in which the expertise of the members of a multidisciplinary design team is used in conjunction with a generative design implementation. An experimental system implements this approach in a model that includes two mechanisms: one mechanism for evolutionary development of new generations, based on traditional genetic programming; the other mechanism is a natural selection mechanism that defines the fitness of the generated solutions, determining the input for the next generation of the population.

Traditional rule-based systems, such as shape grammars or fractals, have been very inspiring and led to surprising results. The main drawback of such computational design systems, however, is that the influence of the designer on the system is limited, indirect, and not intuitive. Avoiding this drawback, *Landreneau, Ozener, Pak, Akleman & Keyser* have developed an approach and prototype system that offers designers a high level of interactivity and control over fractal and L-systems. The system was tested in a graduate level architectural course.

The concept of a semantically rich building information model as a core medium for multidisciplinary design processes is very appealing to the design and construction practice. However, several major issues have remained unsolved that inhibit full exploitation of this powerful medium. One of these issues is related to the limitations imposed by the usage of central databases where such a building information model should reside. For practical reasons mostly, design teams prefer to work with local data, even when collaborating in distributed settings. This implies the need for a working mechanism that can compare and merge different versions of a model, enabling the design team members to exchange modifications and keep their data updated. *Arthaud & Lombardo* present an automatic semantic comparison of product models based on the STEP standard.

Virtual Environments and Augmented Reality

The availability of new technologies for computing and communication, such as mobile computing and wireless networks, offers a challenge to integrate our presence in virtual and real environments. The city of Bath was chosen as an experimental playground for research on interaction spaces, in relation to the urban space, by *Fatah gen. Schieck, Penn, Kostakos, O'Neill, Kindberg, Stanton Fraser & Jones*. This work aims to develop a better understanding of the urban landscape when augmented with a digital landscape, for example how do the physical and digital flow of people through the city interrelate. This research is expected to give insight in the impact of using pervasive computing systems in urban environments on people's relationships, with each other and with, e.g., our cultural heritage.

Motivated by, on the one hand, the need of designers to use their body expressively in design actions and, on the other hand, the need for novice students to get a feeling for the scale and texture of their design, *Chen & Chang* have developed and demonstrate an augmented reality design environment that deploys 1:1 projections of a design and allows for direct manipulation of the design through designer interaction using gestures. The system uses a camera to detect hand gestures made by the designer to indicate design activities, such as drawing or manipulating geometry and adding texture to the design.

To conclude, the editors are proud to present this book, which provides an interesting range of innovations in design & decision support systems in architecture and urban planning. We hope and expect that this book will continue to function, after the DDSS 2006 conference, as a means to bring researchers together and as a valuable resource for our continuous joint effort to improve the design and planning of our environment.

DEVELOPMENT OF DDSS IN URBAN PLANNING

Land Use Simulation and Visualisation

Can Decision Making Processes Benefit from a User Friendly Land Use and Transport Interaction Model?

Guenter Emberger, Nikolaus Ibesich, and Paul Pfaffenbichler
Vienna University of Technology, Institute for Transport Planning and Traffic Engineering

Keywords: Land-use and transport model, Dynamic modelling, Decision-making support, Policy instruments, Flight simulator

Abstract: Urban regions today face serious challenges caused by past and ongoing transport and land use developments. Decision making in this context is a challenging task which was explored in detail in a series of research projects. To support decision making, tools were developed to reduce the risk of inappropriate decisions in the land use and transport context. One of these tools is MARS (Metropolitan Activity Relocation Simulator); an integrated dynamic land use and transport model. The paper presented here focuses therefore on two main issues: 1) the introduction of the decision support tool MARS and of the cause-effect relations between the land-use and the transport system implemented within MARS and 2) the design and application of the MARS flight simulator (MARS FS) as a graphical user interface for MARS especially designed to the needs of decision makers.

1. INTRODUCTION

Urban regions today face serious challenges caused by transport and land use developments. To deal with these tasks decision makers need knowledge about what kind of strategies can contribute to goals like reduction of road congestion, increase quality of life, or how to ensure future economic prosperity. It is common knowledge that transport and land use planning are strongly interrelated and play a key role in the achievement of present and future objectives. It is also well known that the decision making process

Jos P. van Leeuwen and Harry J.P. Timmermans (eds.), Innovations in Design & Decision Support Systems in Architecture and Urban Planning, 3-17.

concerning land use development becomes more and more complex. On one hand the number of involved stakeholders is increasing and with it the number of objectives which have to be met; on the other hand long term feedback loops exist between the land use and the transport system which have to be taken into consideration.

To reduce the risk of inappropriate and publicly unacceptable decisions the use of state of the art decision support tools is essential. One of these tools is MARS (Metropolitan Activity Relocation Simulator).

The paper is structured in the following way: In section 2 the typical decision making process for land use and transport planning policy decisions is introduced. To support this process a series of decision making support tools were developed during the past decades. MARS, as one of these tools – especially designed to follow as close as possible the decision making process, is described in section 2. MARS was designed to support the decision makers at all steps of decision making (objective definitions, policy instrument identifications, assessment of short and long term impacts and appraisal).

In section 3 an overview of existing comparable land use and transport models is given and some detail on the philosophy behind MARS is provided. During our work with MARS and collaboration with decision makers it was realised that the use of MARS was becoming too complex for decision makers.

To overcome this problem a simpler and especially for the user needs adapted graphical user interface was designed and implemented. This user interface allowing an easy handling of the MARS LUTI model will be introduced in section 4.

Finally in section 5 empirical experience of using the MARS graphical user interface gained from a series of workshops with decision makers is given. The section concludes with an outlook of future research needs and lists potential improvements of the software package.

2. DECISION MAKING PROCESS

The "typical" decision making process was explored in detail within the EU-project PROSPECTS (PROSPECTS 2000-2003). Simplified it consists of the following 4 major steps:

1. Identification of objectives/setting targets
2. Identification of possible instruments and combinations of instruments (strategies)

3. Assessing and appraising of the outcome of the instruments/strategies against the objectives/targets including identification of barriers to implementation
4. Can the objectives/targets be met?
   If yes → strategy found – implementation of strategy
   If no → go to step 2

To 1) Identification of objectives/setting targets
Without a clear defined vision of the objectives no goal orientated policy package can be identified. A set of well defined key indicators is necessary to monitor the outcome of the suggested policy packages. Useful indicators are noise and air pollution, accident costs, present value of finance, fossil fuel consumption, etc.

To 2) Identification of possible instruments and combinations of instruments (strategies)
It is proved that only policy packages can deliver objective adequate solutions. Combinations of policies are needed to overcome barriers to implementation or help to generate enough revenue to implement costly instruments (e.g. parking fees to support public frequency increases). A comprehensive list of instruments in combination with a comprehensive description of impacts can be found in the KonSULT-database (ITS 2002).

To 3) Assessment and Appraisal
Presently the Cost-Benefit-Analyses (CBA) approach is the most widely accepted method to appraise the outcome of transport related policy implementations. A well known weak point of this method is the monetarisation of all impacts, which is not always an easy task. Additionally the major share of benefits calculated in CBAs stem from time savings (in most cases more than 70%) - which is problematic taking into consideration the theory of constant travel time budget (for more info on this issue see (Pfaffenbichler, Shepherd et al. 2004)).

Another approach to assess the outcome of policy packages is the use of Multi-Criteria-Analyses (MCA), again the weighting problem is still an issue of ongoing research discussions. However within the MARS FS environment both methods can be applied.

To 4) Can the objectives/targets be met?
The run time of MARS respective MARS FS is a very fast (less than 1 minute for a 30 year simulation), so that users are able to test a wide variety of different policy instruments in one single session. This allows transport planners and decision makers to get a ‘feeling' of the impacts and the

magnitude of impacts of their tested policy combinations. Additionally it has to be mentioned that MARS can be set up to maximize by its own any given objective (-function).

Within the above-described decision making process, the MARS model/flight simulator is used to assess the impacts on land use and the transport system of either a single instrument or a set of instruments combined to a strategy. As can be imagined to assess impacts of a wide set of instruments a complex model (= MARS model) is necessary. On the other hand, transport planners and decision makers want to have a simple tool to test easily and explore the impacts of their strategies. For this reason the MARS FS was developed.

3. LAND USE AND TRANSPORT MODELS

3.1 Overview

Operational land-use modelling was pioneered by Lowry in the 1960ies (Lowry 1964). Initial models drew heavily on analogies to physics, e.g. the law of gravity. Most current models have their foundation in random utility theory, which is based on the principle of utility maximization originating from micro-economics. Typically, LUTI models combine at least two separate components: a land-use and a transport sub-model, which generate dynamic behaviour based on time lags between the two systems. State of the art models feature a modular structure, which entails a flexibility to include further aspects such as imperfect markets (David Simmonds Consultancy 1999). There are, however, concerns that LUTI models focus mainly on the redistribution of activities, neglecting aggregate effects, e.g. on employment, as overall economic activity is usually exogenously specified (SACTRA 1998). Some of the most advanced European LUTI models are IRPUD (Wegener 1998; Wegener 2004), DELTA (Simmonds 1999; Simmonds 2001), MEPLAN (Echenique, Flowerdew et al. 1990) and MARS (Pfaffenbichler 2003). Most of the models listed above can be classified as "academic" models and therefore the focus of these models lies on the functionality and not the user friendliness.

3.2 Introduction to MARS

MARS is a dynamic Land Use and Transport Integrated (LUTI) model. The basic underlying hypothesis of MARS is that settlements and activities within them are self organizing systems. Therefore MARS is based on the principles of systems dynamics (Sterman 2000) and synergetics (Haken

1983). The development of MARS started some 10 years ago partly funded by a series of EU-research projects (OPTIMA, FATIMA, PROSPECTS, SPARKLE). A comprehensive description of MARS can be found in (Minken, Jonsson et al. 2003). To date MARS has been applied to six European cities (Edinburgh - UK, Helsinki - FIN, Leeds - UK, Madrid - ESP, Oslo - NOR, Stockholm - S, and Vienna - A) and 3 Asian cities (Chiang Mai and Ubon Ratchathani - Thailand and Hanoi - Vietnam).

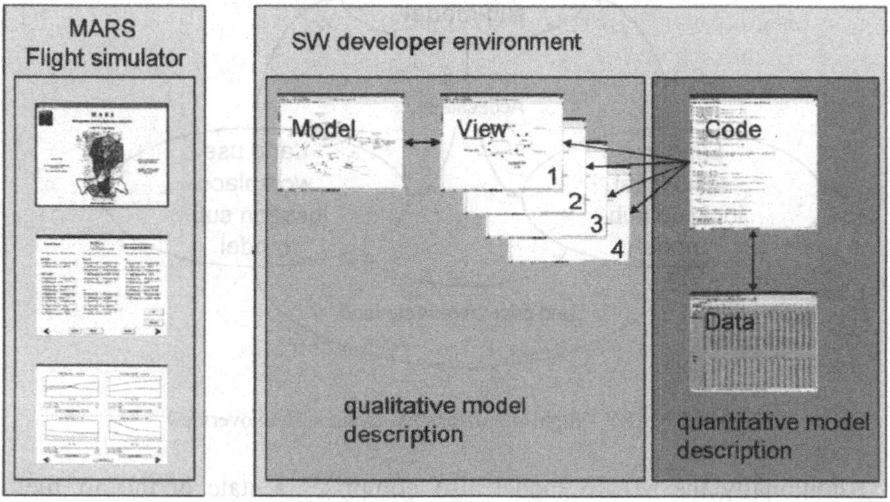

Figure 1. MARS – flight simulator – software system overview.

The present version of MARS is implemented in Vensim®, a System Dynamics programming environment. This environment is specialised for dynamic problems, and is therefore an ideal tool to model dynamic processes.

The MARS model includes a transport model which simulates the travel behaviour of the population related to their housing and workplace location, a housing development model combined with a household location choice model and of a workplace development model combined with a workplace location choice model as shown in *Figure 2*. In the current version of MARS two employment sectors are implemented (service sector and industrial sector). The growth of population (and related to that the growth of work force) is one hand driven by exogenously defined growth rates and on the other hand influenced by the development within the land use model

Accessibility is one of the outputs of the transport model. Accessibility in the year n is used as an input into the location models in the year n+1. Workplace and residential location is an output of the land use model. The number of workplaces and residents in each zone in the year n is used as attraction and potential in the transport model in the year n−1. There are also links between the land use sub-models as they are competing for land and availability of land influences its price. MARS iterates in a time lagged manner between the transport and the land use sub-model over a period of 30 years.

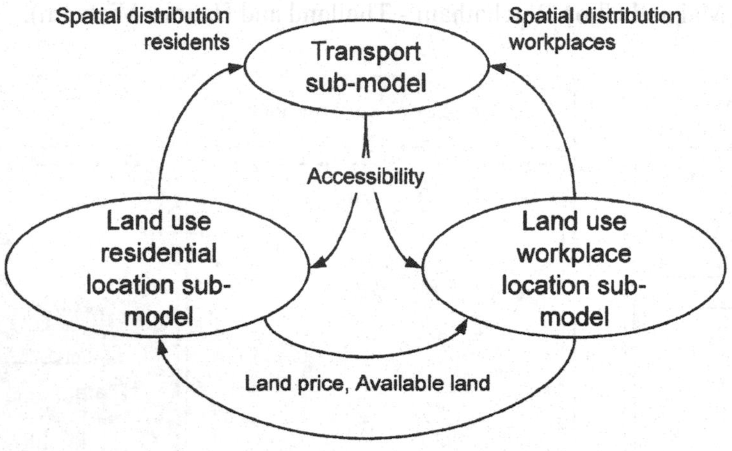

Figure 2. MARS – flight simulator – software system overview.

Additionally the MARS model also comprises a state of the art fuel consumption and emission model which allows the assessment of environmental impacts of transport and land use developments.

4. MARS APPLICATION – FLIGHT SIMULATOR

The MARS FS is a graphical user interface which is specially designed for the needs of decision makers. It enables the user to set up a wide range of transport and land use policy instruments and to simulate their impacts on the development of the underlying case study area. In that way it supports the decision maker to identify their optimal policy mix for their local context. The paper demonstrates the added value generated by the use of the MARS FS within numerous decision making processes and environments.

4.1 Introduction

The MARS FS is a simple to use push button interface for the MARS land use transport model. The application uses a set of commands to give users simplified access to the model. To the user, the MARS FS appears as a series of buttons, menus, or a sequence of screens allowing him or her to use and analyze the MARS model in a straightforward and meaningful way. (Ventana-Systems 2003).

Decision makers can themselves tryout their policies and see the consequences immediately. To navigate through MARS FS you have to operate the appropriate buttons, just like in any other Windows® application.

When opening the MARS FS the first screen you see is the area of the case study with its zones (*Figure 3*).

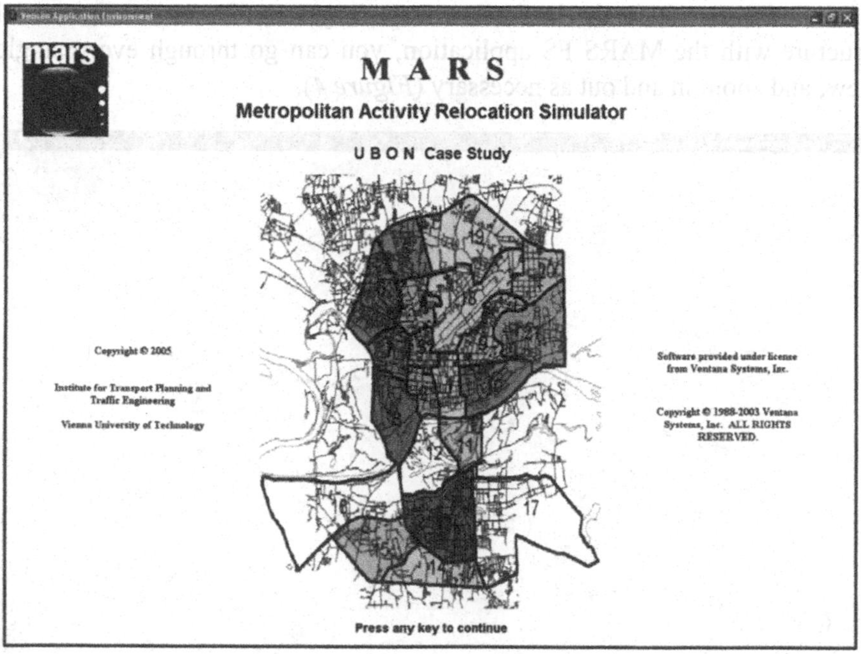

Figure 3. Start screen of MARS FS (case study Ubon, Thailand).

This screen gives you an impression how the underlying city looks like and how the zones, used in the model, are numbered. The main menu lets you choose:

1. to review the model structure;
2. to simulate the model or to;
3. to request help how to operate the MARS FS.

The Help pages cover the most urgent matters, e.g. how to review the model, how to simulate with different policies and how to use the output pages. Also every graph, table or document window in the application can be exported to the clipboard for further use.

4.2 Review of the Model Structure

The button "Review Model Structure" leads the user to a section of MARS FS where he can explore the model inherent cause-effect relations. To be able to do so two different ways are offered:

4.2.1 View Mode

The model is split up into different views, which in turn depict specific thematic issues within the land use transport system. To review the model structure with the MARS FS application, you can go through every single view, and zoom in and out as necessary (*Figure 4*).

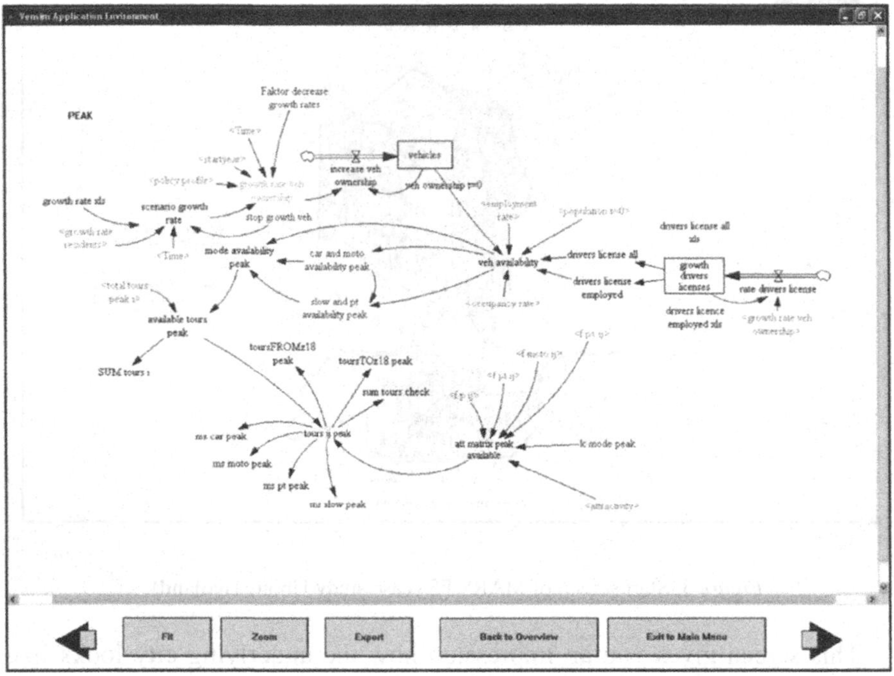

Figure 4. MARS FS – review model structure.

4.2.2 Causes Tree mode

Another way to review the relationships between the model variables is to use the Causes Tree function. Here a tree-type graphical representation is created showing the causes of the chosen model variable (*Figure 5*). By clicking on a variable name, the causes tree can be expanded until the final cause is reached.

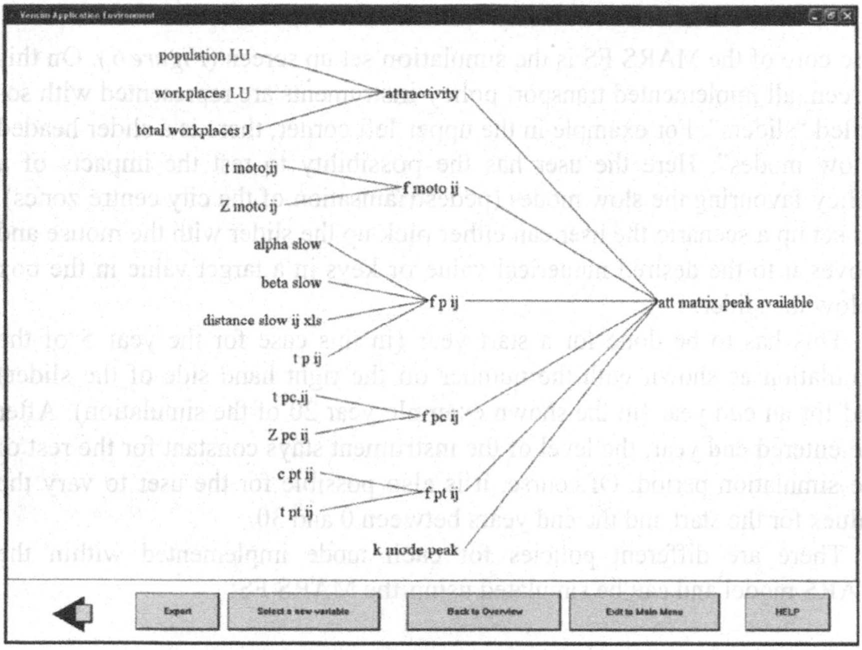

Figure 5. MARS FS – Causes Tree.

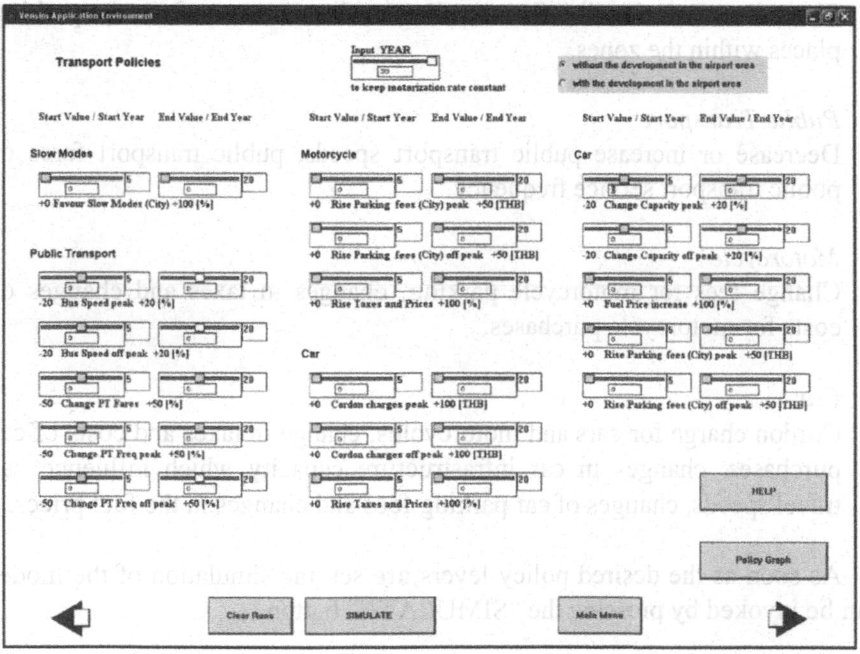

Figure 6. MARS FS – input policies and scenarios for simulation.

4.3 Simulation of the Model

The core of the MARS FS is the simulation set up screen (*Figure 6*). On this screen, all implemented transport policy instruments are represented with so-called "sliders". For example in the upper left corner, there is a slider headed "slow modes". Here the user has the possibility to test the impacts of a policy favouring the slow modes (pedestrianisation of the city centre zones). To set up a scenario the user can either pick up the slider with the mouse and moves it to the desired numerical value or keys in a target value in the box below the slider.

This has to be done for a start year (in this case for the year 5 of the simulation as shown with the number on the right hand side of the slider) and for an end year (in the shown example year 20 of the simulation). After the entered end year, the level of the instrument stays constant for the rest of the simulation period. Of course, it is also possible for the user to vary the values for the start and the end years between 0 and 30.

There are different policies for each mode implemented within the MARS model and can be simulated using the MARS FS:

Slow Mode
Partially pedestrianisation of zones and/or implementation of cycle lanes. These instruments will influence also the distance to or from the parking places within the zones.

Public Transport
Decrease or increase public transport speeds, public transport fares or public transport service frequency.

Motorcycle
Change fees for motorcycle parking, changes in taxes and changes of costs for motorcycle purchases.

Car
Cordon charge for cars and motorcycles, change in taxes and costs of car purchases, changes in car infrastructure capacity which influence car travel speeds, changes of car parking fees and changes in the fuel prices.

As soon as the desired policy levers are set, the simulation of the model can be invoked by pressing the "SIMULATE" button.

4.4 Output

4.4.1 Pre-prepared Indicators

After the simulation (which takes about 20 seconds on a standard PC), the user can immediately switch between different output variables and formats. Within the existing MARS FS, the following variables are shown in graphs for a "do-nothing" scenario and the user defined "do-something" scenario:

- Mode share in peak and off peak
- Population, workplaces
- Total vehicle km motorized
- Average commuting distance motorized and non motorized
- Total CO2 emissions for car and motorcycle
- Average commuting speed per mode.

The simultaneous display of the "do-nothing" scenario and the "do-something" scenario shows the user the impacts of his tested policy instrument in a clear way. Analyzing graphs is a good way to get the overall picture, you can see immediately if, for example CO2, increases or decreases. For the ones interested in absolute numbers, there is also the option to present for every graph an according table (*Figure 7*).

4.4.2 Individual Output

If the MARS FS user wants to go into even more detail of analyses, there is the possibility to choose from a list any variable within the model and display it as graph or as a table. This feature also enables the user to display and export information for a certain range of zones, for specific modes, or for specific trip purposes, etc.

4.4.3 GIS – Output

Until now the GIS - Output was static – it consisted of a series of coloured pictures viewed individually. The newest development of MARS FS includes "AniMap interactive", a tool developed from our colleagues and GIS specialists at the Vienna University of Technology.

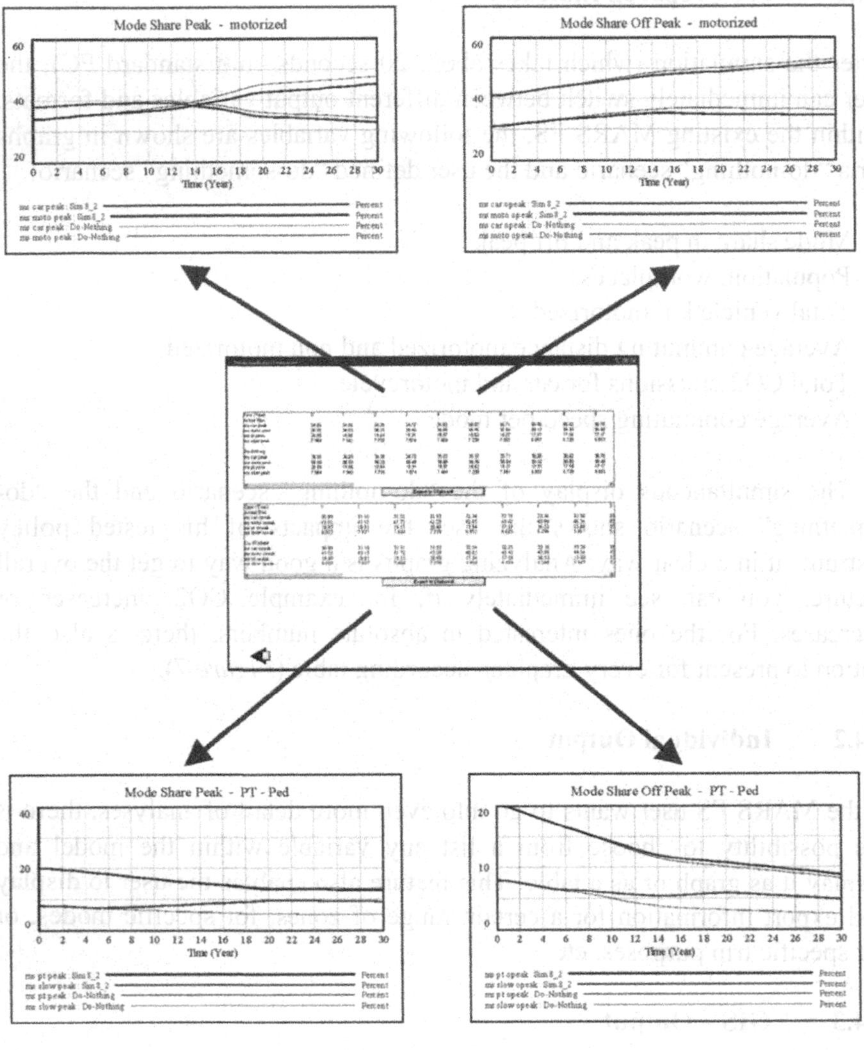

Figure 7. MARS FS – Output Mode Share.

"AniMap interactive" displays animated spatio-temporal information with open-source software and open standards including a Web server, PHP, MySQL, Internet Explorer and XML_SVG (Hocevar, Lunak et al. 2004).

After simulation MARS writes out automatically pre-chosen variables into a MySQL database. The user just has to open Microsoft's Internet Explorer to view the changes in the selected indicators as a short movie over the simulated time period.

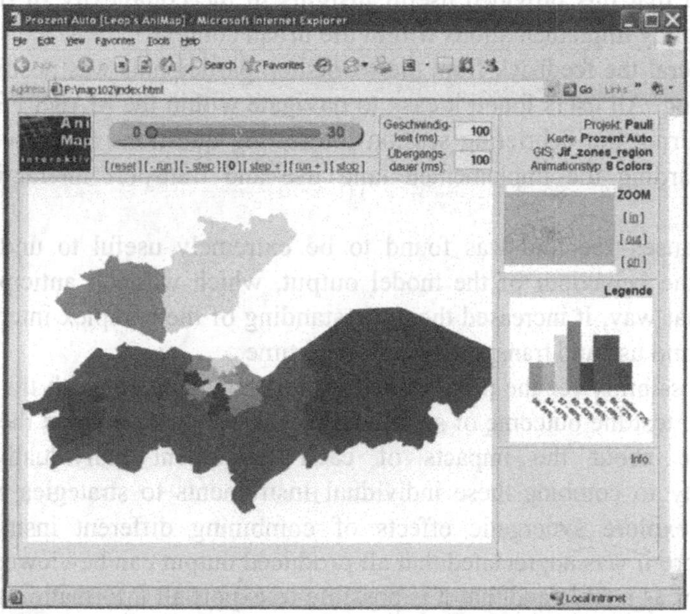

Figure 8. AniMap interactive (case study Edinburgh, UK).

5. OBSERVATIONS

Up to now, the MARS FS has been used as a decision support tool in two workshops in Thailand. More than 40 high and medium level decision makers took part in these workshops. The setting of the workshops was the following: In small working groups (consisting of 4-6 persons) the objectives for the underlying case study were discussed till an agreement regarding the objectives was reached. In a next step the participants had to agree on a set of key-indicators to monitor the outcome of the strategies and to define desired target values, which should be reached within the 30 year simulation period. After an intensive discussion round of 2-3 hours the working groups were asked to present their findings and agreements to all other participants in a plenum session.

After that the participants were introduced to the MARS FS software. In a 30-minutes session the key features of the software were explained and then each working group was put in front of a computer to use MARS FS for the personally. They were asked to use the software to test if their objectives respective target values can be reached within their set of policy packages during the 30 year simulation period. This exercise was repeated until a satisfactory result was obtained. Not in all cases it was possible to meet all their set targets simultaneously. Participants found this especially interesting and stated that this provided useful insights in the complexity of decision making policy implementations within the urban context.

In general the feedback from the workshop participants was positive and encouraging. All users found it easy to navigate within the MARS FS application. During the debriefing session interesting questions and discussions arose regarding the implemented land use and transport interactions of MARS.

The causes-tree tool was found to be extremely useful to understand some of the behaviour of the model output, which was not anticipated at first. In that way, it increased the understanding of the complex interactions between land use and transport system over time.

The possibility for the participants to work interactively with the LUTI-model and test the outcome of single policy instruments provided them with knowledge about the impacts of each instrument individually. The opportunity to combine these individual instruments to strategies allowed them to explore synergetic effects of combining different instruments. Furthermore it was appreciated that all produced output can be viewed either as graph or as a table and that it is possible to export all information to other software packages to carry out more detailed investigations.

A minor point of criticism was the chosen list of pre-prepared indicators depicted in the graphs. Although there is a feature implemented, which enables the user to generate a graph or a table of every individual variable, users preferred to have a fuller set of pre-prepared graphs to click through. This will be improved before our next set of workshops held in Vietnam in March 2006.

6. CONCLUSION

Summing up, the MARS model is a land use transport interaction model, which can be used by experts to explore the interactions between these two systems. The modular and open structure allows adding on additional parts to the MARS model easily.

The MARS FS, as a graphical user interface, was assessed to be useful for easy access to the MARS model. The feedback from users at the workshops proved that the combination of the MARS model and the MARS FS is a useful tool to support the decision making process and encourages us to improve the software in future.

7. ACKNOWLEDEGEMENT

Finally we would like to thank the referees for their useful hints how to improve our work we present here.

8. REFERENCES

David Simmonds Consultancy, 1999, *Review of Land-Use/Transport Interaction Models*, London, Department of the Environment, Transport and the Regions.

Echenique, M., A. D. J. Flowerdew, et al., 1990, "The MEPLAN models of Bilbao, Leeds and Dortmund," Transport Reviews **10**: 309–322.

Haken, H., 1983, *Advanced Synergetics - Instability Hierarchies of Self-Organizing Systems and Devices*, Springer-Verlag.

Hocevar, A., D. Lunak, et al., 2004, *Darstellung von Zeitreihen räumlicher Daten mittels WebMapping*. CORP 2004 & Geomultimedia04. Vienna, Manfred Schrenk.

ITS., 2002, *KonSULT - Knowledgebase on Sustainable Urban Land use and Transport.* from Knowledgebase on Sustainable Urban Land use and Transport.

Lowry, I. S., 1964, *A model of metropolis*. [s.l.], Rand Corp.

Minken, H., D. Jonsson, et al., 2003, *Developing Sustainable Land Use and Transport Strategies - A Methodological Guidebook.* Oslo, Institute of Transport Economics.

Pfaffenbichler, P., S. P. Shepherd, et al., 2004, "The impact of small individual time savings in transport policy appraisal," *World Conference on Transport WCTR04*. Istanbul.

Pfaffenbichler, P. C., 2003, *The strategic, dynamic and integrated urban land use and transport model MARS, Metropolitan Activity Relocation Simulator)*, Institut fuer Verkehrsplanung und Verkehrstechnik - TU Wien. Vienna, Technische Universitaet Wien.

PROSPECTS, 2000-2003, "PROSPECTS - Procedures for Recommending Optimal Sustainable Planning of European City Transport Systems," *http://www.ivv.tuwien.ac.at/projects/prospects.html*

SACTRA, 1998, *Transport and the Economy*, London, DETR.

Simmonds, D. C., 1999, *The design of the DELTA land-use modelling package.*

Simmonds, D. C., 2001, "The objectives and design of a new land use modelling package: DELTA," in G. P. Clark and M. Madden, eds.) *Regional science in business*, Berlin, Springer-Verlag.

Sterman, J. D. 2000, *Business Dynamics - Systems Thinking and Modeling for a Complex World*, McGraw-Hill Higher Education.

Ventana-Systems, 2003, *Vensim DSS Reference Supplement.*

Wegener, M., 1998, "Das IRPUD-Modell: Überblick." Retrieved 27.9.2005, from *http://www.raumplanung.uni-dortmund.de/irpud/pro/mod/mod.htm.*

Wegener, M., 2004, "Overview of Land Use Transport Models," in D. A. Hensher, *Handbook of transport geography and spatial systems*, Elsevier. 5: xxii, 672 p.

Development of a Hierarchical Approach to Assess the Impacts of Transport Policies

The Madrid case study

Paul Pfaffenbichler, Andrés Monzón[1], Ana M. Pardeiro[1], and Paula Vieira[1]
Vienna University of Technology, Institute for Transport Planning and Traffic Engineering
[1] *TRANSyT Transport Research Centre, Madrid University of Technology*

Key words: Simulation, Evaluation, Land-use-transport models, Sustainability, Environmental effects

Abstract: To make our cities sustainable is one of today's major challenges. The complexity of this task requires suitable planning tools. The aim of this paper is to present a hierarchical modelling approach to assess the effects of transport and land use projects and instruments. First a brief definition of the overall objective sustainability is given. This is followed by the description of the suggested hierarchical approach. A strategic, dynamic land use and transport interaction model builds the basis and is linked to models on a different spatial and functional level. A case study covering the Spanish region of Madrid (fiComunidad de Madridfl) was selected to demonstrate the applicability. In particular the effects of the public transport infrastructure projects, the extension of the metro line number 9 and bus lanes on all radial highways, should be assessed. It is demonstrated that the suggested approach is applicable and suitable. The overall effect of the metro line extension and the bus lanes is positive. Nevertheless their contribution to a sustainable urban region is limited. Comprehensive strategies are needed to achieve the objective of sustainability. It could be shown that the projects can even have some negative local effects in the long term.

1. INTRODUCTION

Sustainability is one of today's major challenges. A widely accepted definition is based on intergenerational equity and sub-objectives. Studies provide evidence that our cities do not fulfil the requirements of

Jos P. van Leeuwen and Harry J.P. Timmermans (eds.), Innovations in Design & Decision Support Systems in Architecture and Urban Planning, 19-34.

sustainability. Sustainability requires the consideration of global, regional and local effects. Therefore a hierarchical assessment approach was developed. Basis is a dynamic land use and transport interaction model working on a strategic level. Effects on land use and regional travel patterns are predicted for the whole region. The results, settlement patterns and origin-destination matrices, are disaggregated in the corridors affected by the projects. An assignment model is applied to these areas. This allows the assessment of environmental effects on local and regional level. Finally the modelling results are assessed by applying a modified cost benefit analysis.

The work presented here assesses transport policies of the city of Madrid to determine their environmental effects and their contribution to the objective of sustainability. The case study covers an area of about 8,000 km^2 with about 5 million inhabitants. Land use is characterised by a rapid development of housing and businesses in the outskirts. As a result a high share of people commutes between the periphery and the core city. Although Madrid has an extensive metro line system, this causes high levels of congestion. Indicators show that the land use and transport system are not sustainable. Several instruments to improve the situation have been proposed. The case study covers a selection of them: two projects finalised in the 90ies (extension of a metro line and a bus and high occupancy vehicle lane) and a recent proposal of bus lanes on all radial highways.

The case study delivers methodological advice as well as policy recommendations. The feasibility of the hierarchical approach to assess the contribution of transport projects to sustainability is demonstrated. Potential improvements for future research are highlighted. The contribution of specific projects, the extension of a metro line and bus lanes, towards the overall objective of sustainability is shown.

2. THE OBJECTIVE OF SUSTAINABILITY

2.1 Definition

Sustainability can be defined as equity between today's and future generations (May et al. 2003). I.e. "A sustainable land use and transport system does not endanger the opportunities of future generations to reach at least the same welfare level as those living now, including the welfare they derive from their natural environment and cultural heritage" (May et al. 2003 p. 12). Another definition is that a sustainable system "does not leave any negative impacts or costs for future generations to solve or bear – present builders and users of the system should pay such costs today" (Schipper et al. 2005 p. 621). Both definitions are basically equivalent and stem from the findings of the Brundtland Commission (WCED 1987). Furthermore the use of a set of sub-objectives and indicators is suggested to make these general definitions operational (May et al. 2003; Schipper et al. 2005). Sub-objectives are amongst others "Careful treatment of non-renewable resources", "Protection of the environment" or "Equity and social inclusion" (May et al. 2003 p. 13).

2.2 Indicators

Minken et al. (2003) and Schipper et al. (2005) suggest a wide range of indicators to assess sustainability towards its sub-objectives. Consumption of land, consumption of fossil fuels and atmospheric emissions are amongst them. Land and fossil fuel are non renewable resources. Their consumption might limit the opportunities of future generations and is therefore suitable as an indicator to measure the sub-objective "Careful treatment of non-renewable resources". Local atmospheric emissions endanger the environment and are therefore suitable to represent the sub-objective "Protection of the environment". Accessibility can be interpreted as a proxy for the sub-objective "Equity and social inclusion". The development of these indicators in combination with a modified cost-benefit-analysis as suggested by (Minken et al. 2003) is used to assess the contribution of policies to the objective of sustainability.

3. A HIERARCHICAL ASSESSMENT APPROACH

3.1 Motivation

The objective of sustainability requires a joint consideration of global effects like greenhouse gas emissions, regional effects like changes in land use and local effects like pollutant emissions. No single model is able to cover all these levels simultaneously. Sub-objectives like "*Contribution to economic growth*" (May et al. 2003, p. 13) require the endogenous modelling of macro-economic developments. This is impossible at the regional or local level. On the other end of the spectrum strategic regional models cannot calculate local indicators like noise and pollutant emissions. To overcome these limitations the idea of a hierarchical approach linking models covering different levels of dis-aggregation was borne.

3.2 The Integrated Land use and Transport Model MARS

The starting point of the work presented here was the regional, dynamic, integrated land use and transport model MARS (**M**etropolitan **A**ctivity **R**elocation **S**imulator). The basic underlying hypothesis of MARS is that settlements and the activities within them are self organizing systems. Therefore it is sensible to use the principles of synergetics to describe collective behaviour (Haken 1983).

MARS assumes that land-use is not constant but rather part of a dynamic system that is influenced by transport infrastructure. Therefore at the highest level of aggregation MARS can be divided into two main sub-models: the land-use model and the transport model. The interaction process is implemented through time-lagged feedback loops between the transport and land-use sub-models over a period of 30 years.

Two person groups, one with and one without access to a private car are considered in the transport model part. The transport model is broken down by commuting and non-commuting trips, including travel by non-motorized modes. Car speed is volume and capacity dependent and hence not constant. Energy consumption and emission sub-models utilize speed dependent specific values. The land-use model considers residential and workplace location preferences based on accessibility, available land, average rents and amount of recreational space. Decisions in the land-use sub-model are based on random utility theory. Due to its strategic characteristic MARS uses a rather high level of spatial aggregation. The output of the transport model are accessibility measures by mode for each zone while the land-use model yields workplace and residential location preferences per zone.

MARS is able to estimate the effects of demand and supply-sided instruments whose results can be measured against targets of sustainability. They range from demand-sided measures, such as public transport fare (increases or decreases), parking or road pricing charges to supply-sided measures such as increased transit service or capacity changes for road or non-motorized transport. These measures can be applied to various spatial levels and/or to time-of-day periods (peak or off-peak).

To date MARS was applied to six European cities (Edinburgh, Helsinki, Leeds, Madrid, Oslo, Stockholm and Vienna). Within the project SPARKLE (**S**ustainability **P**lanning for **A**sian cities making use of **R**esearch, **K**now-how and **L**essons from **E**urope) MARS is adopted and applied to the Asian cities Ubon Ratchasthani, Thailand and Hanoi, Vietnam (Emberger et al. 2005). An extensive testing was carried out with historical data of the city Vienna (Pfaffenbichler 2003). A full description of MARS is given in (Pfaffenbichler 2003).

A MARS model of the Madrid region was available from the research project PROSPECTS - **P**rocedures for **R**ecommending **O**ptimal **S**ustainable **P**lanning of **E**uropean **C**ity **T**ransport **S**ystems (May et al. 2002; Pfaffenbichler and Emberger 2003). This model was used as a starting point in (Vieira 2005). A re-calibration of the model was performed. The land use part was calibrated to observed changes in the number of housing units, residents and workplaces in the industrial and the service sector between 1996 and 2001. The transport model part is calibrated to fit the origin-destination matrices by mode as observed in the 1996 household survey (CRTM 1996). As an example for the model testing a comparison of MARS results and the 1996 Madrid travel survey for total daily trips is shown in *Figure 1*. Both, the fit and the slope of the linear regression are satisfying. The resulting model was utilized for the case study presented here.

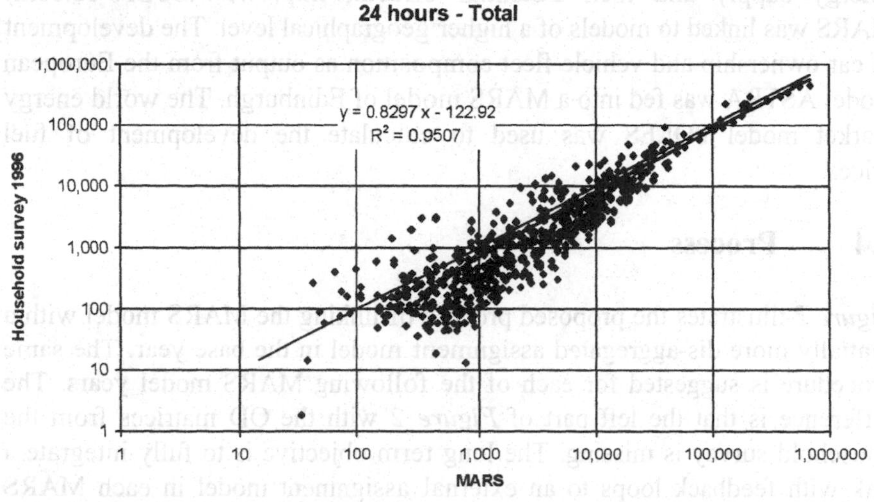

Figure 1. Comparison of the MARS model results with the results of the 1996 household survey (CRTM 1996) – daily trips total per zone.

3.3 First Steps

A first step to realize such a hierarchical approach was to link the MARS model of the region Madrid with a spatially higher dis-aggregated transport demand model of the corridor of the radial highway A3 (Vieira 2005). Five zones of the MARS model, which are directly affected by a metro line extension, were split up into the 28 census zones of the corridor A3. The relevant MARS output (number of residents, number of workplaces, car speed etc.) were dis-aggregated to census zones after achieving a reasonable conformity between empirical and calculated data for the base year 1996. The transport demand model was fed with the dis-aggregated MARS model results for the years 1996, 2005 and 2025. The next step, currently ongoing in a follow up PhD-thesis, is to link the results of the transport demand model to an assignment model. The commercial German software VISUM (www.ptv.de) is used in this process.

Another attempt to combine MARS with an assignment model was made within consultancy work for the Viennese municipal department 18. MARS origin-destination (OD) matrices for the years 2020 and 2030 were dis-aggregated and assigned to a VISUM transport network. Gravity type models were used for the dis-aggregation. The limitation was that there was no feedback of the VISUM results back into the MARS model. In the EU-funded research project STEPS (**S**cenarios for the **T**ransport system and **E**nergy supply and their **P**otential effect**S**, http://www.STEPs-eu.com) MARS was linked to models of a higher geographical level. The development of car ownership and vehicle fleet composition as output from the European model ASTRA was fed into a MARS model of Edinburgh. The world energy market model POLES was used to calculate the development of fuel prices.

3.4 Process

Figure 2 illustrates the proposed process of linking the MARS model with a spatially more dis-aggregated assignment model in the base year. The same procedure is suggested for each of the following MARS model years. The difference is that the left part of *Figure 2* with the OD matrices from the household survey is missing. The long term objective is to fully integrate a link with feedback loops to an external assignment model in each MARS iteration.

3.5 Dis-aggregation

A simple algorithm was used in (Vieira 2005) to dis-aggregate the MARS land use results. New developed housing units and businesses were distributed to the census zones in proportion to the area available for the corresponding type of land use. Gravity type models were used to disaggregate OD matrices were used in the work for the Viennese municipal department 18. The use of more sophisticated algorithms is foreseen for future applications. E.g. the same principles as used within the model MARS could be used in the dis-aggregation.

4. THE MADRID CASE STUDY

4.1 Case Study Area

The region "Comunidad de Madrid" is situated in the centre of Spain (*Figure 3*). It covers an area of about 8,000 km^2 and is populated by about 5.4 million people. Land use is characterized by a rapid development of housing and businesses in the surroundings of the city. The share of population living there steadily increases (*Figure 2*). As a result a high share of people commutes between the outskirts and the core city. The share of car trips is higher during the peak period than during the rest of the day (CRTM 1996). Although Madrid has an efficient metro line system and about 55% of the people commuting into the city use public transport (CRTM 1996), this results in a high level of peak hour congestion. The bus based part of the public transport system is stuck in congestion and therefore not attractive. Neither the land use nor the transport system currently do fulfil the requirements of sustainability.

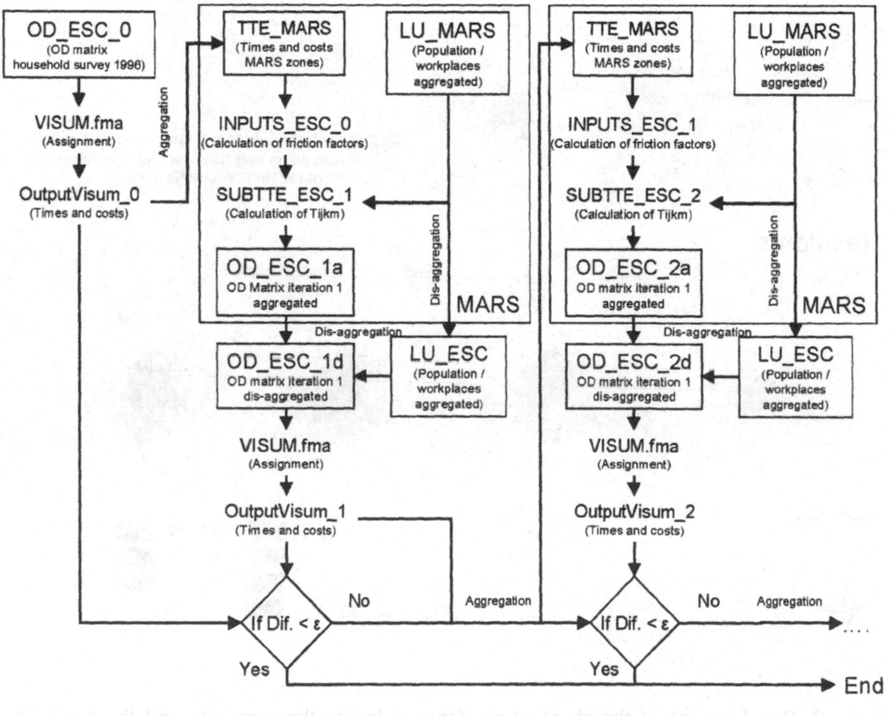

Figure 2. Linking the Madrid MARS model in the base year with an assignment model.

4.2 Extension of the Metro Line Number 9

The municipalities Rivas-Vaciamadrid and Arganda del Rey are situated in the corridor of the highway A3 in the South-East of the city of Madrid (*Figure 4*, detail X). Due to the relative proximity many residents commute into the city of Madrid. When public transport was solely bus based the share of public transport trips was only about 21% in 1996 (CRTM 1996). The situation was worsened by a rapid development of housing, service sector business and industry within the corridor (*Figure 5*). The authorities reacted with the decision to extend the metro line number 9 from its former end station Puerta de Arganda until Arganda del Rey (*Figure 4*). The extension, opened in 1999, consists of four new metro stations: Rivas Urbanizaciones, Rivas Vaciamadrid, Poveda and Arganda del Rey. Details about the planning process and the funding of the project have been published elsewhere (Monzon 2003; Monzon and Gonzalez 2000). A more detailed description is also given in (Vieira 2005).

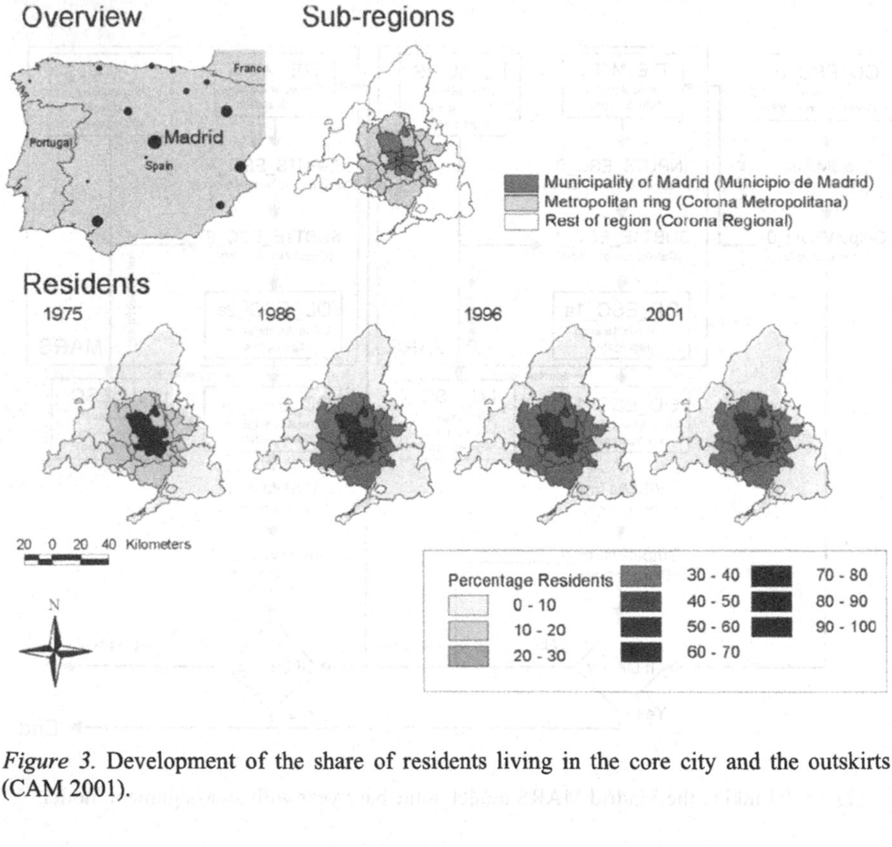

Figure 3. Development of the share of residents living in the core city and the outskirts (CAM 2001).

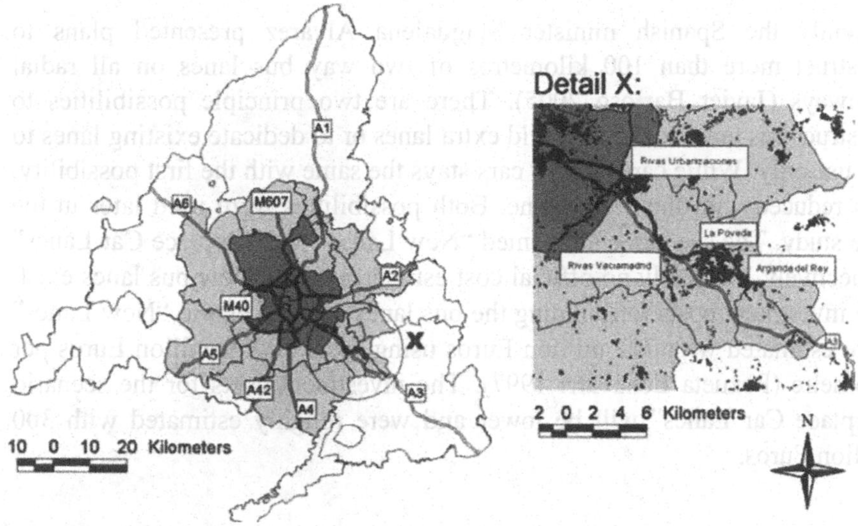

Figure 4. Extension of the metro line number 9 from Puerta de Arganda to Arganda del Rey.

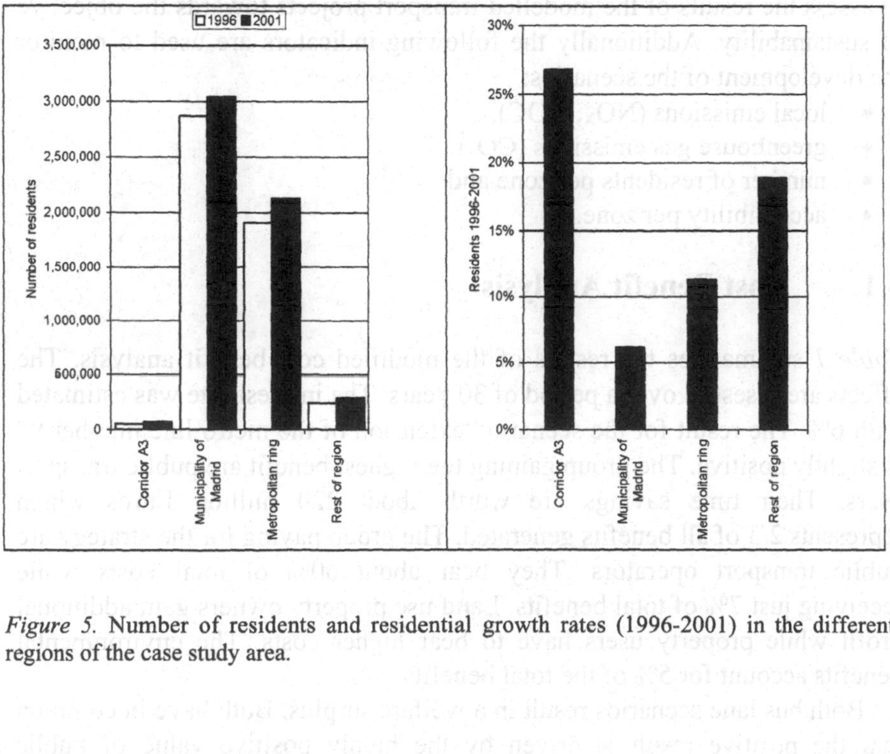

Figure 5. Number of residents and residential growth rates (1996-2001) in the different regions of the case study area.

4.3 Bus Lanes on Radial Highways

Recently the Spanish minister Magdalena Álvarez presented plans to construct more than 100 kilometres of two way bus lanes on all radial highways (Javier Barroso 2005). There are two principle possibilities to construct bus lanes: either to build extra lanes or to dedicate existing lanes to bus use only. While capacity for cars stays the same with the first possibility, it is reduced with the second one. Both possibilities were used later in the case study. The scenarios are named "New Lanes" and "Replace Car Lanes" respectively. Currently no official cost estimates for the new bus lanes exist. The investment costs for building the bus lanes in the scenario "New Lanes" were estimated with 722 million Euros using costs of 3.3 million Euros per kilometre (Pozueta Echavarri 1997). The investment costs for the scenario "Replace Car Lanes" will be lower and were roughly estimated with 300 million Euros.

5. CASE STUDY RESULTS

A modified cost benefit analysis as laid out by (Minken et al. 2003) is used to assess the results of the modelled transport projects towards the objective of sustainability. Additionally the following indicators are used to monitor the development of the scenarios:

- local emissions (NO_X, VOC),
- greenhouse gas emissions (CO_2),
- number of residents per zone and
- accessibility per zone.

5.1 Cost Benefit Analysis

Table 1 summarizes the results of the modified cost benefit analysis. The effects are assessed over a period of 30 years. The interest rate was estimated with 6%. The result for the scenario "extension of the metro line number 9" is slightly positive. The group gaining the highest benefit are public transport users. Their time savings are worth about 220 million Euros which represents 2/3 of all benefits generated. The group paying for the strategy are public transport operators. They bear about 60% of total costs while receiving just 7% of total benefits. Land use property owners gain additional profit while property users have to bear higher costs. The environmental benefits account for 5% of the total benefits.

Both bus lane scenarios result in a welfare surplus. Both have in common that the positive result is driven by the highly positive value of public transport user time savings. Car user time savings are positive in the scenario "New lanes" and negative in the scenario "Replace Car Lanes". The same is true for car user costs. In both scenarios public transport operators create about the same revenues from additional fares. The government finances the investments in both scenarios. The total external costs are negative for "New Lanes" and positive for "Replace Car Lanes". The present value of finance is negative for all three scenarios, i.e. they increase public spending compared to "Do minimum".

Table 1: Results of the cost benefit analysis (million Euros).

Group	Source of costs and benefits		Value (million €)		
			Line 9	New	Replace
User	Public transport	Time savings	221.5	1,181.8	1,140.8
	Car	Time savings	–3.5	112.7	–255.1
		Money	4.0	33.3	–81.2
	Residences (rent, mortgage)		–123.5	–148.0	–340.7
Operator	Public transport	Investment	–113.3	–722.0	–300.0
		Operating costs	–77.6	–5.9	0.0
		Revenues	22.4	141.4	123.8
	Road	Maintenance	–0.4	0.0	0.0
	Residences		71.1	151.1	346.3
Government	Fuel tax, Parking		–8.5	1.3	17.7
Society	Accidents, local emissions		10.2	–36.0	–5.5
(external costs)	Greenhouse gas emissions		5.2	6.9	12.7
Total			7.6	716.6	658.8

5.2 Atmospheric Emissions

The potential of the metro line extension to reduce emissions is rather limited. For the local pollutants NO_X and VOC it is about –0.1% to –0.2%. The potential to reduce CO_2-emissions is of the same order of magnitude. They are reduced by about –0.2% in the years following the implementation but the reduction decreases continuously to about –0.1% in the long run. The detailed analysis of NO_X and VOC emissions at the local level shows a reduction in the short term but a significant increase in the long term (about -–3% for both in 2005 but +23% for NO_X and +38% for VOC in 2025). There are explanations. First, more residents move into the corridor with the metro line extension than without (see section 5.3). Therefore the number of trips and hence car trips is in the long term higher than in the scenario "do minimum" (*Figure 6*). Secondly, average car speed increases in the short run years due to the reduced number of car trips. Thus making travel by car more attractive and stimulating car use.

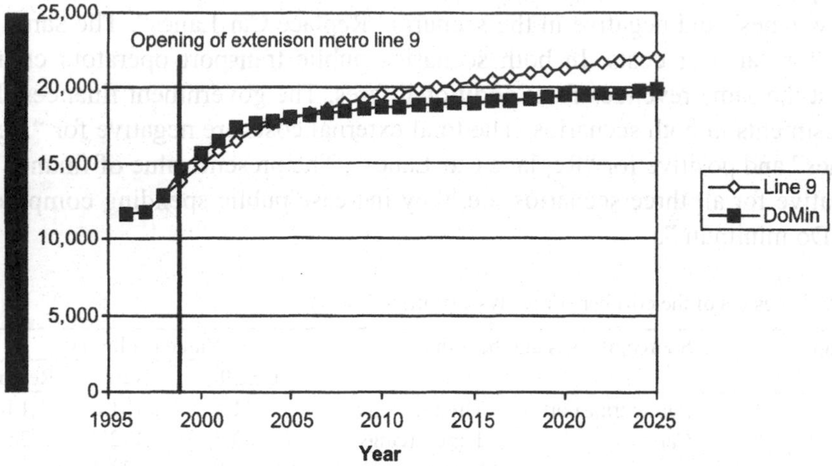

Figure 6. Development of the number of car trips originating in the corridor A3.

In the short run the bus lanes increase the modal share of public transport significantly. In the long run the modal share of public transport goes back to about the value of the base year. This behaviour coincides with the observations in the aftermath of the installation of the bus/HOV lane on the A6 (Monzón et al. 2003). Nevertheless it remains significantly higher than in the scenario "Do Minimum". The scenario "Replace Car Lanes" significantly decreases NO_X and VOC emissions, while the scenario "New Lanes" only has significant effect on VOC emissions. Both scenarios reduce CO_2-emissions. The relative amount is small: about −0.3% for "New Lanes" and about −0.5% for "Replace Car Lanes". Even these small reductions are in danger of being lost in the long term. Especially in the scenario "New Lanes": the additional road capacity is filled up again and CO_2-emissions are above the "Do Minimum" levels in the years 2023 and 2024.

5.3 Residents

The metro line extension increases the attractiveness of Rivas and Arganda as a living place. More people decide to settle within the corridor (*Figure 7*). The location choice within the corridor is determined by two main factors: availability of land and proximity to the metro stations. Overall land consumption increases slightly due to the fact that development densities are lower in the corridor A3 than in the core city.

The pattern of land use changes caused by the bus lane scenarios is not very clear. The location decisions are determined by relative differences between zones. But as the bus lanes are quite evenly distributed around the city centre the differences are not very distinct. A more detailed analysis o clarify this issue is planned.

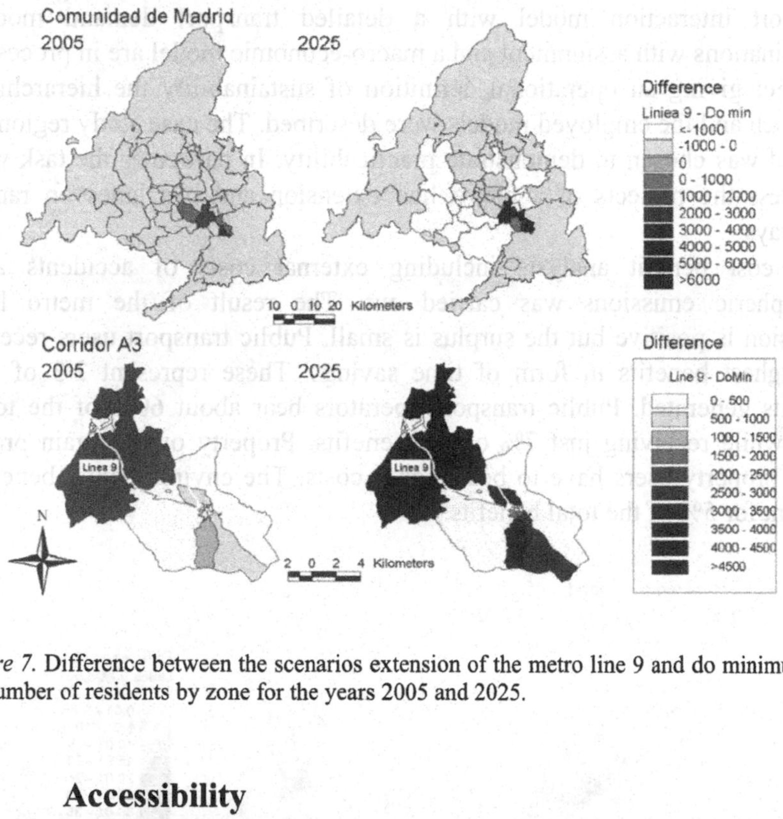

Figure 7. Difference between the scenarios extension of the metro line 9 and do minimum in the number of residents by zone for the years 2005 and 2025.

5.4 Accessibility

Accessibility is measured as the number of working places which can be reached within a weighted by travel time. The difference in the accessibility by public transport in the scenarios "line 9" and "do minimum" is shown in *Figure 8*. It is obvious that the metro line extension increases the public transport accessibility of the corridor significantly.

In both bus lane scenarios, accessibility by public transport increases. It is highest in the bus lane corridors. In "New Lanes", accessibility by car increases slightly in the whole region. The increase is highest in the bus lane corridors. In "Replace Car Lanes", accessibility by car decreases in the bus lane corridors where road capacity for cars is reduced. Due to the overall reduction in car trips it increases in the other zones.

6. SUMMARY AND CONCLUSIONS

This paper presents a hierarchical approach to assess transport and land use projects against the high level objective of sustainability. The proposed approach makes use of simulation models on different spatial and functional levels. The case study presented here links a strategic, dynamic land use and transport interaction model with a detailed transport demand model. Combinations with assignment and a macro-economic model are in process.

After giving an operational definition of sustainability the hierarchical approach and the employed models were described. The case study region of Madrid was chosen to demonstrate practicability. In particular the task was to assess the projects of a metro line extension and bus lanes on radial highways.

A cost benefit analysis including external costs of accidents and atmospheric emissions was carried out. The result of the metro line extension is positive but the surplus is small. Public transport users receive the highest benefits in form of time savings. These represent 2/3 of the benefits generated. Public transport operators bear about 60% of the total costs while receiving just 7% of the benefits. Property owners gain profit while property users have to bear higher costs. The environmental benefits account for 5% of the total benefits.

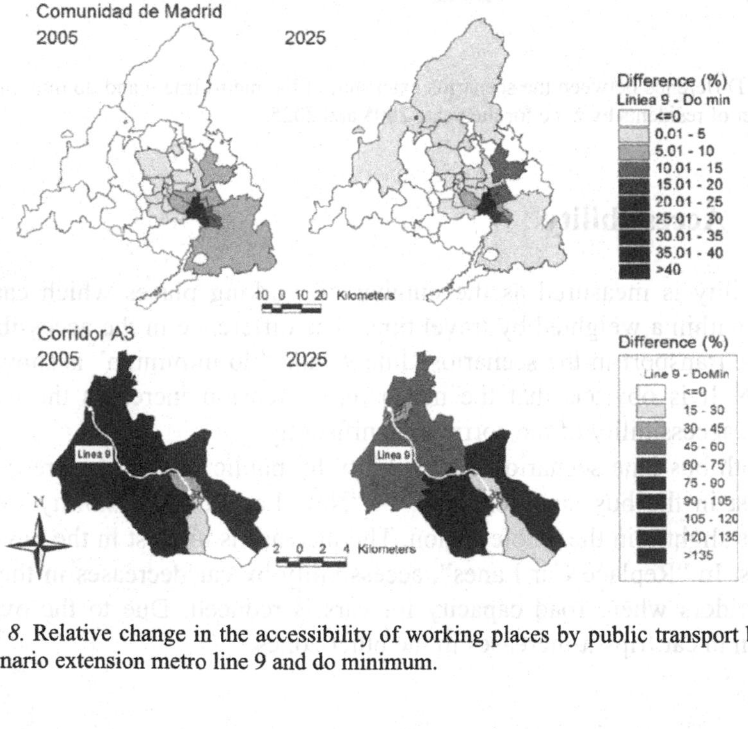

Figure 8. Relative change in the accessibility of working places by public transport between the scenario extension metro line 9 and do minimum.

The welfare surplus created by the bus lanes is higher than that of the metro line extension. As the bus lane scenarios cover a wider area this was expected. About 70% of the benefits are created by time savings. Public transport operators and government are financing the surplus. They bear 30% ("Replace Car Lanes") to 80% ("New Lanes") of all costs. The reduction of greenhouse gas emissions is small for both and makes up less than 1% of the benefits. External costs of the scenario "New Lanes" are even negative due to increased accident costs caused by higher car speed.

Different indicators were used to assess effects on a more local level. The attractiveness of the corridor A3 benefits from the expansion of the metro line. Accessibility by public transport improves. Investors are attracted to develop living space and more people locate in the corridor. The results for NO_X and VOC emissions are ambiguous. Although they decrease in the short term, they increase in the long term. The main reason is the growth in population. This demonstrates that the use of a land use and transport interaction model is essential for assessing sustainability.

Public transport accessibility improves in the bus lane corridors. Modal share of public transport increases in the short term but goes back to initial values in the long term, which is still higher than in the scenario "Do Minimum". "Replace Car Lanes" decreases NO_X and VOC emissions, while "New Lanes" only has a significant effect on VOC. The increase in accidents caused by higher car speed offsets improvements in the category external costs. The effects on land use need further investigation. From a sustainability viewpoint the option "Replace Car Lanes" has to be favoured.

Although the analysed projects are substantially big especially from a financial viewpoint, they affect only a rather small area of the total region. Their potential to reduce negative environmental impacts is limited. Single projects like these are not able to achieve the goal of a sustainable city or urban region. A comprehensive strategy including other complementary instruments like pricing would be necessary.

Summing up, the use of already available tools (detailed assignment models, detailed transport demand models and dynamic land use transport interaction models) and their linking together delivers useful information on different spatial and temporal levels and produce so synergies. Local impacts can so estimated in more detail (e.g. output from detailed assignment model) whereas more regional impacts (e.g. output from LUTI-model) can be taken into consideration at the same time. We think that the here introduced hierarchical approach is an innovative and feasible way forward to generate more accurate information for decision making in an urban and regional context.

7. REFERENCES

CAM., 2001, *Censos y Padrones oficiales*, Consejería de Economía e Innovación Tecnológica, Instituto de Estadística de la Comunidad de Madrid, Madrid.

CRTM., 1996, *Encuesta Domiciliaria de Movilidad en la Comunidad Madrid 1996*, Consorcio Regional de Transportes de Madrid, Madrid.

Emberger, G., Ibesich, N., and Pfaffenbichler, P., 2005, "Die Entwicklung eines integrierten dynamischen Siedlungsentwicklungs- und Verkehrsmodells für Asiatische Städte," *CORP 2005*, Wien.

Haken, H., 1983, Advanced Synergetics - Instability Hierarchies of Self-Organizing Systems and Devices, Springer-Verlag.

Javier Barroso, F., 2005, "Fomento construirá 120 kilómetros de carriles bus en las autovías de entrada en la capital," *El País*, Madrid, p. 4.

May, A. D., Karlstrom, A., Marler, N., Matthews, B., Minken, H., Monzon, A., Page, M., Pfaffenbichler, P., and S., S., 2003, *Developing Sustainable Urban Land Use and Transport Strategies - A Decision Makers' Guidebook*, Institute for Transport Studies, University of Leeds, Leeds.

May, A. D., Shepherd, S., and Pfaffenbichler, P., 2002, "The Development of Optimal Urban Land Use and Transport Strategies," *International Conference on Seamless & Sustainable Transport*, Singapore.

Minken, H., Jonsson, D., Shepherd, S. P., Järvi, T., May, A. D., Page, M., Pearman, A., Pfaffenbichler, P. C., Timms, P., and Vold, A., 2003, *Developing Sustainable Land Use and Transport Strategies - A Methodological Guidebook*, Institute of Transport Economics, Oslo.

Monzon, A., 2003, "Integrated Policies for Improving Modal Split in Urban Areas," *16th ECMT Symposium*, Budapest.

Monzon, A., and Gonzalez, J. D., 2000, "Travel Demand Impacts of a New Privately Operated Suburban Rail in the Madrid N-III Corridor," *Eurpean Transport Conference*, Cambridge, UK.

Monzón, A., Puy, J., Pardillo, J. M., Cascajo, R., and Mateos, M., 2003, "Modelización y evaluación de medidas de gestión en corredores urbanos," *CICYT Spanish National R + D Programme*, Madrid.

Pfaffenbichler, P., 2003, *The strategic, dynamic and integrated urban land use and transport model MARS (Metropolitan Activity Relocation Simulator) - Development, testing and application, Beiträge zu einer ökologisch und sozial verträglichen Verkehrsplanung Nr. 1/2003*, Vienna University of Technology, Vienna.

Pfaffenbichler, P., and Emberger, G., 2003, "Are European cities becoming similar?" *CORP2003, 8. internationales Symposion zur Rolle der IT in der und für die Planung sowie zu den Wechselwirkungen zwischen realem und virtuellem Raum*, Wien, 243-250.

Pozueta Echavarri, J., 1997, "Experiencia Española en carriles de alta ocupación. La Calzada bus/VAO en la N-VI: Balance de un año de funcionamiento," *Cuadernos de Investigación Urbanística*, Escuela Técnica Superior de Arquitectura, Madrid (16).

Schipper, L., Huizenga, C., and Ng, W., 2005, "Indicators: Reliable signposts on the road to sustainable transportation - The partnership for sustainable transport in Asia," *ECEE 2005 Summer Study Proceedings*, Mandelieu, France.

Vieira, P., 2005, *Modelización de la interacción de usos del suelo y transporte. Aplicación al corredor de la A-3 de Madrid*, PhD-thesis, Universidad Politécnica de Madrid, Madrid.

WCED., 1987, *Our Common Future. The Report of the World Commission on Environment and Development*, Oxford University Press, New York.

Development of a Support System for Community-Based Disaster Mitigation Planning Integrated with a Fire Spread Simulation Model Using CA

The results of an experimentation for verification of its usefulness

Y. Gohnai, A. Ohgai, S. Ikaruga[1], T. Kato[2], K. Hitaka[3], M. Murakami[4], and K. Watanabe[5]

Toyohashi University of Technology, Toyohashi, Japan
[1]Yamaguchi University, Yamaguchi, Japan
[2]University of Tokyo, Tokyo, Japan
[3]Kyushu Sangyo University, Fukuoka, Japna
[4]Kogakuin University, Tokyo, Japna
[5]The University of Tokushima, Tokushima, Japan

Keywords: Community-based planning for disaster mitigation, Planning support system, Fire spread simulation, Cellular automata, WebGIS

Abstract: This research carried out an experiment to verify the usefulness of a WebGIS-based system that we have been developing (A support system for community-based disaster mitigation planning integrated with a fire spread simulation model using CA) BEFORE, DURING and AFTER Workshops. Based on the data collected from the experiment, the usefulness and advantages of the system were verified from various angles. As a result, it was found that even though there are still some issues to be solved, the system is useful and effective as a support tool and also in enhancing awareness before and after Workshops.

1. INTRODUCTION

In case of seismic disasters, old wooden structures in Japanese densely built-up areas would suffer immense damage. For that reason, the Japanese government has, among its policies, given a priority to the improvement of these areas. In order to solve this problem, promotion of planning for improvement is needed through community-based planning for disaster

Jos P. van Leeuwen and Harry J.P. Timmermans (eds.), Innovations in Design & Decision Support Systems in Architecture and Urban Planning, 35-51.

mitigation with consensus-building by the residents and the administration of the areas.

One method for planning called "Workshop" has attracted a great deal of attention in Japan. A typical Workshop (hereinafter referred to as WS) is summarized as follows. (i) Understanding; The residents understand and confirms the existence of the problems and useful resources for disaster mitigation in the area. (ii) Summarization; Based on the "Understanding", the residents summarize the problems for the disaster mitigation in the area. (iii) Planning; The residents discuss the measures to solve the disaster mitigation problems of the area clarified by "Summarization". And, the residents summarize the measures through cooperative decision making and consensus-building.

Especially in scene (iii) of densely built-up areas with old wooden structures, improvement for the prevention of fire spread is an important consideration. If there were a system that could offer the residents "the result of fire spread simulations applied to existing circumstances on a digital map and the result of fire spread simulations applied to the improved circumstances" and "geographic information related to the WS", the system would be effective in consensus building and decision making in group discussions of the WS.

Based on above hypothesis, we tried to develop a fire spread simulation model as a community-based planning support tool, which can plainly offer residents the information about the effect of increase in disaster mitigation performance by improvement of the local environment (Ohgai, Gohnai and et al., 2003, 2004). We have developed a WebGIS-based system with a GUI that could apply the model to any city, anywhere, anytime as long as data is available. The developed system was used on trial in the scene of an actual WS for verification of its usefulness (Gohnai, Ohgai and et al., 2005).

As a property of this WebGIS-based system, not only can the system be used DURING the WS scene, but also BEFORE and AFTER the WS. Based on its potential, we have set up the following hypotheses. (I) The residents with experience of using the system BEFORE the WS would effectively use the system DURING the WS. Because of this, there is some possibility of smooth consensus-building and substantial planning. (II) If the residents use the system AFTER the WS, it would be possible to consider the measures. Using these considerations, there could be a possibility of planning more effective measures. (III) The resident's awareness of disaster risk and the need for improvement of the local environment is enhanced through the discussions and considerations done when planning using the system.

Accordingly, this paper tries to carry out a verification experiment for the above hypothesis using the above mentioned system. Based on the result of the experiment, we discuss the verification, effectiveness and problems of the system.

2. OUTLINE OF THE SUPPORT SYSTEM FOR COMMUNITY-BASED DISASTER MITIGATION PLANNING

This chapter explains the outline of four subsystems that make up the support system for community-based disaster mitigation planning; 1) Disaster mitigation related Geographic information subsystem, 2) Fire spread simulation subsystem, 3) Built-up area improvement subsystem and 4) Confirmation of effects of built-up area improvement subsystem. *Figure 1* shows each of the subsystems user interface.

a) Disaster Mitigation Related Geographic Information Subsystem

This subsystem has a function that displays disaster mitigation related geographic information in the area such as data shown in *table 1*. And, it has

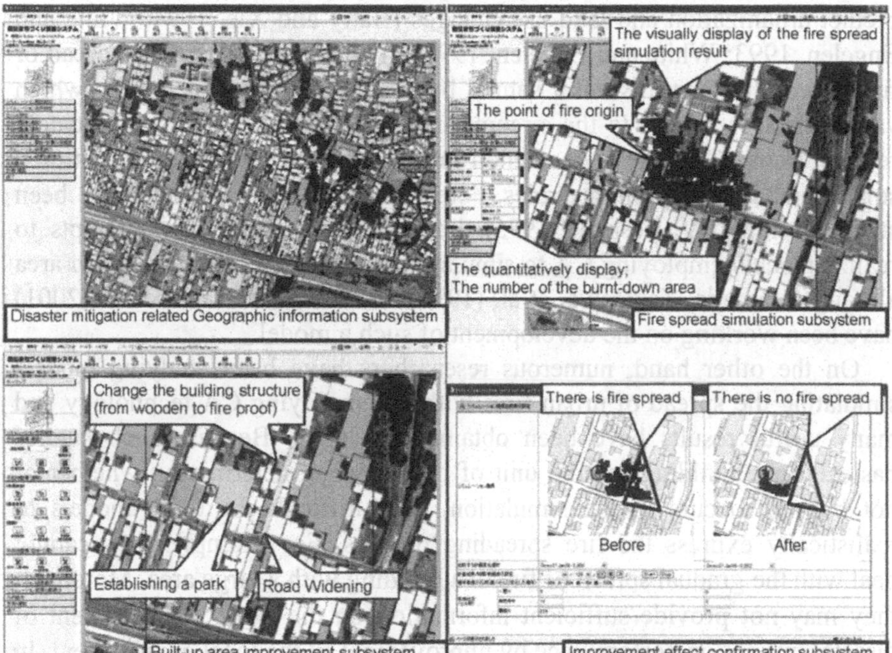

Figure 1. Example of Basic Interface of the System.

Table 1. Data and functions of the disaster mitigation related geographic information subsystem.

Data	Building shape, use, structure, floors and age/ Town block shape/ Ground Shape/ Land use/ Aerial photo/ width of roads/ water utility point for fire fighting/ Refuge, etc.
Function	Overlay, Thematic map, Zoom, Panning, Distance and Area measurement, etc.

the basic GIS functions as shown on *table 1*. This subsystem's screenshot (*Figure 1*) shows aerial photo, building structure, road width and position of refuge place and water utilization point for fire fighting using overlay and thematic map of this subsystem.

b) Fire Spread Simulation Subsystem

This subsystem can simulate fire spread on the digital map, and can visually and quantitatively display the simulation results. The system user can freely set simulation end time, wind direction, wind velocity, point of fire origin and number of origin points. The subsystem has two methods of displaying the result; 1) displaying the result of at any given point of time, 2) displaying the result with animation. Additionally, the system not only shows the result visually, but can also confirm the size of burnt-down area (*Figure 1*). As a simulation model, this subsystem uses a fire spread simulation model with Cellular automata (CA) that we had developed earlier. (Ohgai, Gohnai and et al., 2004)

Research using CA has been advancing in various fields including transportation, economics, and chemistry. In the field of city planning, CA is used to understand dynamic and complicated urban phenomena and to predict urbanization and land use changes (Batty and Xie, 1994; White and Engelen, 1993; White and Engelen, 1997; Li and Yeh, 2000). The spread of fire in a built-up area is not a natural but a physical phenomenon, in which various factors related to the circumstances of built-up areas are intricately related. Using CA, this phenomenon must be reproduced by constructing a small number of appropriate rules. Forest fire simulation models have been studied using CA (Resnick, 1995), but there have been few attempts to realize models employing CA to simulate the spread of fire in an urban area although Yamada, Takizawa, et al. (1999) and Xie, Sakamoto, et al. (2001) have been working on the development of such a model.

On the other hand, numerous researchers have been studying models simulating the spread of urban fires without employing CA technology and many useful results have been obtained in Japan. Because such studies basically use buildings as the unit of output when determining whether or not burning occurs in the simulation, it is difficult for these models to realistically express the fire spreading process. For example, they cannot deal with the gradual spread of fire in a building with a large area. Moreover, they may not provide sufficient information on the likely enhancement of disaster mitigation performance by improvement of the local environment. In our study, using the results of previous research on fire spread simulation models and the characteristics of CA, we have developed a model that can simulate the spread of fire using a smaller unit than that in the previous research and can visually show the detailed fire spreading process in built-up areas containing many wooden buildings.

c) Built-up Area Improvement Subsystem

This subsystem can edit the digital map to change the local environment (*table 2* shows the edit/ improvement contents). This subsystem's screenshot (*Figure 1*) shows an improvement example; building structure change (from wooden to fireproof), road widening, establishment a park by the user.

d) Improvement Effect Confirmation Subsystem

This subsystem can visually and quantitatively compare fire spread simulation result of before the improvement and after the improvement. By displaying the difference of before and after the improvement, a user can confirm the effect of the improvement.

Table 2. Improvement object and contents.

Object	contents
Road	Widening roads (can be set to any width by the user)
Building	Changing building structures (Wooden, Fire prevention wooden Fireproof), Removing/Adding buildings
Open space	Establishing parks, Deleting parks

3. EXPERIMENTATION OUTLINE

This chapter explains the outline of the experimentation for the verification of the hypotheses (i) to (iii) (in chapter 1) using the above mentioned system.

Two experiments were carried out for verification of the hypotheses; Case 1 (Akumi District): Use of the system by residents "BEFORE the WS",

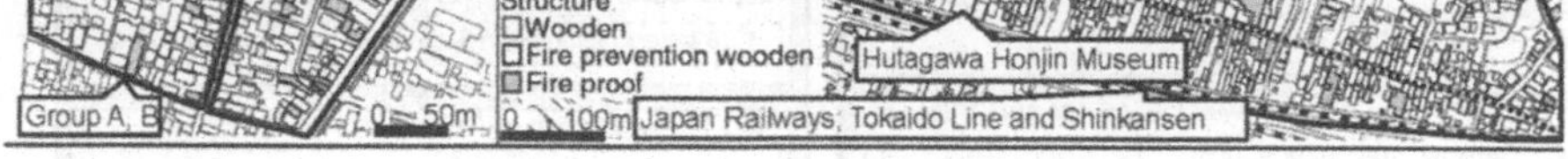

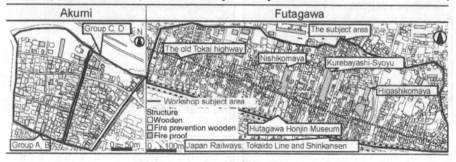

Figure 2. The subject area of the workshop.

Table 3. Subject area conditions.

Akumi	Futagawa
Escaped war damage and therefore has been no major renewal of the urban infrastructure and is characterized by: - Densely built old wooden houses, many narrow roads, no large open-spaces and fire-proof buildings for prevention of fire spread. - no effective water utility points, and refuges	A Historical built-up area that also escaped war damage. It is characterized by: - old historical stores called Nishikomaya, Higasikomaya, and Kurebayashi-Syoyu - There is also a historical road that was called the Tokai highway in the Edo era. - densely built old wooden houses, many narrow roads, no large open-spaces and fire-proof buildings for prevention of fire spread.

"During the Workshop" and "AFTER the WS". Case 2 (Futagawa District): Use of the system by the residents "During the WS" and "After the WS".

The subject areas for the experimentation were Futagawa and Akumi district, old built-up areas, located in Toyohashi city of Aichi Prefecture, Japan. Both districts have densely built old wooden houses with many roads of which the width is no wider than 4-5 meters and therefore would suffer immense damage in case of a disasters such as fires or earthquakes. Because of this, development and improvement projects of the built-up area for disaster mitigation by the Central and Local Governments is in progress in both districts. *Figure 2* and *Table 3* show subject areas and their conditions.

Figure 3 shows the experimentation outline and flow.

a) Experimentation Outline at Akumi District

The ouline of the experimentation at Akumi District (on Dec. 2005) is as shown below; **Step-1**: for educating the selected test subjects on the use of the system. **Step-2**: in order to improve their skills, the test subjects could use the system anywhere they liked (home, work place, etc). During this period, the system use log data were stored in the system server. **Step-3**: Explained in Chapter 4. **Step-4**: the system use skills of test subjects' were examined and interviewed to find out if there was any change in their awareness of the need for improvement. **Step-5**: data of the created improvement draft plan were collected, used video and voice recorders to collect data about the system use, and issued a questionnaire on the

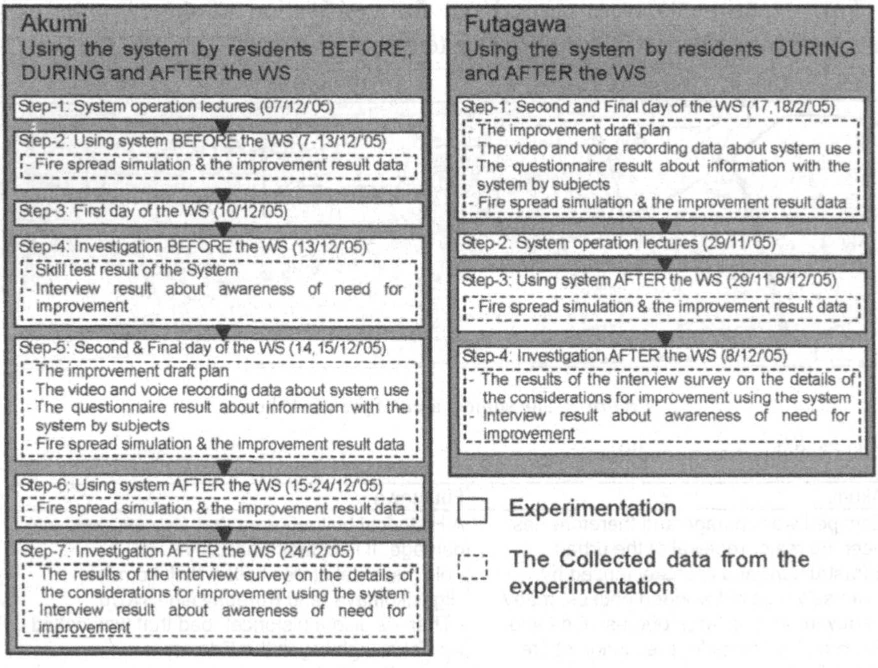

Figure 3. Experimentation flow and outline and collected data.

usefulness of both the system and the information provided by subjects. The fire spread simulation and the improvement result data were logged in the system server. **Step-6**: AFTER the WS, the test subjects used the system anywhere they liked to deliberate on urban improvement options putting into consideration the contents of the WS discussion. The system use log data were stored in the system server. **Step-7**: investigation of the test subjects' use of the system and the change in their disaster mitigation awareness due to system use AFTER the WS. Six test subjects were chosen based on their being residents of the Akumi district, were to participate in the WS and had some experience in using computers.

b) Experimentation Outline at Futagawa District

The experiment outline as shown on *Figure 3*, is similar to Step-1,5,6,7 of Akumi district. Seven test subjects were chosen based on their being residents of Futagawa district, had participate in a WS done the year BEFORE and had some experience in using computers.

4. EXPERIMENTATION DETAILS AND RESULT ANALYSIS

4.1 Results of system use BEFORE the workshop

a) The Examination of System Operation Skill

The system use skills of test subjects' were examined BEFORE the workshop. The examiner asked the examinees to use the functions of each subsystem and they were given 2 points if they performed the request smoothly, 1 point if it took time, and a no point if they did not succeed. The results of this test are shown on *Table 4*. From these results it is confirmed that most of the Test Subjects had acquired enough skills to operate the system through use of the system BEFORE the workshop.

Table 4. The test result about operation skill of the system.

	A1	A2	A3	A4	A5	A6
Map subsystem (6 functions)	10	2	10	8	12	9
Fire spread simulation subsystem (5 functions)	10	10	8	8	10	8
Improvement subsystem (6 functions)	10	3	1	6	12	9
Improvement effect confirmation subsystem (4 functions)	6	4	5	4	8	7
Total point (perfect score: 42 point)	36	19	24	26	42	33

Table 5. The interview survey contents of system use BEFORE the WS.

(1) The setting reason of the fire spread simulation area **(2)** The setting reason of fire origin point **(3)** The impression by confirmation of fire spread simulation result **(4)** The setting reason of improvement area **(5)** The setting reason of improvement contents (change the building structure, widening roads, newly establishing park) **(6)** The impression by improvement effect confirmation **(7)** The consideration of effective improvement **(8)** The recognition of need for improvement **(9)** The enhancing the awareness of disaster mitigation **(10)** The recognition of disaster risk by using the system **(11)** The change in awareness for disaster mitigation through system use

b) The Change in Awareness for Disaster Mitigation

An interview survey as shown on *Table 5* was carried out to investigate the change in the Test Subjects' disaster mitigation awareness due to their use of the system. From the results it became clear that the following 2 points were common among all the Test subjects. (1) Through the use of the Geographic information subsystem, there was a renewed recognition of presence of densely built old wooden houses, many narrow roads, etc. (2) From the fire spread simulation results, there was a realization of the risks in their district under the existing circumstances. Although this was not a quantitative analysis, it may be gathered that using the system BEFORE the workshop enhanced the Test Subjects' awareness of the need for prevention countermeasure in their district.

4.2 Results of system use DURING workshop

a) The Workshop Outline

The outline of the 3 day WS held in Akumi and Futagawa Districts where the experimentation for the verification of the hypotheses (i) to (iii) (mentioned in chapter 1) was done is as shown in *Table 6* and had the

Table 6. Outline of the workshop in Akumi and Futagawa.

	First day		Second day		Third day	
	Akumi	Futagawa	Akumi	Futagawa	Akumi	Futagawa
Date	10 Dec.	12 Feb.	14 Dec.	17 Feb.	15 Dec.	18 Dec.
Time	13:00-18:00		19:00-21:00		19:00-21:00	
Residents	22	41	33	30	22	26
Organizers	13	14	12	16	12	12
Theme	1) Understanding existing circumstances through exploration, 2) Summarizing the disaster mitigation problems in the area based on 1)		Discussing the measures to solve the disaster mitigation problems based on the results of First day		Summarizing the improvement plan with consensus-building based on the results of Second day	

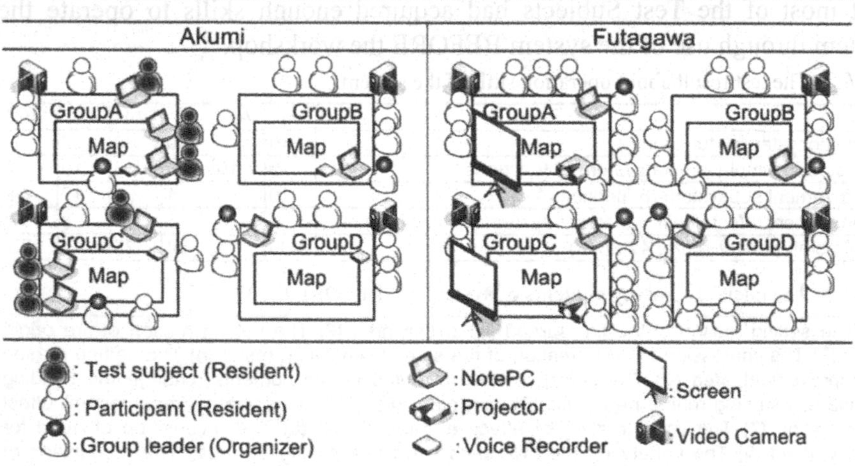

Figure 4. The rough sketch plan of the venue and the system set up.

following 2 aims: (1) Making a concrete draft plan for the improvement of disaster mitigation. (2) Enhancement of the residents' awareness of the need for improvement of the local environment.

Figure 4 shows the venue layout. There was a total of 4 groups in which 2 groups (A and C) had Test Subjects included while the other 2 (B and D) had none. Also, as shown on *Figure 2,* groups A and B members are from the same area (Western) and those of groups C and D are from the Eastern area and each had a discussion on their respective areas. The system was used on the 2nd and 3rd day of the WS to discuss measures for solving the disaster mitigation problems raised on the 1st day. *Figure 5* shows two scenes from the group discussion using the system.

The system was used in the second day and the third day of the group discussion. In the scene of discussion for the measures to solve disaster mitigation problems, especially when discussing fire spread, the system was

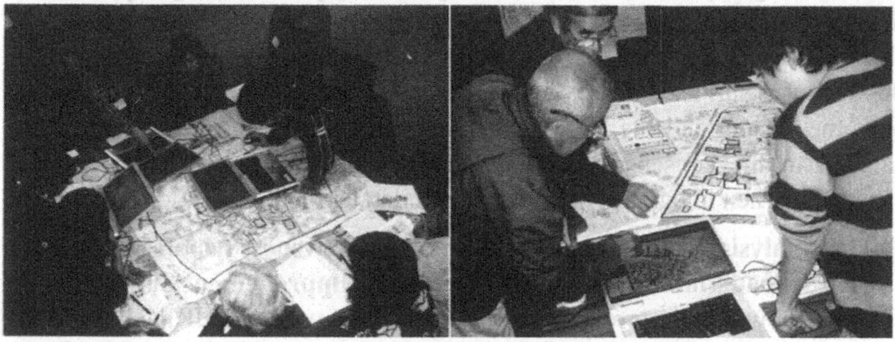

Figure 5. The scene of the group discussion using the system.

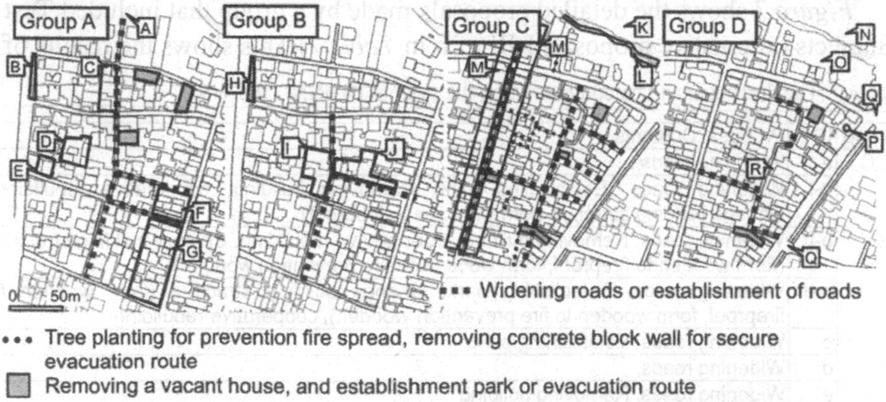

A: Establishment of a riverside park to be used as a temporary refuge and to draw water for firefighting from the River, B,H,L: Reinforcement of dangerous slopes, C,G,J: Cooperative rebuilding, D,F: Park establishment for prevention of fire spread and as a refuge, E,M: Structure change from wooden to fireproof, I: Structure change from wooden to fireproof, refuge and firefighting water point establishment, K,O: Refuge and firefighting water point establishment, N,P: Firefighting water point establishment, Q: Bridge reinforcement, R: Emergency Store

Figure 6. The improvement draft plan.

utilized by test subject or group leader. In case of the contents of unconcerned with fire spread problem, the measures were put together by discussion without system use.

The role of the group leader (Organizer) is assist and support of the group discussion. When it was necessary to use the system for the assist/support, the organizer sometime encouraged the participants to use the system.

b) The Analysis of the System and Subject Effectiveness from the Draft Plan made in the Workshop

There were both physical measures (such as road widening, park establishment, etc) and non-physical measures (such as disaster drills, awareness enhancement, etc) were included in the draft plan made in the WS. Only physical measures that had a definite location and detailed contents were picked up and organized as shown on *Figure 6*. Compared to the draft plan of the groups that had no Test Subject (Group B,D), the group that had the Test Subjects (Group A, C) had a greater number of proposals that showed measures with more detailed contents and definite locations. From this fact, it is surmised that by effectively using the system DURING the WS, both the system and the Test Subjects provided support and also acted as catalyst to the consensus building and decision making in their groups especially when integrating proposals with details such as location.

c) The Analysis of the System and Subject Effectiveness from Log Data of Fire Spread Simulation and Built-up Area Improvement Subsystem

Table 7 shows all the proposals made using the system DURING the WS. From the table, it can be confirmed that the groups with the Test Subjects actively used the system.

Figure 7 shows the detailed proposals made by a group that included Test Subjects (alternative proposals I-III of b in *Table 7*).This shows the results of

Table 7. The draft plan and its contents using the system DURING the WS.

Group	ID	Proposal Contents
A, C	a	Widening roads
	b-I	Widening roads, Establishing park, Removing building, cooperative rebuilding, Change building structure (form wooden to fireproof)
	b-II	Widening roads, Removing building, cooperative rebuilding, Change building structure (form wooden to fireproof, form wooden to fire prevention wooden)
	b-III	Widening roads, Establishing park, Change building structure (form wooden to fireproof, form wooden to fire prevention wooden), cooperative rebuilding
	c	Widening roads, Establishing park,
	d	Widening roads,
	e	Widening roads, Removing building
	f	Widening roads, Change building structure (form wooden to fireproof), Removing building
	g	Widening roads, Establishing park,
	h	Widening roads, Change building structure (form wooden to fireproof)Removing building
B,D	i	Widening roads, Change building structure (form wooden to fireproof),
	j	Change building structure (form wooden to fireproof),
	k	Establishing park, Removing building

the fire spread simulation and the improvement discussion process. From this figure, it is clear that the fire spread area decreases with accumulated discussion. Moreover the proposals become more detailed as discussions are repeated. This could be because the discussions done were aided by visual and quantitative information for prevention of fire spread provided by the Test Subjects. This implies that the information provided by the Test Subjects had a degree of effect on decision making and consensus building in the group discussions.

c) The Analysis of the System and Subject Effectiveness from Video and Voice Record Data

In order to clarify the utilization of the System and the actions of the Test Subjects, Video and Voice Recorders were used at each group to collect data DURING WS as shown in *Figure 4*. From these data, from the series of discussions, we picked up each unique discussion and summed it up as 1 discussion.

Table 8 shows the total number of discussions, the number of discussions in which the system was used and their ratio. From this table, the following about the system use in Akumi district is evident: (1) 1/4 of the discussions in the group with Test Subjects utilized the system. (2) Compared to the group without Test Subjects, the system use in the group with Test Subjects is about 2.5 times higher.

From the above, it is clear that there was an active use of the system in the group with Test Subjects. Aside from this, the system use in Akumi by the group with the Test Subjects when compared to the the system use in Futagawa shows a ratio higher by about 10%.This also suggests that there was an active use of the system by the Test Subjects.

Likewise, in order to clarify the actual system utilization, the discussion in which the system was used were picked up, classified and summed up into

Table 8. The draft plan and its contents using the system DURING the WS.

Group	ID	Proposal Contents
A, C	a	Widening roads
	b-I	Widening roads, Establishing park, Removing building, cooperative rebuilding, Change building structure (form wooden to fireproof)
	b-II	Widening roads, Removing building, cooperative rebuilding, Change building structure (form wooden to fireproof, form wooden to fire prevention wooden)
	b-III	Widening roads, Establishing park, Change building structure (form wooden to fireproof, form wooden to fire prevention wooden), cooperative rebuilding
	c	Widening roads, Establishing park,
	d	Widening roads,
	e	Widening roads, Removing building
	f	Widening roads, Change building structure (form wooden to fireproof), Removing building
	g	Widening roads, Establishing park,
	h	Widening roads, Change building structure (form wooden to fireproof) Removing building
B,D	i	Widening roads, Change building structure (form wooden to fireproof),
	j	Change building structure (form wooden to fireproof),
	k	Establishing park, Removing building

the following 3 items:(1) The origin of discussion and target (2) The subsystem used during the discussion. (3) The discussion scene and its contents.

Table 9 shows the results of this classification. From this table, in the groups with Test Subjects, discussions originating from the Test Subjects are about 4 times more than those originating from the other participants. The total number of discussions using the system by group A and C was 166 (*Table 8*) while the number of times of use of the built-up improvement subsystem by group A and C was 72 (*Table 9*). This shows that about 43% of the system use by group A and C was utilizing the built-up area improvement subsystem. Also, the discussion scene and contents of group A and C shows that the use of the system was mainly in the Planning Scene (this is where the residents discuss measures to solve disaster mitigation problems of the area).

From the above, in the scene of Planning, the Test Subjects made effective use of the built-up area improvement subsystem to provided the information which was used to support detailed proposals and decision making.

In addition, when the Akumi and the Futagawa cases are compared, the Futagawa case shows that only 20% of the discussions originated from the

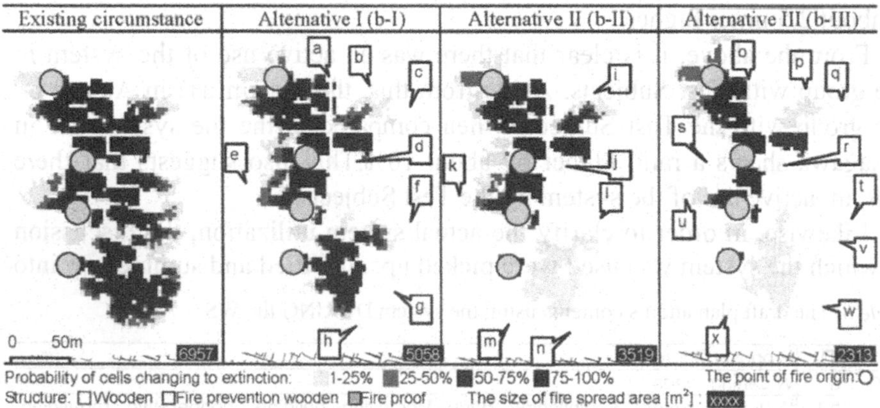

Probability of cells changing to extinction: 1-25% 25-50% 50-75% 75-100% The point of fire origin:○

Structure: □Wooden □Fire prevention wooden ■Fire proof The size of fire spread area [m^2]: xxxx

All simulation parameter settings (wind direction, velocity, simulation end time, fire origin point) are the same.

a,f,l,m,q,t,x: Widening roads, b,e,o: Vacant house to park, c: Removing a vacant house and establishment evacuation route, d,i,k,p,u: Vacant house to Fireproof building, g: Removing a vacant house, h,n,w: Cooperative rebuilding, j,r: Changing building structure from wooden to fire prevention wooden, s,v: Park establishment for prevention of fire spread and as a refuge

Figure 7. A example of log data of fire spread simulation and improvement.

Table 9. The total number of discussion and system use discussion.

Group	Akumi		Futagawa
	A, C	B, D	
(a): The total number of discussion	643	500	530
(b): The total number of system use discussion	166	53	79
(c): The ratio of (b) to (a) [%]	25.8	10.6	14.9

Group A, C: including subjects / Group B, D: not including subjects participating residents and the remaining 80% originated from the group organizer a result similar with the result of the Akumi groups that had no Test Subjects. On the other hand, 45% of all the discussions originated from the participating residents in the groups that had Test Subjects. One reason for this could be that the Test Subjects were able to acquire the skills needed to operate the system through using the system BEFORE the WS and therefore were able to use it effectively to convey and share their thinking during the discussions. Also, the enhanced awareness of the Test Subjects could have been a contributing factor.

Looking at the System functions that were used, built-up area improvement and the improvement effect confirmation subsystems were mainly used by the groups that had Test Subjects while in the Futagawa groups and in the Akumi groups that had no Test Subjects, the use of the above subsystems was minimal. It can be concluded that the use of the system promoted efficient use of time in the limited WS time since more and better detailed opinions came from the groups with the Test Subjects.

In any case, the bottom line would seem to be that the groups with the Test Subject had a more lively discussion with the participating residents as the main players and with not only more opinions but also with more details which were reflected in the final draft plan.

d) The Analysis of the System and Subject Effectiveness from Questionnaire

In order to find out the effectiveness and the problems of the system and the information provided by the Test subject, the results of a questionnaire survey given to the WS Participants was analyzed. The questionnaire survey

Table 10. The classification of the discussion and collected results.

		Group	Akumi		Futagawa
			A,C	B,D	
The origin of discussion and target	From the organizer to residents or test subjects		91	41	61
	From a residents to the organizer or other residents or subjects		16	12	18
	From subjects to the organizer or residents or other resident		59	-	-
The subsystem used during the discussion	Geographic information related Community-based planning for disaster mitigation providing subsystem		30	10	22
	Fire spread simulation subsystem		22	11	31
	Built-up area improvement subsystem		72	5	10
	Improvement effect confirmation subsystem		33	16	16
	Combined use of subsystems		9	11	-
The discussion scene and its contents	Understanding	Question/Providing topics	22	7	17
		Explanation/Answer	8	2	4
		Confirmation/Agreement	14	6	6
	Summarization	Question/Providing topics	1	0	0
		Explanation/Answer	0	0	0
		Confirmation/Agreement	0	0	0
	Planning	Question/Providing topics	65	9	28
		Explanation/Answer	37	18	13
		Confirmation/Agreement	19	11	11

Group A, C: including subjects/Group B, D: not including subjects outline is as shown on *Table 9*. Also, part of the contents of the questions and the summed up result are shown on *Figure 8*. Hereinafter, the analysis and discussion of the results of the questionnaire survey will be done.

The Survey answer sheets was divided into those from groups with the Test Subject, without the Test Subject and Futagawa and the proportion of the response to each question in the Survey summed up as shown on the graphs in *Figure 9* . This figure shows that the percentage of participants who answered "Extremely useful" and "Useful" was about 90% in each question Q1-10. Judging form the result, it is conceivable that the Map subsystem contributed to the understanding of the existing circumstances and the livening up of the discussion to some degree. In addition, it can be presumed that Fire Spread Simulation Subsystem and built-up area improvement Subsystem were useful in enhancing the awareness of fire spread risk under the existing circumstances, enhancing the awareness of the need for improvement of the local environment, verifying the restraining effect by the improvement and in making consensus-building during the discussions.

When compared to the group without Test Subjects and the Futagawa groups, the system rating of the group with Test Subjects was highly favourable of the system. This could be as a result of the effective use of the system by the Test Subjects. On top of that, the questions Q12-14 shown on *Table 10* were given to the group with Test Subjects. From the response, it is clear that the information provided by the Test Subjects promoted a more lively discussion, more detailed opinions and generally supported a smoother group consensus building.

4.3 Results of System Use AFTER Workshop

As with the interview Survey conducted BEFORE the WS, an interview survey with the items showed on *Table 5* was conducted with the aim of better understand how the System was used AFTER the WS and to find out the change in the Test Subjects disaster mitigation awareness due to system use. In addition to, they were also asked whether they used the results from the WS in consideration of the Proposals.

Table 11. Outline of the Questionnaire Survey.

	Akumi	Futagawa
Distribution/collection	The second day, the third day of the Workshop	
Respondent	Participants of the group discussion (residents)	
Respondent's sex	Male: 12, Female: 11	Male: 29, Female: 10
Respondent's age [person]	Under 20: [2], 20 - 60: [11], Over 60: [10]	Under 20: [5], 20 - 60: [12], Over 60: [22]
Question	1) Attribute of the respondent, 2) Evaluation of Map subsystem, 3) Evaluation of Fire Spread Simulation Subsystem and Improvement Subsystem, 4) Overall evaluation of the system	

All Test Subjects from both Akumi and Futagawa districts gave the following responses: (1) the results from the WS were used in reconfirmation of the draft plan. (2) from the fire spread simulation results, there was a realization of the risks in their district under the existing circumstances (3) the need for improvement of the local environment to make it more prepared for disaster was strongly felt.

From this result, even though there was a difference in the degree of awareness, the results from the WS were used in reconfirmation of the draft plan and that the awareness of Test Subjects were heightened through use of the system.

In this experiment, when the system was used AFTER the WS, there was no notable proposal as seen from the interview survey results and the Fire spread simulation and the improvement result data logs. This is because the

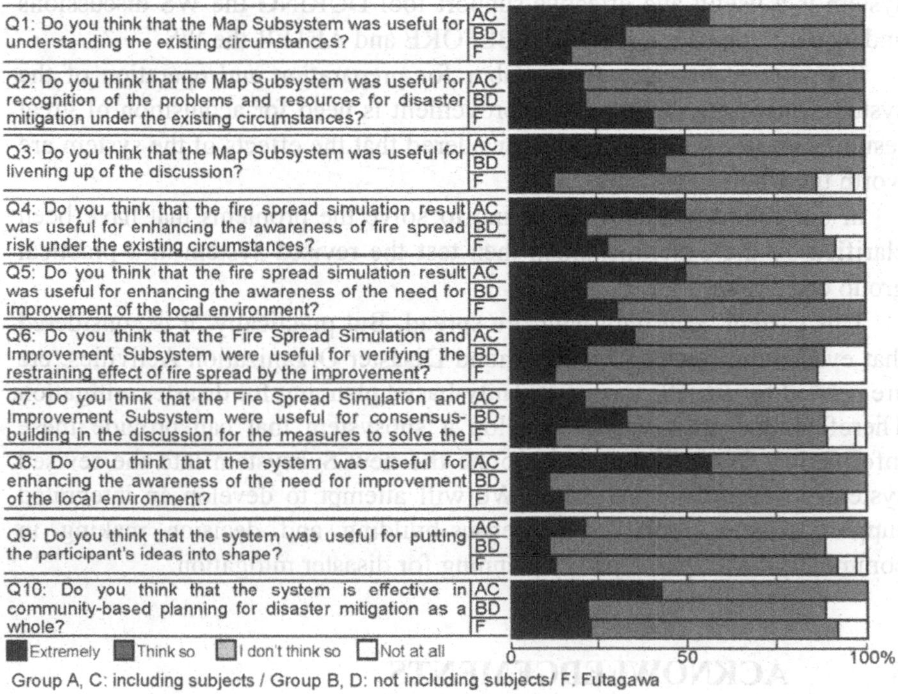

Figure 8. A part of the questionnaires and the summed up results.

Table 12. A part of the questionnaires and the summed up results.

	Extremely useful	Useful	Not very useful	Not at all useful
Q11: Do you think that providing information by subject was useful for living up of the discussion?	3(23)	10(77)	0(0)	0(0)
Q12: Do you think that providing information by subject was useful for concrete discussion?	4(31)	9(69)	0(0)	0(0)
Q13: Do you think that providing information by subject was useful for community-based planning for disaster mitigation as a whole?	3(23)	10(77)	0(0)	0(0)

<Note> Number: Answers. ( ): Component ratio

draft plan proposal made DURING the WS was of high quality with little room for modification according to the residents.

Nevertheless, each Test Subject reviewed the draft plan proposed DURING the WS and acknowledged its reasonableness.

5. CONCLUSION

This research carried out an experiment to verify the usefulness BEFORE, DURING and AFTER the WS of a WebGIS-based system that we have been developing. Based on the data collected from the experiment, the usefulness and advantages of the system were verified from various angles. As a result, it was found that even though there are still some issues to be solved, the system is a useful and effective support tool DURING the WS discussions and also for enhancing awareness BEFORE and AFTER the WS.

It was need to some degree labor for preparation and operation of the system. Therefore, the system improvement is need for acquisition of great results with less work. But, it is considered that the effects of the system are worth the labor.

In our future work, we are going to solve the problems that have been clarified by the experimentation then test the revised system in a practical group discussion.

This system dealt only with fire spread. But practically, it is considered that evacuation routes, Neighbourhood Disaster Organization Activities, etc are needed for an effective community-based planning for disaster mitigation. Therefore, we are going to develop a subsystem that can provide these information. We will then incorporate the new Subsystem into the revised system developed in this paper. We will attempt to develop an integrated support system (tool) for consensus-building and decision making in community-based participatory planning for disaster mitigation.

6. ACKNOWLEDGEMENTS

This research was partially supported by the Ministry of Education, Science, Sports and Culture, Grant-in-Aid for JSPS Fellows.

7. REFERENCES

Batty M and Y Xie, 1994, "Form cells to cities", *Environment and Planning B, Planning and Design*, **21**(5): 531-548.

Gohnai Y., A. Ohgai and et al., 2005, "Development of a WebGIS supporting community-based planning for disaster mitigation integrated with a fire spread simulation model using CA", *9th International Conference, Computer in Urban Planning and Urban Management*, University College London, London.

Jirou K., K. Kobayashi, 1997, "Large area fire", in: Fire Institute of Japan (eds.) *Fire Handbook third edition*, Kyoritsu Publication Co., Ltd., Tokyo, p. 508-573 (in Japanese).

Matubara M. and K. Suzuki, 1996, "Actual conditions of fire", in: Fire Institute of Japan (eds.) *Inspection report concerning a fire in southern part of Hyogo Prefecture earthquake in 1995*, Tokyo, p. 32-39 (in Japanese).

Murosaki Y. and et al., 1984, "Research on spreading factor of an urban expansion fire", *Journal of The City Planning Institute of Japan*, p. 373-378 (in Japanese).

Ohgai A., Y. Gohnai, 2003, "A Fire spread model to support community-based planning for disaster mitigation using cellular automata: application to historical built-up area, Japan" *Proceedings of 4th International Symposium on City Planning and Environmental Management in Asian Countries*, Korea Environment Institute, Soul, Korea, p.259-268.

Ohgai A., Y. Gohnai and et al., 2004, " Simulation and Agent Technology" in: Jos P. Leeuwen and P. Timmermans (eds.), *Recent Advances in Design and Decision Support Systems in Architecture and Urban Planning*, Kluwer Academic Publishers, p. 193-209.

Resnick M., 1995, *Turtles, Termites, and Traffic Jams -Explorations in Massively Parallel Microworlds-*, A Bradford Book, MIT Press.

Sugawara S., T. Naruse, 1997, "Building fire", in: Fire Institute of Japan (eds.) *Fire Handbook third edition*, Kyoritsu Publication Co., Ltd., Tokyo, p. 368-447 (in Japanese).

Wakamatsu T., 1978, "The risk of fire", in: Editorial committee of architectural outline (eds.) *Architecture outline No. 21 theory of architectural fire prevention*, Syokokusya, Tokyo, p. 279-298 (in Japanese).

White R. and G. Engelen, 1993, "Cellular automata and fractal urban form: a cellular modeling approach to the evolution of urban land-use patterns", *Environment and Planning A*, **25**(8): 1175-1193.

White R. and G. Engelen, 1997, "Cellular automata as the basis of integrated dynamic regional modeling", *Environment and Planning B*, Planning Design **24**(2): 235-246.

Xie L. and A. G. Yeh, 2000, "Modelling sustainable urban development by the integration of constrained cellular automata and GIS", *International Journal of Geographical Information Science*, **14**(2): 131-152.

Xie M. and et al., 2001, "Application of cellular automata to the city fire simulation" *Memoirs of Fukui National College of Technology*, **35**: 19-24 (in Japanese).

Yamada A, A. Takizawa and et al., 1999, "The city fire simulation using the cell automaton", *Journal of Architecture Institute of Japan Tokai Chapter*, p. 657-660 (in Japanese).

Yasuno K., Y. Nanba, 1999, "City fire prevention design", in: Horiuchi (eds.) *New edition architectural fire prevention*, Asakurasyoten, Tokyo, p. 178-235 (in Japanese).

Transition Rule Elicitation Methods for Urban Cellular Automata Models

Junfeng Jiao[1] and Luc Boerboom[2]
[1] *Texas A&M University, USA, hkujjf@gmail.com*
[2] *ITC, Enschede, the Netherlands, boerboom@itc.nl*

Keywords: Cellular Automata (CA), Simulation, Modelling, Transition rule, Elicitation

Abstract: In this chapter, transition rules used in urban CA models are reviewed and classified into two categories: transition potential rules and conflict resolution rules. Then, four widely used rule elicitation methods: Regression analysis, Artificial Neural network (ANN), Visual calibration, and Analytical Hierarchy Processing – Multi Criteria Evaluation (AHP-MCE) are discussed. Most of these methods are data driven methods and can be used to elicit the transition potential rules in the urban CA models. In the following, three possible rule elicitation methods: Interview, Document analysis, and Card sorting are explained and demonstrated. These three methods are driven by knowledge and can be used to elicit conflict resolution rules as well as transition potential rules in urban CA models.

1. INTRODUCTION

1.1 Cellular Automata Model

Cellular Automata (CA) were firstly devised by John von Neumann and Stanislaw Ulam in the 1940s as a framework to investigate the logical underpinnings of life. One can say that the "cellular" comes from Ulam and the "automata" comes from von Neumann (Rucker, 1999). CA can generate complex behaviors based on a relatively simple set of rules. It makes them suitable to be applied in complex system simulation, such as urban development, fire, disease spreading, traffic simulation etc. Basic CA, as

Jos P. van Leeuwen and Harry J.P. Timmermans (eds.), Innovations in Design & Decision Support Systems in Architecture and Urban Planning, 53-68.

defined by Ulam, von Neumann, and Wolfram (Wolfram, 1984; Rucker, 1999) have five components: Lattice, Neighborhood, Cell State, Transition rules, and Time. The lattice is the space where CA exist and evolve over time. The neighborhood is the place, where the cells are exactly located. A neighborhood consists of an examined cell itself and any number of cells in a given configuration around the examined cell (Torrens, 2000). Cell state is the status or value, which a cell can take. Basic CA models often have a Boolean cell state, 0 or 1. The transition rule is the control component in the CA model that determines the future cell sate as a function of the current state of a cell and the states of its surrounding cells. Time is the temporal scale in the CA model and is represented in discrete time steps.

1.2 Urban Cellular Automata Model

Since CA models are good at generating different scenarios based on pre-defined criteria or constraints, many researchers have employed CA models to simulate urban development and try to answer different "what-if" questions. These questions include land use dynamics (White and Engelen, 1993, White, Engelen, et al., 1997), regional scale urbanization (Semboloni, 1997; White and Engelen, 1997), poly centricity (Wu 1998), urban spatial development (Wu and Webster, 1998), and urban growth and sprawl (Batty, Xie, et al., 1999; Clarke, Hoppen, et al., 1997).

CA models have demonstrated their ability in urban research, especially for academic study. However, the use of CA in urban simulations often entails substantial departures from the original formal structure of CA described by von Neumann, Ulam, and Wolfram. Although the application of CA to urban systems seems natural and intuitive, this is not in itself sufficient justification for their use (Couclelis, 1985). Basic CA, as defined by Ulam, von Neumann, and Wolfram (Wolfram, 1984; Rucker, 1999), is not well suited to urban applications; since the framework is too simplified and constrained to represent real cities (Torrens, 2000).

In order to simulate an urban system successfully, it is necessary to make some modifications to the basic CA model. In the five components of the basic CA model, transition rules are the most important part. They serve as the algorithms that code real-world behavior into the artificial CA world. In fact, in the context of urban CA, transition rules are responsible for explaining *how cities work* (Torrens, 2000). Different transition rules will generate different simulation results and the simulation precision is mainly determined by transition rules. Therefore accurate elicitation and understanding of transition rules is at the heart of CA modeling.

To that end, we see the need to explicitly differentiate transition rules and consider conflict resolution rules, which we discuss in section 2. Then, in section 3, we review a number of methods to establish transition rules, most of which are data driven. In section 4 we review methods to fire knowledge driven rule elicitation, which are of particular interest to establish conflict resolution rules. Conclusions are drawn in section 5.

2. TRANSITION RULES IN URBAN CA MODELS

2.1 Transition Rules in Basic CA Models

Transition rules specify the behavior of cells between time-steps, deciding the future states of cells. In urban models, states mostly are land-uses. In a strict CA, transition rules are applied uniformly across cells in a synchronous fashion (Torrens, 2000).As argued by Batty (1997) the basic CA rules are formulated with IF, THEN, ELSE sentences and rely on the input from a neighborhood template to evaluate cell state changes. For example, one typical CA transition rule can take the following form:

$$TP_{T+1} = f(S_T, NB) \qquad \text{(Equation 1)}$$

TP_{T+1}— Transition Potential of tested cell in time T + 1

S_T — Tested cell state in time T.

NB—Neighborhood states

In this formula, each cell's Transition Potential in time T + 1 is determined by neighborhood states and its own state in time T.

2.2 Transition Rules in Urban CA Models

To make CA models applicable to urban environments we need to subject transition rules to conflict resolution rules. When applying the CA model to urban systems, many influential factors need to be considered as well as many states a cell can take. The state of a tested cell will not only be affected by the neighborhood effect (e.g. neighboring land uses), but also be affected by other influential factors in the urban system. For example, in most urban CA models, factors such as accessibility and suitability will also be included in transition rules. Thus, the cell's transition potential for instance from rural to urban or from one land use type to another can be calculated as follows.

$$TP_{T+1} = f(S_T, NB. AC, SU...) \qquad \text{(Equation 2)}$$

TP_{T+1}, S_T and NB have the same meaning as Equation 1.

AC—Accessibility effect

SU—Suitability effect

From the above model, people can calculate the transition potential for different states in each of the tested cells. In this paper, the above formula will be called the ***transition potential rule.*** In most available urban CA models, the above rule will be regarded as the transition rule. In these models, the cell's state will only be determined by its transition potential, where the state for which the highest potential was calculated, is the state that the cell will take in the next time step.

In some urban CA models, which aim to simulate the change from non-urban to urban area, such simplicity is accepted, since we only have one objective cell state: urban area. However, if we want to model more precise urban development process, such as the change between different land-use types, the above simplicity is not sufficient because in such a model, a cell's transition potential to different land use types will be different and because the final cell state will not only be determined by its transition potential, but also be affected by other factors.

For example, we have four cells: A, B, C and D. Each of them has two possible cell states, namely residential or industrial land use. Suppose each land use type will require two cells. In this case, the cells' transition potential can be calculated from above equation resulting in for example values found in Table 1.

Table 1. Changing Potentiality of four tested cells.

A	C	Changing Potential to Residential Area	**0.9**	**0.8**	Changing Potentialto Industrial Area	**0.8**	**0.9**
B	**D**		**0.7**	**0.5**		**0.6**	**0.6**

However, to which state the cell A, B, C and D will change is still unknown. If the modeler changes the cell states according to the ranking of **its transition potentiality**, then the above cells will evolve into the following states. See the following table 2.

Table 2. One possible changing results.

A	**C**	Changing states according to their Potential Ranking	**R**	**I**
B	**D**		**R**	**I**

If the modeler uses a different rule, the result of these four cells will be different. For example, if the modeler decides that the residential land use type has changing priority, then the final cell states will be the following.

Table 3. Another possible changing results.

A	**C**	Changing states according to the priority of land use types.	**R**	**R**
B	**D**		**I**	**I**

In this case, although cell C has a higher potential to change into industrial area, yet since the residential land use type has the priority to select the cells and cell C has the second highest changing potential for residential land use type, thus its final state will be residential area. And if two land use types have different land demands, the case will be more complicated.

With the two different approaches the modeler implies that each cell changes to the state with the highest potential, or that each cell changes according to an outside power e.g. in the form of a policy, or financial power, etc.

From this discussion, we can see a cell's final state is not only determined by a cell transition potential value, but also affected by other transition rules. In this paper, these rules which will deal with possible land use conflict will be named *conflict resolution rules*. The whole structure of transition rules of urban CA models can be represented as in figure1. In most urban CA models, the conflict resolution rule has been neglected. The cell's status is only be determined by its transition potential.

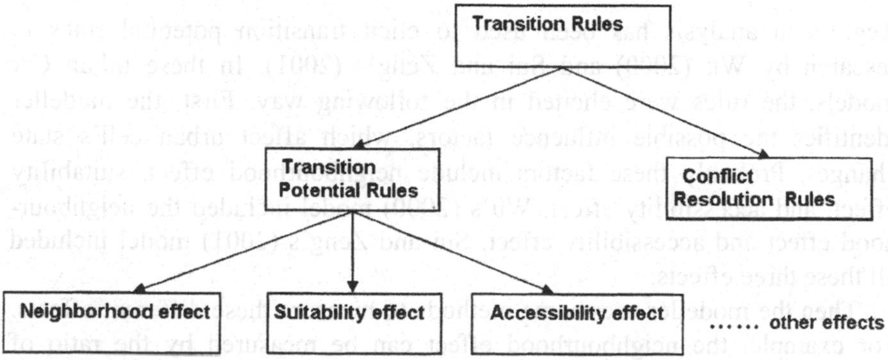

Figure 1. Transition rule structure in urban CA models.

Although in many cases transition potential and conflict resolution rules can be clearly differentiated, situations do exist where they cannot, for instance, if a cell with current state A can possibly change to state B or C, dependent on the concentration of state A cells or concentration of state B or C cells. To illustrate, the potential of a cell of poor quality residential state, to be converted to institutional state or to be converted to higher quality residential state, could be a function of the current concentration of poor quality residential cells, in relation to the current concentrations of institutional and higher quality neighbouring cells. Is that an expression of a neighbourhood effect or is it a conflict resolution rule? Without the distinction it would have been considered a neighbourhood effect, but if powers of lobbies of developers are behind these transitions, conflict rules would be more meaningful. Obviously, with such explicit definition of conflict resolution rules, we seem to treat cells as agents, although we are certainly not talking about agent-based systems.

3. TRANSITION POTENTIAL RULE ELICITATION METHODS, MOSTLY DATA DRIVE, SOME KNOWLEDGE DRIVEN

In this part, four available rule elicitation methods: regression analysis, artificial neural network, visual observation (trial-error) and Analytical Hierarchy Process and Multi-Criteria Evaluation (AHP-MCE) method will be reviewed.

3.1 Regression Analysis

Regression analysis has been used to elicit transition potential rules in research by Wu (2000) and Sui and Zeng's (2001). In these urban CA models, the rules were elicited in the following way. First, the modeller identifies the possible influence factors, which affect urban cell's state changes. Probably these factors include neighbourhood effect, suitability effect, and accessibility effect. Wu's (2000) model included the neighbourhood effect and accessibility effect. Sui and Zeng's (2001) model included all these three effects.

Then the modeller uses some methods to measure these different effects. For example, the neighbourhood effect can be measured by the ratio of developed cells to all neighbourhood cells. The suitability of cells can be calculated by traditional land suitability analysis. And the accessibility of cells can be measured by the distance to different areas (urban centre, main road…) with GIS software.

In the third step, the modeller overlays different land use maps and identified changing areas. Then random samples will be selected from these changing areas. Next, modellers use multiple regression analysis to find explanatory coefficients to the different influence factors. These coefficients will be input in the transition potential rule to calculate the changing potentiality of different cells.

Finally the modeller uses regression analysis to calculate future land demand based on past urban development. The predicted land demand will be used as the threshold to determine how many cells will change states in the simulation process. After the above analysis, the transition potential rule can be represented by the following formula.

$$TP = \sum_{i=1}^{n} CiRi \qquad \text{(Equation 3)}$$

Here, *TP* is the cell's transition potentiality. *Ri* is the influence factor which has been identified and regressed in the above process. *Ci* is its coefficient. This rule elicitation method is a data-driven method without consideration and understanding of people's decision making process, which will affect urban development directly.

3.2 Artificial Neural Network: ANN

In some research, Artificial Neural Networks have been used to elicit the transition rules in urban CA models. Li and Yeh (2001, 2002) have developed their models based on ANN. In their research, a Back-Propagation (BP) neural network was employed, which is good at capturing non-linear characteristics and strong in prediction. Firstly, the authors identified the possible factors which will affect land use change. In Li and Yeh's (2001, 2002) research neighbourhood effects, accessibility effects and suitability effects were identified. Then, some methods were used to measure the above three effects. In the third step, a neural network was formed. Generally, the modeller will select a three-layer network (input layer, hidden layer, and output layer). In such network, each layer will include some neurons. For the input layer, the amount of neurons will be equal to the selected variables. To the output layer, the number of neurons will be same as the predicted land use types. Suppose we want to predict the land use change from rural to urban area, we will have two neurons in the output layer. Neurons in hidden layer are selected based on experience.

After forming the network structure, the modeller still needs to select some functions to link the neurons. In Li and Yeh's case, since the authors used a Back-Propagation (BP) neural network to elicit the transition rules, they selected the functions according to BP network regulation. In the fourth step, a land use change map will be formed by overlaying historical data. Generally, the network will require at least two land use change maps, one will be used to select some random samples to train the network; the other change map will be used to test the formed network. After that, a mature neural network will be formed and can act as the transition potential rule in urban CA models. Finally, the total land consumption in a given period will be calculated from the historical data and be used to control the whole iteration times of the formed network.

However, the above rule elicitation method also inherits some shortcomings, e.g. the black box operation. People don't know what the exact transition potential rules are. Only some data go in and some results come out. This drawback restricts its application in rule elicitation, in which a high interpretative ability is required. Furthermore, how to determine the neural network structure is still under discussion. Finally, such rule elicitation method is also a data-driven method, although good at generating urban configurations, yet lacking understanding of the real urban development process.

3.3 Visual Observation (Trial and Error) Method

Although many researchers have used visual observation to calibrate their models and elicit the transition rules, yet such a method is best demonstrated in the RIKS CA models, which are developed by the Research Institute of Knowledge System (RIKS), in the Netherlands. Generally, RIKS models include two parts: the micro model and the macro model. The former part is developed from CA model and will calculate the transition potential of different cells. The latter part will control the detailed cell state change based on the outside land demand. Here, RIKS models used the visual observation method to elicit the interactions of the different land use types within the neighbourhood, which can be represented by distance curves like the following figure 2. The horizontal value: 1…7 denotes the distance between tested cell and central cell. The vertical value: −100∼+100 reflects the interaction between tested cell and central cell. Based on these interactions, the modeller can describe the neighbourhood effect easily and then develop out the transition potential rules for the models.

In order to get these curves, firstly, the modeller will use some coarse curves to reflect the interaction of different land use types in the case study area. Then, some simulation results will be generated based on these curves. The modeller will compare the simulation results with the real urban land use map and try to adjust the parameter values. After rough adjustment, a new simulation result will be generated, which is then compared with the real land use map. If the accuracy was increased then the parameter will be modified in a subtler step and in the same direction as the first modification. If the simulation accuracy was decreased, it may suggest the first adjustment is in a wrong direction. Clearly, this process is a time consuming process and full of uncertainty, which limits its application greatly.

3.4 Analytical Hierarchy Process and Multi-Criteria Evaluation (AHP-MCE) Method

From the above introductions, we can see these rule elicitation methods are driven by data and lack of the consideration about people’s decision making process, which in the end is the determining force in urban development process. Comparing to these methods, Wu and Webster's (1998) method as it tries to elicit behaviour-driven transition rules. In their model, analytical hierarchy process (AHP) and multi-criteria evaluation (MCE) have been used to capture the characteristics in people's decision making process and the elicited rule can be quantified by pairwise comparison and weighting process.

In detail, the transition rule can be elicited in the following way. Firstly, the modeller identifies some factors which will affect the land development in a case study area. Then he will form a criteria hierarchy to represent the relationship between these factors and the simulation objective. Thirdly, the factors' importance will be determined by pairwise comparison, a weighting process to establish importance of the different factors. This step is very important in the whole rule elicitation process, since in this step, the decision makers' opinions will be represented by their different weight sets. Wu and Webster (1998) simplified this step and determined the factors' weights according to possible planning policies and their own understanding of the urban development.

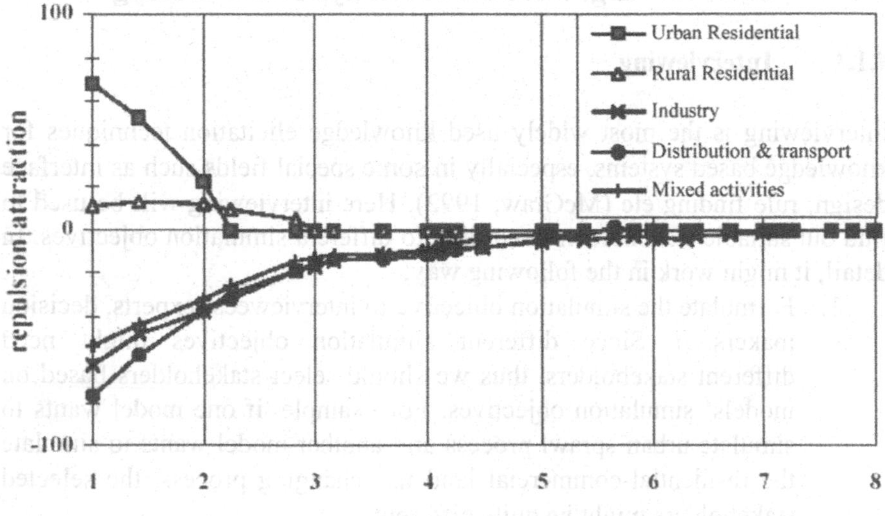

Figure 2. Distance Curves in RIKS models (Engelen, Geertman, et al., 1999).

Comparing with the above data-driven methods, this method is easier to reflect peoples' decision making priorities. And it is possible to generate different urban development scenarios based on different decision making processes. However, similar to the other methods, this method also only focuses on the transition potential rule in the CA models. Some questions such as how to select suitable decision makers and deal with their conflicts about urban developments should be well considered in the future modelling process.

4. POSSIBLE KNOWLEDGE DRIVEN RULE ELICITATION METHODS

Generally speaking, rule elicitation is one kind of knowledge acquisition form, which aims to develop some transition rules based on available knowledge. Burge discussed (1998) many knowledge acquisition methods such as: *Interviewing, Case Study, Protocols, Critiquing, Role Playing, Simulation, Prototyping, Teach back, Observation, List Related, Construct Elicitation, Sorting, Laddering, and Document Analysis*. In the following the authors will use some of the above knowledge acquisition methods to elicit and represent conflict-resolution rules, which are derived from stakeholders' different opinions.

4.1 Interviewing, Document Analysis and Sorting

4.1.1 Interviewing

Interviewing is the most widely used knowledge elicitation techniques for knowledge based systems, especially in some special fields such as interface design, rule finding etc (McGraw, 1992). Here interviewing will be used to find out suitable stakeholders according to different simulation objectives. In detail, it might work in the following way.

1. Formulate the simulation objective to interviewees (experts, decision makers…). Since different simulation objectives might need different stakeholders, thus we should select stakeholders based on models' simulation objectives. For example, if one model wants to simulate urban sprawl process and another model wants to simulate the residential-commercial land use changing process, the selected stakeholders might be quite different.
2. Encourage the interviewees to list some possible stakeholders which are related to simulation objectives.

By this method, possible stakeholders to a given simulation objective might be identified.

4.1.2 Document Analysis

In this step, document analysis will be used to rank and validate the different stakeholders which have been identified in the above step. In detail it might work in the following way.

1. Select corresponding documents, which recorded the recent land use changes in the case study area. The source may come from official or unofficial records.
2. Analyze the land use changes process and try to find out the roles of different stakeholders in such process. The focus might be on those documents, which recorded the happened land use conflict in case study area. For example, in some period, a commercial investor and an industrial investor requested a same land parcel. If finally, the commercial investor won the conflict, it may suggest that the commercial investor had priority in the land conflict process. Based on these historical records, modeler can assign them with different weights or rank them in a sequence order to reflect such difference.

Following the above steps, possible stakeholders related to simulation objective and their relative importance can be identified.

4.1.3 Sorting

In sorting method, domain entities are sorted to determine how the expert classifies their knowledge. The domain expert is presented with a list of entities to be sorted. They are then asked to sort them either using pre-defined dimensions or along any dimension they feel is important. Subjects may be asked to perform multiple sorts, each using a different dimension (Burge 1998). Here, sorting will be used to elicit transition potential rule. The detailed working processes will be the following.

1. Write down the possible measurement factors such as, soil types, terrain, slope, and distance to urban center…on the cards and present them to the selected stakeholders. Make them clear and understandable. Also suggest the stakeholders list some new measurement factors.
2. Formulate some sub-effects which will be included in the transition potential rule, such as the neighborhood effect, accessibility effect, suitability effect, planning influence, and social-economic influence ...
3. Let the selected stakeholders sort the measurement factors according to these sub-effects. For example, soil type might be sorted to reflect the suitability effect. Distance factors will be sorted to reflect the accessibility effect…In this step the stakeholders can also form new

sub-effect. Require these stakeholders to sort the cards as possible as they can until they cannot find some new categorizing methods. Finally require the stakeholders to explain why they sort the cards in this way.

By card sorting, modeler can identify suitable factors to measure the transition potential rule in urban CA models.

4.2 Demonstrating the Rule Elicitation Process

In the following part, based on different simulation objectives, the above rule elicitation methods will be demonstrated.

4.2.1 Simulate the Change from Non-Urban to Urban Area

If the modelers want to simulate urban expansion in case study area, the transition rule might be elicited in the following way.

1. By interviewing, a modeler can identify the important stakeholders in the past urban expansion process. Suppose the modeler found a residential investor and a commercial investor are the main driving forces in the urban expansion process. Then some document analysis will be done, which aims to find out the investors' rankings in the urban development process. Here, the focus will be put on those documents, which recorded the past land use conflict in the case study area. By analysis of these conflicts, the investors' relative importance can be identified. Ideally this importance should be quantified quantified in the form of a weight: ***WR*** and ***WC***. If more than 2 important stakeholders have been identified in such process, their weights can also be attained through this method.
2. Let the identified stakeholder select suitable factors to represent the transition potential rules. Here card sorting will be used. And finally different stakeholder probably will form different transition potential rules.
3. Use the Analytical of Hierarchy Process (AHP) method to form two different hierarchy structures of transition potential rules: ***TPr, TPc***. ***TPr*** comes from residential investor and ***TPc*** comes from commercial investor.
4. Use the pairwise comparison and weighting method to quantify the different factors in these two hierarchical transition rules; and use the following formula to calculate the transition potential rules: ***TPr and TPc***. Here take ***TPr*** as an example.

TPr = w1*neighborhood effect + w2*suitability effect + w3*accessibility effect... (Equation 4)

= w1*deve-density + w2*(wa*slope + wb*terrain + wc*soil) + w3*(wd*Dist-Road + we*Dist-Center)... (Equation 5)

5. Use the firstly assigned weights of residential investor and commercial investor ***WR*** and ***WC*** to assemble these two hierarchical transition rules (***TPr*** and ***TPc***) as ***TPwhole***. Its structure can be found in the following figure 3.

TPwhole = WR* TPr + WC*TPc (Equation 6)

(Here Deve-density means the development density, which can be measured by the ratio of developed cells to the total cells in the neighbourhood. We can use this method to reflect neighbourhood effect.)

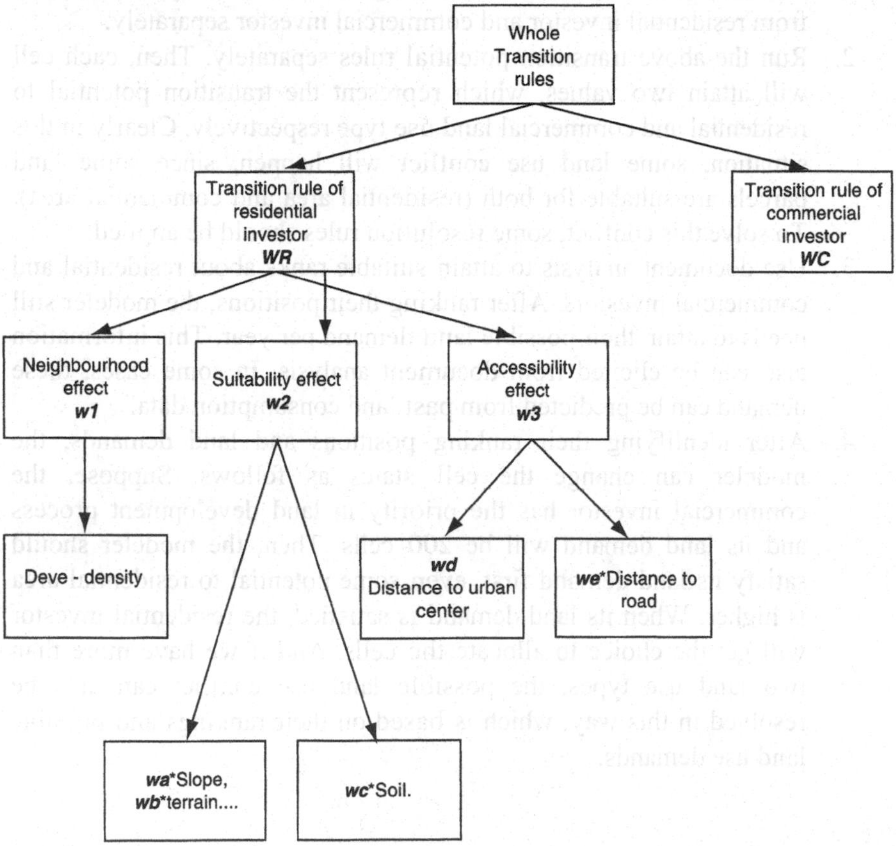

Figure 3. Combined Transition Rule elicited from residential and commercial investors.

In this model, the possible land use conflict between residential investor and commercial investor has been modeled in the above formula by their different weights ***WR*** and ***WC***. The cell's changing potential from non-urban to urban area can be calculated from the above formula. The total amount of cells, which developed in urban area, will be controlled by the predicted land consumption.

4.2.2 Simulate the Inter-Changing Among Different Land Use Types

In the above example, the modeler only has one final cell state: urban area, thus it's possible to combine the transition potential rule and conflict resolution rule together. If modelers want to simulate the change among different land use types, the conflict resolution rule should be considered separately. The rule elicitation process might be the following.

1. Use the above step 2 and step 3 to get the transition potential rules from residential investor and commercial investor separately.
2. Run the above transition potential rules separately. Then, each cell will attain two values, which represent the transition potential to residential and commercial land use type respectively. Clearly in this situation, some land use conflict will happen, since some land parcels are suitable for both (residential area and commercial area). To solve this conflict, some resolution rules should be applied.
3. Use document analysis to attain suitable ranks about residential and commercial investors. After ranking their positions, the modeler still needs to attain their possible land demand per year. This information also can be elicited from document analysis. In some cases, these demand can be predicted from past land consumption data.
4. After identifying their ranking positions and land demands, the modeler can change the cell states as follows. Suppose, the commercial investor has the priority in land development process and its land demand will be 200 cells. Then, the modeler should satisfy its land demand first, even some potential to residential area is higher. When its land demand is satisfied, the residential investor will get the choice to allocate the cells. And if we have more than two land use types, the possible land use conflict can also be resolved in this way, which is based on their rankings and possible land use demands.

5. CONCLUSION AND DISCUSSION

In this paper, we have argued that a distinction should be made between transition potential rules and conflict resolution rules. The transition potential rules can calculate the transition potential of different cells. The conflict-resolution rules will determine the cell's final state in CA models, which is of particular interest for those cases where a cell has high potential for multiple purposes. In most CA models, conflict resolution rules have been ignored and the cell's state is determined by its transition potential only.

After defining the concept of transition rule in urban CA models, four rule elicitation methods were discussed. They are Regression analysis, ANN, Visual comparison (trial and error) and AHP-MCE method. The former two methods are data driven methods, which try to elicit the transition rules from historical data. The visual comparison method is very time consuming and full of uncertainty. Comparing with these three methods, AHP-MCE is more hopeful, and aims to develop the transition potential rule based on people's decision making process. However, all these 4 rule elicitation methods focus on the transition potent rules.

Therefore, we also discussed three possible rule elicitation methods: interview, document analysis and sorting. These three methods are knowledge driven methods and intended to elicit the conflict resolution rule as well as the transition potential rule. Finally, these rule elicitation methods have been demonstrated according to different simulation objectives.

Acknowledgment: The authors would like to thank three anonymous reviewers for their valuable suggestions on this paper.

6. REFERENCES

Batty, M.,1997, "CA and Urban Form: A primer", *Journal of the American Planning Association* **63**(2): 266-274.

Batty, M., Y. Xie, and Z. Sun,1999. "Modelling urban dynamics through GIS-based cellular automata", *Computers, Environment and Urban Systems* **23**(3): 205-233.

Burge, J. E., 1998, *Knowledge Elicitation for Design Task Sequencing Knowledge*, Master thesis, Worcester Polytechnic Institute, Worcester, Mass, USA.

Clarke, K. C., S. Hoppen, and Gaydos, L., 1997, "A self-modifying cellular automaton model of historical urbanization in the San Francisco Bay area." *Environment and Planning B* **24**(2): 247-261.

Couclelis, H., 1985, "Cellular worlds: a framework for modelling micro-macro dynamics", *Environment and Planning A* **17**(5): 585-596.

Engelen, G., S. Geertman, P. Smits, and C. Wessels, 1999, "Dynamic GIS and Strategic Physical Planning Support: a practical application to the IJmond/Zuid-Kennemerland region", in: Stillwell, Geertman, and Openshaw (eds.) *Geographical Information and Planning*, Sringer, p. 87-111.

Li, X. and A. G.-O. Yeh, 2001, "Calibration of cellular automata by using neural networks for the simulation of complex urban systems." *Environment and Planning A* **33**(8): 1445-1462.

Li, X. and A. G.-O. Yeh, 2002. "Neural-network-based cellular automata for simulating multiple land use changes using GIS." *International Journal of Geographical Information Science*, **16**(4): 323-343.

McGraw, K. L., 1992, "Review: Knowledge Elicitation for User Interface Design", *Designing and Evaluating User Interfaces for Knowledge Based Systems-* Chapter 4: Ellis Horwood Limited, p. 45-62.

Rucker, R. 1999, *Seek! Selected Nonfiction*, New York. Four Walls Eight Windows, New York.

Semboloni, F., 1997, "An urban and regional model based on cellular automata", *Environment and Planning B*, **24**(4): 589-612.

Sui, D. Z. and H. Zeng, 2001, "Modelling the dynamics of landscape structure in Asia's emerging desakota regions: a case study in Shenzhen." *Landscape and Urban Planning* **53**(1-4): 37-52.

Torrens, P. M., 2000, "How cellular automata models of urban system work. (1.theory)" *Centre for Advanced Spatial Analysis Working Paper Series 28*, University of College London, London.

White, R. and G. Engelen, 1993, "Cellular automata and fractal urban form: a cellular modelling approach to the evolution of urban land use patterns", *Environment and Planning A* **25**(8): 1175-1199.

White, R. and G. Engelen, 1997, "Cellular automata as the basis of integrated dynamic regional modelling," *Environment and Planning B* **24**(2): 235-246.

White, R., G. Engelen, and I. Uijee, 1997, "The use of constrained cellular automata for high-resolution modelling of urban land use dynamics", *Environment and Planning B* **24**(3): 323-343.

Wolfram, S. (1984) "Cellular automata as models of complexity," *Nature*, 31(4), p. 419-424.

Wu, F., 1998, "Simulating urban encroachment on rural land with fuzzy-logic-controlled cellular automata in a geographical information system." *Journal of Environmental Management* **53**(4): 293-308.

Wu, F., 2000, "A Parameterised Urban Cellular Model Combining Spontaneous and Self-Organising Growth", in: P. Atkinson and D. Martin (eds.) *Geocomputation: Innovation in GIS 7*, Taylor & Francis, London, p. 73-85.

Wu, F. and C. J. Webster, 1998, "Simulation of land development through the integration of cellular automata and multi-criteria evaluation." *Environment and Planning B* **25**(1): 103-126.

A Method for Estimating Land Use Transition Probability Using Raster Data

Considerations about apatial unit of transition, fixed state of locations, and time-varying probability

Toshihiro Osaragi and Yoshitsugu Aoki
Tokyo Institute of Technology

Keywords: Markov chain, Land use, Transition probability, Building lot, Fixed state, Time-varying probability

Abstract: In the field of urban and regional planning, several Markov chain models for land use conversion have been proposed. However, some problems have been encountered when estimating land use transition probabilities. In this paper, a new estimation method to determine land use transition probability is proposed by taking into account spatial units of land use transition, fixed state of locations, and varying transition probabilities. The effectiveness of the proposed methods and some new findings on land use conversion are presented using numerical examples.

1. INTRODUCTION

In recent years, the use of models based on multi-agent systems (MAS) and cellular automata (CA) are in vogue in many interdisciplinary research fields, including urban planning and urban management, and built-in decision support systems (McClean and Watson, 1995; Torrens, 2002; Arentze and Timmermans, 2003; Saarloos, Arentze, et al. 2005). The strengths and weaknesses of myriad land use and land cover change (LUCC) modeling approaches have been examined, and MAS/CA models have been offered as a means of complementing other techniques. One promising class of models designed to simulate and analyze LUCC might be the MAS/CA models. However, some of the major challenges encountered in these models were also indicated (Parker, Manson, et al. 2003). Most MAS/CA models are, by

Jos P. van Leeuwen and Harry J.P. Timmermans (eds.), Innovations in Design & Decision Support Systems in Architecture and Urban Planning, 69-84.

nature, interdisciplinary. Therefore, it is obvious that we should draw on and combine knowledge from many disciplines in order to develop creative new tools for empirical analysis.

A large body of basic research using Markov chain models exists in LUCC studies (Drewett, 1969; Bourne, 1971; Bell, 1974; Bell and Hinojosa, 1977; Robinson, 1978; Jahan, 1986; Ishizaka, 1992; Muller and Middleton, 1994; Theobald and Hobbs, 1998; Qihao, 2000). Markov models may be combined with CA for LUCC modeling, as evidenced by joint CA-Markov models (Li and Reynolds, 1997; Balzter, Braun, et al. 1998). Although this paper addresses fundamental problems hidden in the Markov chain models, the challenges shown in this paper are common and partly overlap those of MAS/CA models. In particular, the question has evolved from "Which is the best?" to "What are the conditions under which it is the best, and how can we flexibly combine the appropriate approaches on a case-by-case basis?"

Employing Markov chain models to predict the distribution of land use is always plagued with several types of errors. One type stems from the uncertainty, which is inevitably inherent in the transition matrix. Since each coefficient in the matrix refers to one sub-area, an error of this type is directly related to the manner in which the sub-areas are formulated. Therefore, the authors traced the manner in which such error was generated and developed methods for estimating and decreasing it. The latter involved the division of a total area into appropriate sub-areas by using a simulated neural network (Aoki, Osaragi, et al. 1993, 1996).

Furthermore, Markov chain models have predominantly focused on providing information regarding the amount, location, and type of LUCC that has occurred. Only a few models have been developed to address the manner in which and the reasons for change (Qihao, 2002). Then, the authors described land use transition probabilities by using the concept of land use utility, and measured and evaluated the effects of socioeconomic factors on land use utility (Osaragi and Kurisaki, 2000).

Thus, several methods have been proposed to sharpen Markov chain models. However, there still remain several types of errors in estimating transition probabilities. In the previously proposed models, land use transition probability was typically estimated using time series raster data by counting the number of cells. Although this method is very simple, there are several problems associated with it, as described below.

First, it has been widely recognized in geography that there exist common scale-related problems related to the verification and validation of MAS/CA models. Nevertheless, we should take into consideration that the basic unit of land use change is not a cell but a building lot. Adopting the conventional method entails the risk of overlooking the true transition structure (Yoshikawa, 1994).

Second, in CA, the system is homogeneous such that the set of possible states is the same for each cell and the same transition rule applies to each cell (Parker, Manson, et al. 2003). This feature corresponds to the assumption underlying Markov chain models (Stewart, 1994). One basic assumption is the consideration of LUCC as a stochastic process. However, there are some locations where land use is in a fixed state. In other words, land use transition at such locations should not be considered as a stochastic process. Hence, the conventional model, which assumes that land use transition at all locations will vary stochastically, might yield an incorrect estimate.

Finally, while cellular modeling techniques offer greater flexibility for representing spatial and temporal dynamics, these dynamics are based on stationary transition probabilities (Parker, Manson, et al. 2003). The models proposed thus far assume that the transition probability will remains Table in the future, except in a few instances where it has been tested (Bourne, 1971; Bell, 1974; Bell and Hinojosa, 1977). Therefore, it is necessary to consider the variations in the transition probability in terms of time.

Hence, this study addresses the abovementioned issues and improves upon previous models by proposing the following methods:

(1) A method for estimating land use transition probabilities based on building lots using raster data.

(2) A method that takes into account the existence of building lots in a fixed state.

(3) A method for expressing changes in transition probability based on the concept of land use utility.

2. LAND USE TRANSITION PROBABILITY BASED ON BUILDING LOTS

2.1 The Order of the Markov Chain

The order of the Markov chain has only been formally tested in a few studies (Bell, 1974; Robinson, 1978). In several Markov chain models, the "first-order Markov chain" property is simply assumed to simplify the theoretical discussion. In other words, it is assumed that "the state at time t+1" is dependent only on "the state at time t." In order to ensure the first-order Markov chain property, an investigation was carried out using the actual land use raster data that has been used in this study (Osaragi, 2005). Since the result showed that the land use transition process could be adequately described by a first-order Markov model, this simple property is assumed in the following discussion.

2.2 Transition Probability Based on Building Lots

In conventional land use transition models, the probability P_{ij} for a land use change from j to i is estimated using the following equation:

$$P_{ij} = \frac{m_{ij}}{\sum_k m_{kj}}, \tag{1}$$

where m_{ij} is the total area of all locations where the land use category changes from j to i during a certain time interval. In other words, the transition probability was estimated with the assumption that all the spots (cells of raster data) varied independently. However, the actual spots do not change independently; rather, all the spots included within the same building lot generally changed together. This implies that the spatial unit of land use change is a building lot. A simple example of the difference between the transition probability matrices based on cells and building lots is shown in Figure 1.

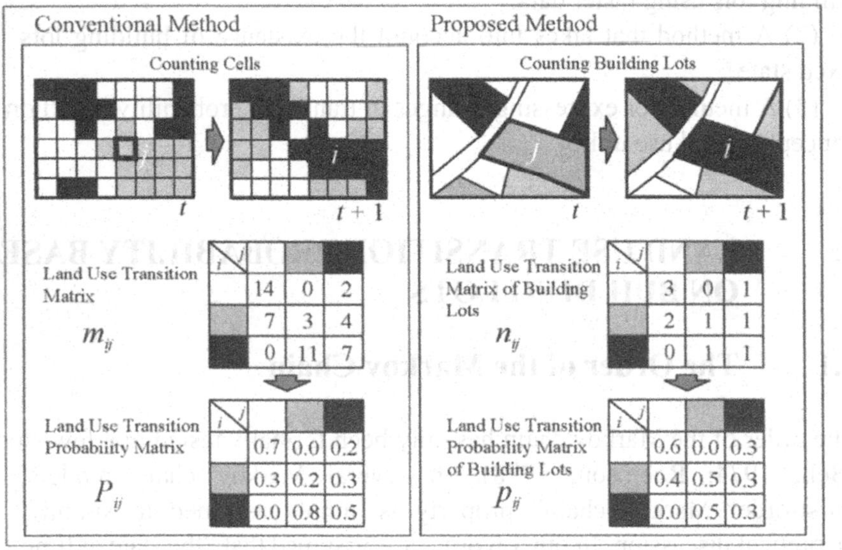

Figure 1. Example of difference in transition probability estimated by counting cells and building lots.

The transition probability matrices obtained are evidently distinct. Therefore, in order to accurately ascertain the structure of the transition, it is necessary to estimate the transition probability by a method based on building lots. Thus, the above-mentioned conventional model (equation (1)) should be transformed into the following enhanced equation.

$$p_{ij} = \frac{n_{ij}}{\sum_k n_{kj}}, \tag{2}$$

where n_{ij} denotes the total number of building lots where the land use category changes from j to i. In this study, a transition matrix composed of n_{ij}, which is called the "transition matrix of building lots," is utilized, and its probabilistic expression is referred to as the "transition probability matrix of building lots."

2.3 Estimation Method of Transition Probability of Building Lots

The transition matrix of building lots should be estimated using vector data of building lots. However, it is relatively simple to synthesize raster data, and large amounts of such data have been collected to date. Assuming efficient and effective use of existing raster data, a method for estimating the transition probabilities of building lots using raster data is proposed. The number of building lots where the land use category changes from j to i, denoted by n_{ij}, can be estimated using the following equation.

$$n_{ij} = \frac{m_{ij}}{a_{ij}}, \tag{3}$$

where m_{ij} denotes the total area of the building lots where the land use category changes from j to i, and a_{ij} denotes the average area. The average area a_{ij} of building lots in the transition from j to i can be estimated by the method shown in Figure 2. If adjacent cells of the raster data display identical transitions, they are considered to constitute one building lot. Therefore, the average area a_{ij} of the building lots in transition from j to i is estimated by the following equation.

$$a_{ij} = \frac{m_{ij}}{n'_{ij}}, \tag{4}$$

where n'_{ij} is the number of building lots estimated by the method shown in Figure 2. However, the number of building lots where the land use category does not change cannot be determined by this method. Therefore, the value of the area a_{jj} is assumed to be equal to the average area of all the building lots that change from j to i ($i = 1,\ldots, n$).

If the values of a_{ij} and a_{jj} are calculated from the raster data using the above-mentioned method, the transition matrix of the building lots can be estimated by equations (2) and (3).

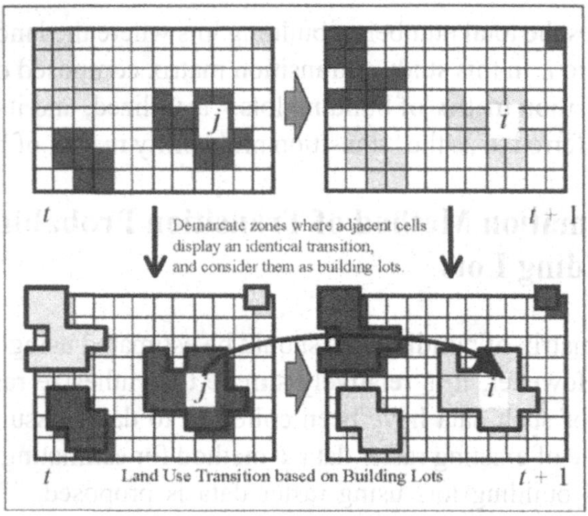

Figure 2. Estimation method of average area of building lots.

2.4 Features of the average area of building lots

The average area a_{ij} of the building lots is estimated using actual land use data (Table 1) of the Tokyo metropolitan area. The estimated values of the average area are presented in Table 2. The average area is comparatively large in forests, land under development, vacant lots, and industrial lots. On the other hand, the average area of private residential lots is relatively small. Furthermore, in order to examine the characteristics of the average area of the building lots in detail, the average area is calculated based on the time distance from the center of Tokyo (CBD). The result is shown in Figure 3. The average area based on pre-conversion is shown in Figure 3 (a). The average area of land under development exhibits a high value at 20 minutes from the CBD. This indicates that large-scale development is in progress in the littoral district. The average area of forests/fields increases with distance from the CBD. On the other hand, the average area of residential and commercial lots does not depend on the distance from the CBD, and it remains almost constant. Additionally, this value is considerably smaller than that of the other land use categories. The average area post-conversion is shown in Figure 3 (b). The average area under each category tends to increase slightly depending on the distance from the CBD. In other words, although the average area based on pre-conversion in the CBD and the suburbs is almost equal, larger building lots tend to be formed only after land use conversion. Therefore, the average area of the building lots is closely related to the pattern of land use conversion and varies depending on the distance from the CBD.

Table 1. Classification of land use data.

No.	Detailed Classification	Classification
1	Forest, Wasteland	Forest/Field
2	Rice field	
3	Field	
4	Land under development	Vacant lot
5	Vacant lot	
6	Industrial lot	Industrial lot
7	Regular residential lot	Residential lot
8	Densely populated residential lot	
9	High-rise residential lot	
10	Commercial lot	Commercial lot
11	Road	Public lot
12	Park, Green area	
13	Public facility lot	

The Detailed Digital Land Use Data (1974, 1979, 1984, 1989)
Cell Size: 10 m by 10 m

Table 2. Average area of building lots in the process of land use transition.

j \ i	1	2	3	4	5	6	7	8	9	10	11	12	13
1	9.9	9.7	9.0	26.3	4.9	6.0	2.8	1.4	1.9	3.6	1.7	4.3	5.6
2	7.6	7.1	11.6	12.8	4.2	11.6	2.1	1.3	1.1	2.8	1.6	2.1	3.5
3	8.5	9.1	5.2	16.7	5.4	4.1	2.0	1.4	1.7	2.0	1.5	3.6	1.8
4	61.6	40.3	20.0	18.0	28.9	24.9	5.7	2.8	6.3	6.2	2.6	19.2	14.2
5	9.9	8.5	5.3	37.7	4.9	6.6	2.5	1.9	2.8	3.3	1.5	4.3	3.9
6	10.7	6.7	5.8	27.9	7.3	7.4	3.1	1.8	2.0	6.4	1.5	5.7	8.1
7	4.8	3.4	4.1	13.0	3.8	3.5	2.6	2.1	3.7	2.2	1.4	2.7	1.8
8	4.4	5.4	4.6	4.7	1.8	6.9	2.6	1.9	2.9	1.9	1.5	1.5	1.3
9	12.3	6.1	6.3	44.8	9.3	14.2	5.6	2.6	3.0	4.5	1.7	7.4	3.9
10	8.2	5.7	5.3	30.6	5.8	10.5	2.8	2.0	2.7	3.4	1.5	6.1	5.3
11	2.8	2.4	2.4	4.2	2.2	2.6	1.9	1.6	2.0	2.5	1.6	3.7	2.7
12	15.3	11.5	9.1	61.2	15.5	14.6	2.6	1.8	3.8	5.6	2.0	5.2	12.7
13	19.5	13.4	10.3	85.0	8.5	13.6	3.3	2.0	2.9	5.7	2.2	11.9	4.4

(x 100 square meters)
ID numbers of land use categories of *i* and *j* are corresponding to ID numbers in Table 1.

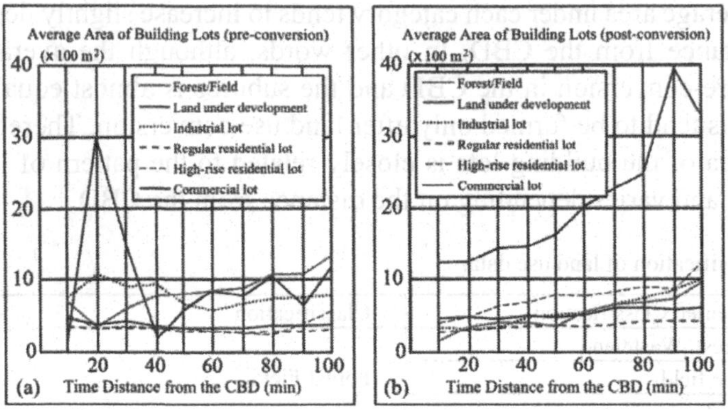

Figure 3. Average area of building lots in the process of land use transition based on time distance from the CBD.

3. LAND USE TRANSITION MODEL CONSIDERING A FIXED STATE

3.1 Formulation of the Model

The necessity and validity of a land use transition model that considers a fixed state of land use has already been examined (Osaragi and Masuda, 1996). However, in the previous model, the transition probability was assumed to be constant and was estimated by counting the cells with raster data. Therefore, the model is improved by discarding the assumption of constancy (time-invariance) and adopting an estimation method based on building lots. According to the previously used method, the number of building lots of land use i at time $t + 1$ is denoted by $x_i(t + 1)$ and is expressed as follows:

$$x_i(t+1) = \sum_j p_{ij} x_j(t), \tag{5}$$

where p_{ij} denotes the transition probability shown in equation (2). However, as described earlier, there may be certain locations where land use does not vary as a stochastic process. A national park with strict development restrictions is a typical example of such a location. Moreover, although forests adjacent to housing sites may be stochastically developed into residential lots, the interior mountainous regions are likely to remain in a fixed state for a long time. This implies that if the building lots in the study area are also considered in this study, the above-mentioned model should be expressed as follows:

$$x_i(t+1) = s_i + \sum_j q_{ij}(t)\left(x_j(t) - s_j\right), \tag{6}$$

where s_j denotes the total number of building lots in the fixed state, and $q_{ij}(t)$ is the transition probability of building lots estimated by excluding the building lots in the fixed state. Moreover, since this transition probability is expected to vary with time, the description of the transition probability should include a suffix for time t. Thus, the estimator of the transition probability $q_{ij}(t)$ is expressed by the following equations:

$$q_{ij}(t) = \frac{n_{ij}(t)}{\sum_k n_{kj}(t) - s_j} \quad (i \neq j), \tag{7}$$

$$q_{jj}(t) = \frac{n_{jj}(t) - s_j}{\sum_k n_{kj}(t) - s_j}, \tag{8}$$

where $n_{ij}(t)$ is the total number of building lots where the land use category changes from j to i during the time interval t to $t + 1$. In addition, the total number of building lots where the land use category j did not change during the time interval t to $t + 2$ is denoted by $r_j(t)$. The estimator of $r_j(t)$ can be expressed using $q_{jj}(t)$ as follows:

$$r_j(t) = s_j + \left(\sum_k n_{kj}(t) - s_j\right) q_{jj}(t) q_{jj}(t+1). \tag{9}$$

The following equation is obtained when $q_{jj}(t)$ and $q_{jj}(t + 1)$ are eliminated using equations (8) and (9):

$$s_j = \frac{r_j(t)\sum_k n_{kj}(t+1) - n_{jj}(t) n_{jj}(t+1)}{r_j(t) + \sum_k n_{kj}(t+1) - n_{jj}(t) - n_{jj}(t+1)}. \tag{10}$$

Therefore, if the values of $n_{ij}(t)$, $n_{ij}(t + 1)$, and $r_j(t)$ are estimated using time series data, the number of building lots in the fixed state s_j can be estimated. Furthermore, the transition probability of building lots $q_{ij}(t)$ and $q_{jj}(t)$ can be estimated using equations (7) and (8). The framework of the estimation method is shown in Figure 4.

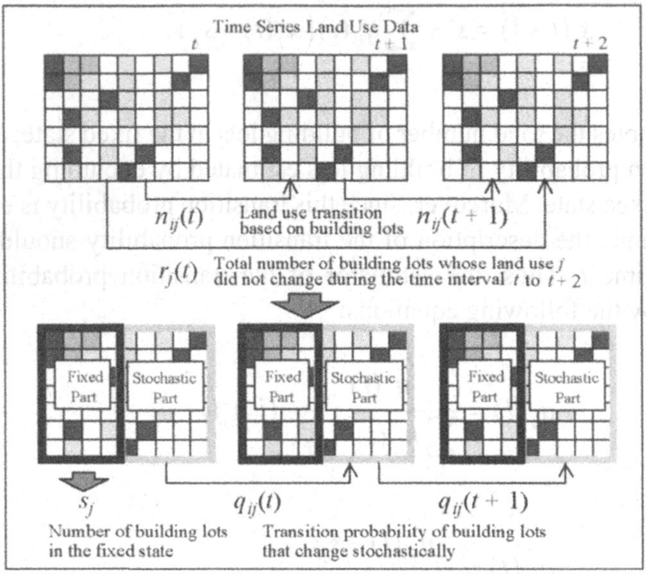

Figure 4. Framework of estimation method for transition probability considering building lots in the fixed state.

3.2 Building Lots in the Fixed State

The actual land use data are analyzed using the land use transition model described earlier and the number of building lots in the fixed state is estimated. If the areas undergoing large-scale development are considered within the study area, the estimated value of the average area becomes unstable. Therefore, the building lots that change by 1 *ha* or more are excluded from the study.

To begin with, the proportion of building lots in the fixed state is estimated, and the results are presented in Table 3. The areas of the lands under development and the vacant lots are unlikely to remain in the fixed state, in contrast with the other land use categories. Next, in order to examine the spatial distribution of building lots in the fixed state, the proportion of building lots in the fixed state is estimated based on the time distance from the CBD. A part of the result obtained is shown in Figure 5. The area under the vacant lot land use category does not depend on the time distance but has a low, almost constant value. On the other hand, the proportion of residential and commercial lots is high in the CBD and low in the suburbs. Although changes in the land use category are frequent in the CBD, the possibility of land use conversion is actually low. In other words, this type of land use is particularly stable in the CBD.

Table 3. Proportion of building lots in the fixed state.

No.	Detailed Classification	Proportion of building lots in the fixed state
1	Forest, Wasteland	0.811
2	Rice field	0.785
3	Field	0.668
4	Land under development	0.175
5	Vacant lot	0.381
6	Industrial lot	0.693
7	Regular residential lot	0.719
8	Densely populated residential lot	0.791
9	High-rise residential lot	0.689
10	Commercial lot	0.530
11	Road	0.694
12	Park, Green area	0.439
13	Public facility lot	-

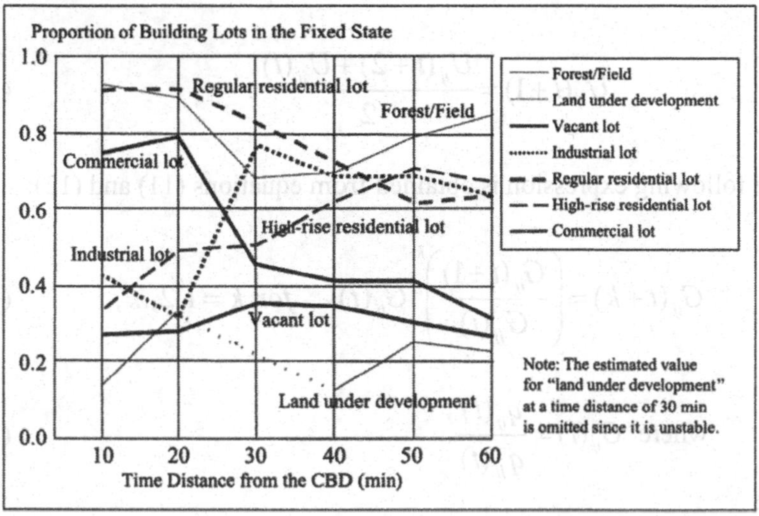

Figure 5. Proportion of building lots in the fixed state based on time distance from the CBD.

4. TIME SERIES CHANGE IN THE TRANSITION PROBABILITY

4.1 Formulation of the Model

The conventional Markov chain model assumes time-invariance transition probabilities. However, this assumption may not be realistic. Therefore, a method for describing time series variations in the transition probability that is based on the concept of land use utility is proposed. The utility, which is to be obtained by changing the land use category from j to i at time t, is denoted by $U'_{ij}(t)$. The expression $U'_{ij}(t)$ is divided into the deterministic term $U_{ij}(t)$ and the probabilistic term $e_{ij}(t)$. If it is assumed that the distribution of $e_{ij}(t)$ follows the Gumbel distribution, the following logit model is obtained:

$$q_{ij}(t) = \frac{\exp[U_{ij}(t)]}{\sum_k \exp[U_{kj}(t)]}, \tag{11}$$

where $q_{ij}(t)$ is the transition probability at time t. If the land use utility is assumed to vary linearly with time, the relationship between land use utility at three consecutive instances of time, i.e., $U_{ij}(t)$, $U_{ij}(t+1)$, and $U_{ij}(t+2)$ can be expressed as follows:

$$U_{ij}(t+1) = \frac{U_{ij}(t+2) + U_{ij}(t)}{2}. \tag{12}$$

The following expression is obtained from equations (11) and (12):

$$G_{ij}(t+k) = \left(\frac{G_{ij}(t+1)}{G_{jj}(t)} \right)^k G_{ij}(t) \quad \textit{for } k = 1, 2, \ldots, \tag{13}$$

$$\text{where } G_{ij}(t) = \frac{q_{ij}(t)}{q_{jj}(t)}. \tag{14}$$

Since the sum of all terms of $q_{ij}(t)$ in equation (14) is equal to 1, the diagonal component of the transition probability matrix at time $t + k$ is expressed by the following equation using $G_{ij}(t + k)$:

$$q_{jj}(t+k) = \frac{1}{\sum_k G_{kj}(t+k)}. \tag{15}$$

Similarly, the non-diagonal component can be obtained from equation (14) as follows:

$$q_{ij}(t+k) = \frac{G_{ij}(t+k)}{\sum_k G_{kj}(t+k)}. \tag{16}$$

In other words, if two sets of transition probability matrices are estimated using time series data, and $G_{ij}(t)$ and $G_{ij}(t+1)$ are expressed based on equation (14), the land use transition matrix $q_{ij}(t+k)$ at time $t+k$ $(k = 1, 2,\ldots)$ can be estimated from equations (13), (15), and (16).

4.2 Validation of the Model

The proposed model is validated using actual data. The validation method of the model is shown in Figure 6. The transition probabilities $q_{ij}(1)$ and $q_{ij}(2)$ of the building lots that change stochastically can be estimated using the time series data for 1974, 1979, and 1984 by the methods described in the previous section. The transition $q_{ij}(3)$ for the period 1984–1989 is estimated by the method presented above. The value of the transition probability $q_{ij}(3)$ is compared with the corresponding value directly estimated from the time series data of 1984 and 1989. The result is shown in Figure 7(b). In addition, in order to examine the time series variation in the transition probability, the transition probability $q_{ij}(3)$ is compared with $q_{ij}(1)$, and the result is shown in Figure 7(a). As shown in Figure 7(a), the variations in the transition probability occur within a certain time interval. This implies that the assumption of the constancy of the transition probability is not a realistic one. If the transition probability is estimated by the proposed method, the values estimated would be more accurate. According to the above-mentioned model, it is possible to examine the time series change in the transition probability. However, the model is not sufficiently robust: some large deviations can be observed in Figure 7(b).

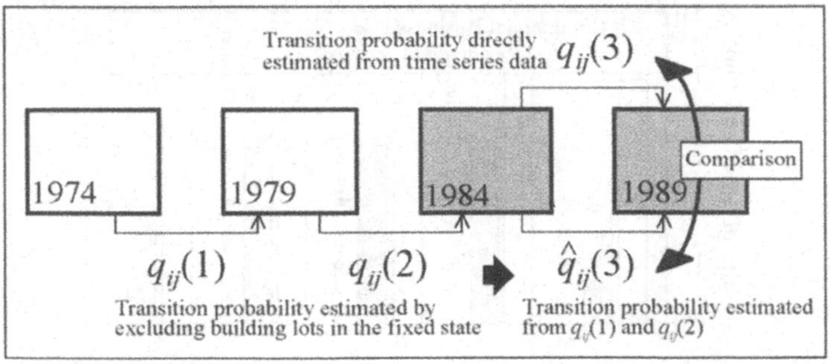

Figure 6. Validation method of model describing change in transition probability.

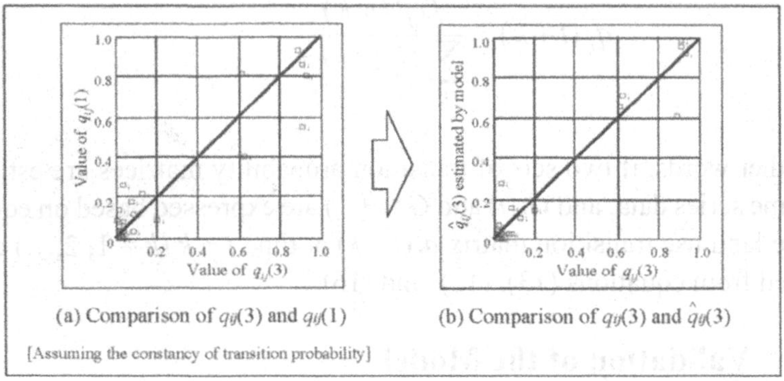

Figure 7. Validation of model describing change in transition probability.

This shortfall seems to result from not having considered the characteristics of the location, such as land use zoning. Land use utility can hardly be considered homogeneous over the entire city. Since land use utility varies depending on the characteristics of different locations, transition probability also varies similar. We therefore examine the variation in land use utility based on the characteristics of different locations. The model that measures land use utility is applied (Osaragi and Kurisaki, 2000), and the influence of land use zoning on land-use utility is examined. The results are shown in Figure 8. In the first and second residential zoning, the utility for residential use is high and that for commercial or business use is rather low. On the other hand, in the commercial zoning, this tendency is reversed. Land use utility varies significantly with the characteristics of locations, which in turn affect the land use transition probability, as indicated in equation (11). It is, therefore, necessary to classify the building lots based on the characteristics of the locations and estimate the value of the transition probability of these building lots (Aoki, Osaragi, et al. 1996).

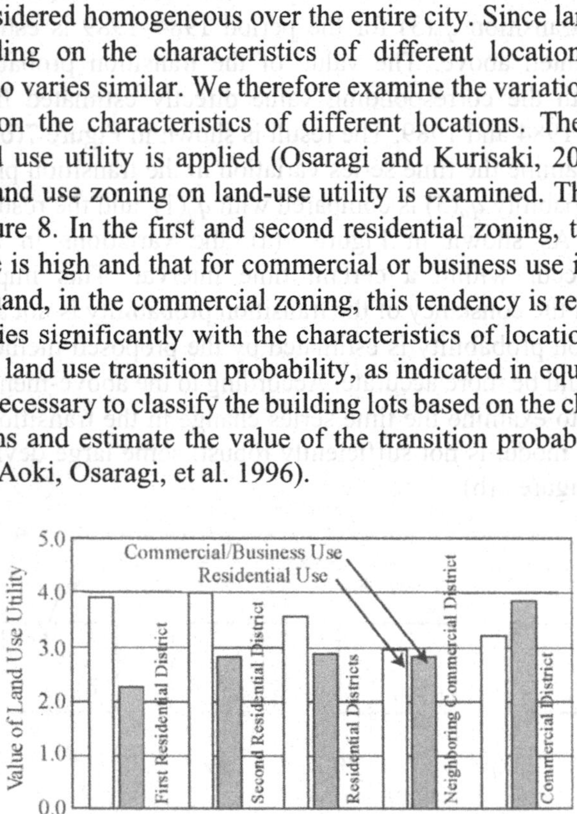

Figure 8. Difference of land use utility according to land use zoning.

5. SUMMARY AND CONCLUSIONS

In this study, certain inherent problems in the conventional method of estimating land use transition probability were discussed and new estimation methods were proposed. First, the need to estimate the transition probability based on building lots was explained. Assuming efficient use of the existing raster data, we developed a method to estimate transition probabilities, based on building lots, from raster data. We then proved that the transition probability should be estimated by excluding the locations in a fixed state in order to obtain an accurate transition structure. We proposed a method by which the number of locations in the fixed state could be estimated, and then discussed the spatial distribution of these locations. To conclude, we proposed a method to estimate time series changes in land use transition probability that is based on the concept of land use utility. The value of land use utility that determines the transition probability is influenced by the characteristics of the locations, such as land use zoning.

6. REFERENCES

Aoki, Y., A. Nagai, and T. Osaragi, 1993, "Area dividing method by neural network", in: R. E. Klosterman and S. French (eds.) *Proceedings of Third International Conference on Computers in Urban Planning and Urban Management, Georgia Institute of Technology*, 2, p. 379-392.

Aoki, Y., T. Osaragi, and A. Nagai, 1996, "Using of the area-dividing method to minimize expected error in land use forecasts", *Environment and Planning B: Planning and Design*, **23**: 655-666.

Arentze, T. and H. Timmermans, 2003, "A multiagent model of negotiation processes between multiple actors in urban developments: a framework for and results of numerical experiments", *Environment & Planning B: Planning & Design*, **30**(3): 391-410.

Balzter, H., P. W. Braun, and W. Kohler, 1998, "Cellular automata models for vegetation dynamics", *Ecological Modelling*, **107**(2/3): 113-25.

Bell, E. J., 1974, "Markov analysis of land use change: an application of stochastic processes to remotely sensed data", *Socio-Economic Planning Sciences*, **8**: 311-316.

Bell, E. J. and R. C. Hinojosa, 1977, "Markov analysis of landuse change: continuous time and stationary processes", *Socio-Economic Planning Sciences*, **11**: 13-17.

Bourne, L. S., 1971, "Physical adjustment processes and land use succession: a review and central city example", *Economic Geography*, **47**: 1-15.

Drewett, J. R., 1969, "A Stochastic Model of the Land Conversion Process", *Regional Studies*, **3**: 269-280.

Ishizaka, K., 1992, "A consideration on analyzing method of land use transition matrix", *Transaction of Architectural Institute of Japan*, **436**: 59-69 (in Japanese).

Jahan, S., 1986, "The determination of stability and similarity of Markovian land use change processes: a theoretical and empirical analysis", *Socio-Economic Planning Sciences*, **20**: 243-251

Li, H. and J. F. Reynolds, 1997, "Modeling effects of spatial pattern, drought, and grazing on rates of range land degradation: A combined Markov and cellular automaton approach", in: D. A. Quattrochi and M. F. Goodchild (eds.) *Scale in remote sensing and GIS, NewYork*, Lewis Publishers, p. 211-230.

McClean, C. J. and P. M. Watson, 1995, "Land use planning: A decision support system", *Journal of Environmental Planning & Management*, **38**(1): 77-92.

Muller, R. M. and J. Middleton, 1994, "A Markov model of land-use change dynamics in the Niagara region, Ontario, Canada", *Landscape Ecology*, **9**: 151-157.

Osaragi, T. and K. Masuda, 1996, "Land use forecasts considering the existence of steady state places", *Theory and Applications of GIS*, **4**(2): 1-6.

Osaragi, T. and N. Kurisaki, 2000, "Modeling of Land use Transition and Its Application", *Geographical and Environmental Modelling*, **4**(2): 203-218.

Osaragi, T., 2005, "Land-use transition model based on building lots considering steady state of locations and inconstancy of transition", in: C. A. Brebbia et al. (eds.) *Sustainable Planning 2005*, p. 77-86.

Parker, D., S. Manson, M. Janssen, M. Hoffmann, and P. Deadman, 2003, "Multi-agent systems for the simulation of land-use and land-cover change: A review", *Annals of the Association of American Geographers*, **93**(2): 314-337.

Qihao, W., 2002, "Land use change analysis in the Zhujiang Delta of China using satellite remote sensing, GIS and stochastic modelling", *Journal of Environmental Management*, **64**(3): 273-284.

Robinson, V. B., 1978, "Information theory and sequences of land use: an application", *The Professional Geographer*, **30**: 174-179.

Saarloos, D., T. Arentze, A. Borgers, and H. Timmermans, 2005, "A multiagent model for alternative plan generation", *Environment & Planning B: Planning & Design*, **32**(4): 505-522.

Stewart, W. J., 1994, "*Introduction to the numerical solution of Markov chains*", Princeton, NJ, Princeton University

Theobald, D. M. and N. T. Hobbs, 1998, "Forecasting rural land-use change: a comparison of regression and spatial transition-based models", *Geographical and Environmental Modelling*, **2**: 65-82.

Torrens, P. M., 2002, "Cellular automata and multi-agent systems as planning support tools", in: S. S. Geertman and J. Stillwell (eds.) *Planning support systems in practice, London*, Springer-Verlag, p. 205-222.

Yoshikawa, T., 1994, "A stochastic model of lot land use transition based on multinomial probability distribution", *Comprehensive Urban Studies*, **53**: 113-121 (in Japanese).

Linking Land Use Modelling and 3D Visualisation

A mission impossible?

J.A.M. Borsboom-van Beurden[1], R.J.A. van Lammeren[2], T. Hoogerwerf[2], and A.A. Bouwman[1]

[1] *Netherlands Environmental Assessment Agency,* [2] *Wageningen University*

Keywords: Land use models, 3D visualisation, Policy-making

Abstract: Additional to the traditional land use maps 3D visualisation could provide valuable information for applications in the field of spatial planning, related to ecological and agricultural policy issues. Maps of future land use do not always reveal the appearance of the physical environment (the perceived landscape) as a result of land use changes. This means that 3D visualisations might shed light on other aspects of changed land use, such as expected differences in height or densities of new volume objects, or the compatibility of these changes with particular characteristics of the landscape or urban built environment. The Land Use Scanner model was applied for the Netherlands Environmental Assessment Agency's "Sustainability Outlook" to explore land use changes, followed by GIS analyses to asses both the development of nature areas and the degree of urbanisation within protected national landscapes. Since it was felt that 3D visualisation could complement the resulting land use maps, the land use model output was coupled to 3D visualisation software in two different ways: 1) through Studio Max software in combination with iconic representation of the concerned land use types and 2) through 3D components of GIS software. However, the use of these techniques on a national scale level for the generation of semi-realistic 3D animations raised a number of conceptual and technical problems. These could be partly ascribed to the particular format and of the Land Use Scanner output. This paper discusses the methods and techniques which have been used to couple the output of the land use model to 3D software, the results of both approaches, and possible solutions for these problems.

Jos P. van Leeuwen and Harry J.P. Timmermans (eds.), Innovations in Design & Decision Support Systems in Architecture and Urban Planning, 85-101.

1. LAND USE MODELLING AND 3D VISUALISATION FOR POLICY APPLICATIONS

A number of current policy issues in The Netherlands benefit from supporting information from land use models and 3D visualisation. Urbanisation pressure, high in the central part of The Netherlands, is not expected to drop in the coming decades. Besides the persistent quantitative housing shortage that can be ascribed to population growth and household fission, there is also a qualitative shortage, since the demand for low density residences in a rural or semi-rural environment has not been met at all. Besides, a number of traffic problems in this congested part of the Netherlands have to be addressed. At the same time further scale enlargement in Dutch agriculture in the countryside is common since revenues are increasingly under pressure by further liberalisation of world trade and reform of the EU Common Agricultural Policy. In traditionally strong sectors such as greenhouse horticulture, cattle breeding and dairy farming, this scale enlargement is regarded as a precondition for continuing one's business in the long term, which requires further rationalisation of agricultural practices. Other policy issues of importance are the extension of existing nature areas to a national ecological network, the introduction of explicit spatial measures in water management, the restructuring of intensive cattle breeding areas and the designation of national landscapes by the National Spatial Strategy of the Ministry of Housing, Spatial Planning and the Environment (VROM et al. 2004).

It is clear that the developments sketched above all bear consequences for the physical environment. For this reason, insight in the effects of long term demographical, societal, technological and economic developments is needed to formulate robust policies anticipating potential problems in the future. Maps of future land use are one of the means used by the Netherlands Environmental Assessment Agency to support policy-making. These maps are usually based on the geodata output of land use simulation. A number of environmental, ecological and spatial (EES) effects however are hardly interpretable from these maps: for example the effects of low density residential development on open landscapes or the effects of scale enlargement in agriculture. A change of colour in such a map only depicts the transition to a different type of land use. For this reason 3D visualisation has been explored in this study as an additional means to communicate these EES effects to policy-makers.

But what are the requirements for such an application? Most existing 3D applications in the field of operational policy-making, for example for urban lay-outs or infrastructural constructs give very precise information about projects. They show highly detailed designs at well defined locations and are often meant for giving information on short term changes. The output of land use models, in favour of strategic planning, however consists usually of rather coarse grid cells and it concerns the long term. This implies that the level of detail and time scale of the output of the land use model and its 3D visualisation should be congruent. It has mainly to give an impression of changes, to show new built-up areas, but not necessarily newly built houses in detail. In addition, it is important to know the geographical location of the changes. And finally, current land use and land use in future should be shown in one view so they can be compared.

From the first part of this section it follows that 3D visualisation can offer information that cannot be provided solely by land use maps. Therefore 3D visualisation can be helpful in stakeholder discussions on environmental and ecological effects of land use changes. A number of scholars have discussed the potential contribution of maps to a comparable policy domain, that of participatory spatial planning (PSP), for example Ball (2002), Appleton and Lovett (2003, Tress and Tress (2003) and Hoogerwerf (2005b). Geertman (2002) considers geo-visualisation in general as helpful for understanding and communicating future planning initiatives and changes. Appleton and Lovett (2005) and Bloemmen et al. (2005) draw attention to the connecting function of maps and 3D visualisation in heterogeneous stake holder groups. These often consist of participants who are familiar with conceptual visualisations during the different stages of the planning procedure and participants who are not trained in working with such visualisations and have no skills to apply these. Often the interpretation of map information demands too high a cognitive level to read and to understand the meaning and impact of the conceptual cartographic visualisation (see Zube et al. 1987, Krause 2001). An efficient flow of information, discussion and communication between these two main groups could be supported by new means, methods and techniques.

For PSP, it is believed that visualisation of geodata by means of 3D computer techniques can contribute to more efficiency, as it adheres closely to the normal human perception of real world objects which makes it easier to read and understand, see for example the Virtual London project of Hudson-Smith et al. (2005) aiming to improve public participation. It is obvious that in other policy domains than PSP the representation and recognition of 2D maps can benefit as well from 3D visualisation. Dijkstra et al. (2003) discuss the incorporation of panoramic views in the evaluation of office space design alternatives and found that the use of virtual reality technology improved the reliability and validity of modelling consumer preference and choice. In this paper we will discuss a particular example where geo-referenced grid based land use model output is processed in 3D visualisation software in order to support environmental, ecological and landscape policy-making.

2. 2D BY 3D

In geo-information science different types of geodata are usually distinguished from different types of visual representation (MacEachren 1999 and Chen 1999). Table 1 shows how geodata types and their visual representations can be combined in various ways. Each combination requires a number of varying transformations to come from geodata to visual representation.

Table 1. Combinations of geodata types and visual representation types, after Lammeren et al. (2004) and Bishop et al. (2005).

	Visual representation			
Geodata	**2D**	**2D + ΔT**	**3D**	**3D + ΔT**
2D	*Land Use Scanner*		*Visual Scan application*	
2D + ΔT				
2.5D			*Landscape Feature approach*	
2.5D + ΔT				
3D				
3D + ΔT				

In the first column the available geodata structures are shown. The ΔT points at time series that are delivered by spatio-temporal simulation models. The 2D, 2.5D and 3D refer to subsequently 2-dimensional referenced data (2D), digital elevation models (so-called 2.5D) and three-dimensional referenced, including three-dimensional topology, as known from computer aided design software (3D). The next four columns represent the various types of visual representations of geodata. Two-dimensional geodata may be visualised in a two- (2D – traditional cartography) or a three-dimensional (3D) way. The ΔT in the second row points at the implicit or explicit animation of the visualisation, an extension of the concept of dynamic visualisation by Zube et al. (1987). The adjective explicit is used to show that the end user of the visualisation cannot influence interactively the temporal and projection parameters of the animation (e.g. a video). An implicit animation offers the user tools to animate the application interactively via adaptations of temporal and projection parameters (e.g. the user can decide upon the view path and the speed of movement).

Most of the Dutch land use simulation models as Environmental Explorer, CLUE and Land Use Scanner provide an output as 2D and 2D + ΔT geodata in the form of grid cells, for example Groen et al. (2004), Klijn et al. (2004), Nijs et al. (2005). Maps of current and future land use show the location of changes but not in which way they have changed. As said in the previous section, if a representation of these 2D changes can be put in a virtual world, it will significantly add to the power of imagination of changing land use because they can convey more information about the landscape comparing to 2D presentations and thereby needs less effort from the user to interpret (DiBiase 1990, Bishop 1994, Lange 2001).

When 2D and 2D + ΔT geodata is transformed to a 3D visual representation, the requested level of detail in the 3D visualisation has to be chosen, depending upon the purpose of the visualisation. Details of objects visible in the real world can have useful representation in the virtual 3D world but only if the distance to the objects is not too far. If the distance view is larger, the details of the objects are not noticeable anymore. Networks of roads, patterns of fields, shapes of towns, structure of the land then serve mainly as information for orientation. It has to be decided which objects to show in detail or how to simplify them depending on the purpose of visualisation. Enhancing some details, which are easily recognizable by users, such as landmarks, will improve orientation and recognition. Farther distance views give the possibility of showing a larger part of the terrain at once, but at a certain viewpoint altitude the representation reaches the point where the proportion of the heights of 3D terrain objects is too small and objects will not be visible in 3D anymore (Momot 2004). The next section will discuss in more detail the chosen approaches in this particular research.

3. LAND USE MODEL OUTPUT

The main goal of the Spatial Impressions project at the Netherlands Environmental Assessment Agency was to explore possible patterns of land use in the year 2030 and to portray these changes with maps and other means of visualisation as illustrations and 3D animations. The underlying spatial processes were derived from scenarios developed for the Sustainability Outlook. In each scenario a different quality of life is desired, and different means are used to realise this quality. In line with the IPCC scenario (IPCC 2000) scenarios were constructed by using the scenario axes technique which at first selects the two most important sources of uncertainties in future. The trend towards globalisation or regionalisation and the trend towards solidarity or efficiency were selected. The resulting quadrants, named Global Economy (A1), Global Solidarity (B1), Safe Region (A2) and

Caring Region (B2), were filled in with qualitative storylines describing societal values, demographical and economic development, technology, governance, energy and food supply and mobility (MNP 2004).

The construction of accompanying maps of future land use consisted of a number of consecutive steps. First Delphi rounds were organized to question thematic specialists about the supposed underlying spatial processes. From specialised models, e.g. on housing, firm location and agriculture, information about the future demand for space was derived. This information was then translated to land use claims, allocation rules and suitability maps which were put in the Land Use Scanner model (see Borsboom-van Beurden 2005) This is a multi-nomial logit model that estimates the probability that a certain type of land use is allocated in a particular grid cell. Land use has been modelled in 28 classes and grid cells of 500 by 500 meter. In the Land Use Scanner version that was applied here, one grid cell may contain several types of land use. Figure 1 shows an example of the model results for the central part of The Netherlands.

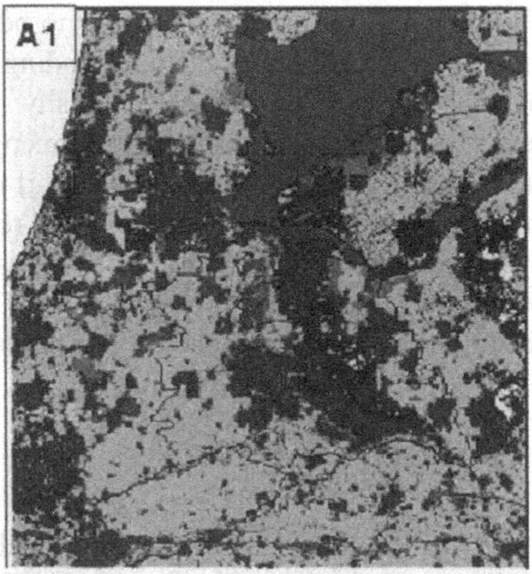

Figure 1. Land use in 2030 according to Global Economy (A1) scenario (Borsboom-van Beurden et al. 2005).

4. 3D SOLUTIONS

To create a 3D-visualisation of the 2D scenario output different options do appear. In fact the options include a transformation of grid-based 2D geodata into a 3D semi-realistic visual representation like presented by figure 2. In this study semi-realism was chosen because it was expected that this level would suit the policy discussions the best. This is in line with Hoogerwerf (2005b) who studied professionals′ preference for realism of visualisations for participatory planning by an internet survey and found that especially the semi-realistic visualisation type is highly suitable for most of the use types and participation level.

Figure 2. Left semi-realism, upper-right conceptual, lower right photo-realism (Hoogerwerf 2003).

Transformation involves the creation of various 3D-objects and linkage to the original grid data. So the geo-referenced data, still two-dimensional, are visualised by help of related 3D objects. The original Land Use Scanner output consisted of grid cells of 500 by 500 metres containing a mix of several land use types. Because grid cells of this size were considered too coarse for 3D visualisation, it was decided to disaggregate the data to cells of 100 meters by 100 meters for a reference data set. It has to be mentioned that the spatial configuration of 3D-objects within one grid of 100 by 100 meters for future situations was not known. For that reason 3D-objects have been designed and by use of these 3D-objects so-called 3D-models or Land Use Icons (LUI) of the 28 land use types are constructed. Each LUI represents a landscape 'stamp' which represents the imagined landscape for a specific type of land use. To conclude, these LUI are the basis for a semi-realistic representation and are intentionally meant to be stand alone 3D-models.

Originally the LUI have been constructed by using 3D Studio Max 2. Every 3D-model has its own geometrical definition. The geometrical definition can range from very simple to complex. The bitmaps used for creating a semi-realism are also ranging from one simple to more combined and complicated bitmaps (figure 3). The last ones are used when semi-realism could not be visualised by the geometry as such.

All the LUI have been designed. It means that each of them is based on a certain imagination of the (future) configuration of the physical space taking into consideration the expression of the land use type adapting to certain social and technological developments. Developing icons means a design task; a land use type in a particular scenario could be visually expressed by a specific LUI. In this example the rural icons have been based on contemporary expressions of nature and agricultural activities, the infrastructural icons too.

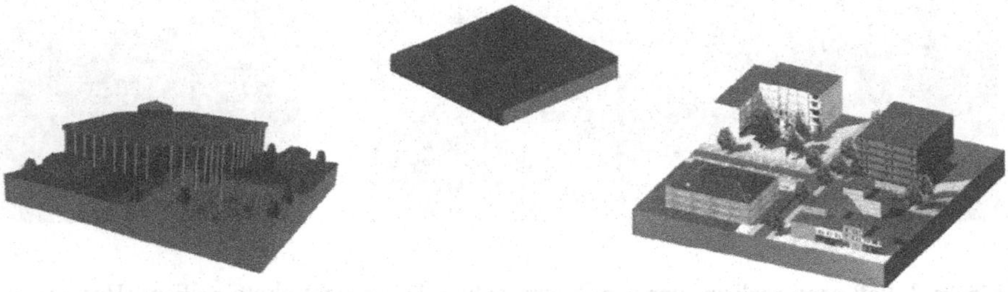

Figure 3. LUI: Simple geometry and bitmap (upper), complex geometry and simple bitmaps (left under), complex geometry and bitmaps (right under) (Lammeren et al. 2004).

The land use model output has been coupled to 3D visualisation software in two distinctive ways, namely the Visual Scan application and the Landscape Feature approach. This first option applies data base principles and operates via modified game-tools technology to translate the two-dimensional grid data into a three-dimensional visualisation. Via the developed application each of the different grid cell values can be connected to a dedicated 3-dimensional land use icon (LUI) by a conversion of the dominant land use class into an RGB-colour value. This RGB-colour value served as a unique key value to the Virtual Tools viewer. The transformation of the scenario output grid delivers a so-called LUI-link map. The LUI-link map is a plain BMP-file. First the application links the LUI-link map (e.g. a red coloured grid cell representing urban area in figure 4) to the related LUI, which represents a kind of land use. Afterwards the application renders the composition of LUI into a 3D-scene (a virtual reality model). This scene we call the VisualScan Scene.

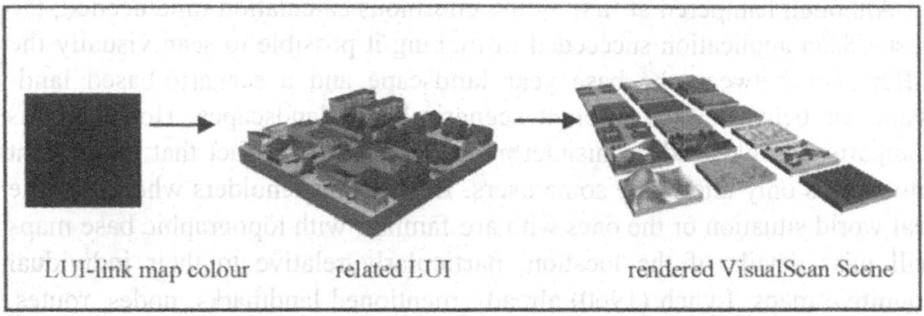

Figure 4. Phases from 2D data into 3D visualisation (Lammeren et al. 2004).

The application developer can set the initial state of all objects in the 3D scene, but Virtools allows also to dynamically change object properties by writing scripts using Building Blocks (BB's) (figure 5). There are more than 450 different BB's included in the programme, but advanced developers may also write their own using Visual Scripting Language (VSL), a C++ based scripting language. The BB's have *behavioural Inputs* (bIn's), which are used to receive a link, *and behavioural Output* (bOut's) to start a link to another BB. This type of link is therefore also called a *behavioural link* (bLink). A Building Block generally receives or transmits specific data, which is provided by the *parameter links (pLink)*, which connects to the top (called pIn) and connect from the bottom (pOut) of the Building Block respectively. There are many parameter data types available, and one can switch between many of them using parameter operations. A script is thus a collection of ordered and linked Building Blocks that is attached to either the scene, to a 3D object, or to any other element in the developer, such as camera's, lights and avatars, and is used to change the properties of these elements at run-time.

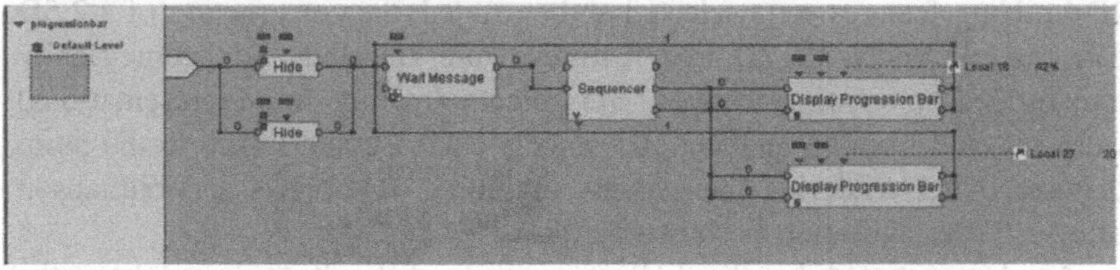

Figure 5. A Virtools script The Building Blocks (BB's) are represented by the squares, the *order of* processing by the behavioural links between them. The numbers depicted with these links indicate the delay at which the link is processed, in this case all BB's are calculated in the same frame (link delay = 0), and in all subsequent frames.

Although hampered at first by the enormous calculation time needed, the VisualScan application succeeded in making it possible to scan visually the differences between the base year landscape and a scenario-based landscape, or between two different scenario-based landscapes. However this comparison could lead to misinterpretations due to the fact that the current situation is only known by some users. Besides, stakeholders who know the real world situation or the ones who are familiar with topographic base maps will miss details of the location, particularly relative to their individual cognitive maps. Lynch (1960) already mentioned landmarks, nodes, routes, edges and areas as primary features of relevance of which a cognitive (mental) map consists of. In other words, the visualisation model needs to meet the users' expectations and knowledge about an area. When the 3D scene would have a too abstract 3D object configuration and for that reason differs too much from the cognitive maps, it creates difficulties for recognition and interpretation.

In the LUI approach the scenario based output and the current land use situation are all based on grid data. All these grid data are transformed into 3D scenes by using the same land use icons. It means that known landmarks, nodes, etcetera are not represented. For simple interpretation of a visualisation, especially taking into account land use characteristics, it seems to be necessary to apply objects that visualise the features mentioned by Lynch, as has been done by Al-Kodmany (2002), who created cognitive maps for community planning based on Lynch' theoretical framework.

For that reason another option has been worked out that tried to link the scenario based grid data with the topographical data set that describes the current landscape (Momot 2004). For this option, called the Landscape Feature approach, we started to use GIS software with 3D visualisation capabilities (ArcGIS 3D Analyst en 3D Scene) and the Top10Vec data. The creation of this visualisation involved two different procedures. Firstly all 2D-building features were selected and extruded: the construction of a 2.5D data set. Secondly all 2D tree features were selected and geometrically transformed into point features. 3D-objects (mainly tree-representations) were selected from an included 3D-objects database and linked to the point features. After rendering the geodata including 3D-objects and bitmaps it offers a 3D visualisation as shown by figure 6.

As demonstrated by the LUI it is quite difficult to know how the projected land use classes will change the landscape in 30 years (Appleton and Lovett 2003). For that reason the existing landscape data (Top10vec) is analysed in such a way that features that will not change within the next 30 years could be selected. A number of assumptions have been made here. The main road network and the existing network of canals and rivers will not change. Existing trees in rows will remain the same through 30 years, because most of these trees are located along the main road network and canals. The areas of which the land use class will not change, keep the same landscape characteristics.

Based on these assumptions an overlay of the current topographical data and the new land use classes (the grid cell output) has been made. The overlay delivers a new geodata set that represents the area that will change including the features as mentioned by the assumptions. The areas that the land use classes will not change are kept the same. After visualisation, like described before, the land use classes that are anticipated to change (in grid cells of 500 by 500 meter) are shown by white squared underground with on top topographical features (infrastructure, water and trees). The unchanged land use classes are visualised by all topographic features from the original geodata set (top10Vec). Finally the white squares have been filled in by a colour that represents the dominant land use class. In fact the original goal of the project was to place the 3D icons of the LUI approach on top of the changed grid cells, but a successful technical solution was not found.

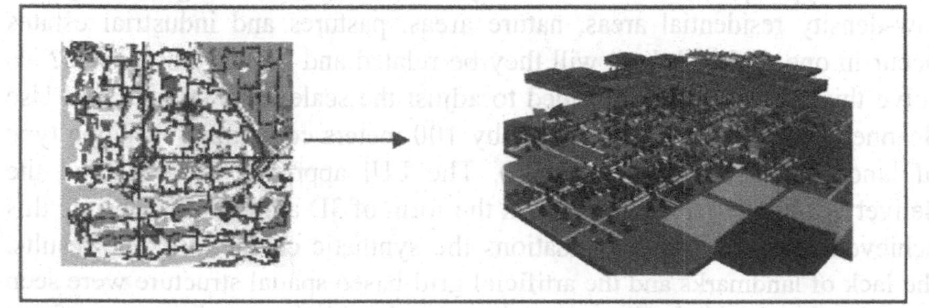

Figure 6. From 2D raster data to a VisualScan Scene (Lammeren et al. 2004).

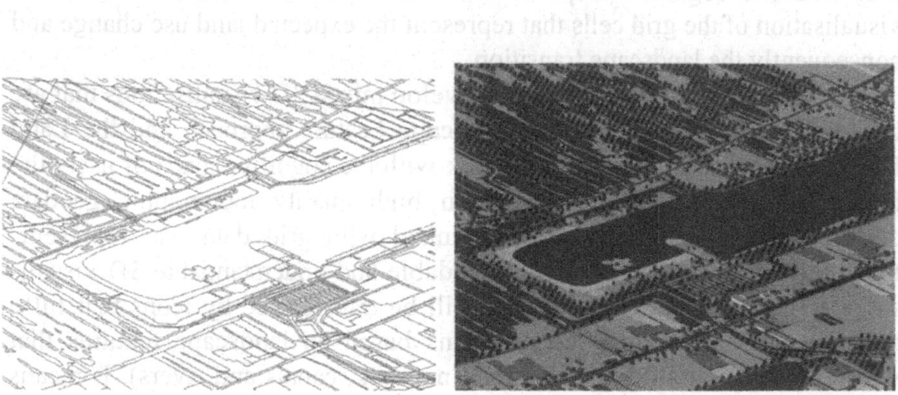

Figure 7. Part of Top10 vector and the visual representation (Momot 2004).

Both approaches appear to have pros and cons. Technical problems due to the enormous amount of time needed to render the visualisations plagued both approaches. Fortunately, a solution could be found for the LUI approach but the Landscape Feature approach was not completed satisfactorily as a result of these technical problems.

In addition to this, conceptual problems occurred too. During the design of the icons it was increasingly realised that information lacked about a number of spatial aspects, such as the assumed spatial structure in future, changes in densities and spatial relations between land use types. These conceptual problems are difficult to solve because there are caused by the fact that the Land Use Scanner model output is already an abstraction of reality in the form grid cells. Other applications, for example the Virtual London project by Hudson-Smith et al. (2005), use the current spatial structure and make only partial changes to this structure by adding, changing or replacing building and other man-made constructions. The Land Use Scanner output used in the Visual Scan approach contains a blend of several land use types per grid-cell of 500 by 500 meters, but if high-density and low-density residential areas, nature areas, pastures and industrial estates occur in one grid-cell, how will they be related and spatially organised? To solve this problem it was decided to adjust the scale level of the Land Use Scanner output to grid cells of 100 by 100 meters containing only one type of land use (Tijbosch et al. 2006). The LUI approach succeeded in the delivery of highly realistic results in the form of 3D animations. Despite this achievement, in informal evaluations the synthetic character of the results, the lack of landmarks and the artificial grid-based spatial structure were seen as major drawbacks, leading to a missing link with the cognitive map of end users. The benefit of the Landscape Feature approach seems to be the better link with the cognitive maps of the users. The drawback is the limited visualisation of the grid cells that represent the expected land use change and consequently the landscape transition.

For the near future we expect to develop mixed 3D visualisations that are based on a combination of the Landscape Feature approach and the Land Use Icon approach. To support the link with the cognitive maps of users the landscape feature approach based on high quality topographic geodata describes the existing landscapes. Linked with grid data that represents future oriented land use it offers a readable and understandable 3D visualisation. The areas that could change will be transformed by icons that fulfil the constraints according the decisions based on landscape features that will not change at all (e.g. main road network, canals and rivers). It means that the spatial configuration of the 3D-objects in LUI has to be arranged by these constraints. To find out if this mixed 3D visualisation will be effective, a number of user tests should explore the possible added value of understanding the impact of land use changes on the landscape. These tests should focus on the requirements as proposed in the first section.

5. CONCLUSION

Participation and consultation of stakeholders have become far more common in various policy fields. An example is development planning in the field of spatial planning, where stakeholders are involved in the drafting of new spatial plans for regions or urban areas. In spite of this involvement, 3D visualisation hitherto has not become a common tool in spatial planning. An exception might the current restructuring of rural areas for environmental reasons, further agricultural rationalisation and the development of new nature areas. The Dutch Ministry of Agriculture, Nature and Food has funded a number of research projects where 3D visualisation plays a central part. Nevertheless, this technology is not yet applied widely in spatial, environmental and ecological policy, for example in designation of locations for further urbanisation and their effects on valuable landscapes and ecological quality, or the spatial effects of water management measures fighting climate change.

Although 3D visualisation seems to offer many advantages, in this research it was concluded that its potential was not fully realised and it might even be a mission impossible if a number of difficulties are not solved. A major difficulty is posed by the nature of the land use model output data. These are rather coarse square grid cells and concern long term changes. They do not include more information about spatial structure and coherence than is stored in the 2D map, what reduces the possibilities for 3D visualisation. This point has been addressed in the Landscape Feature approach, where topographical information and landmarks were included. Even then additional assumptions about geographical inertia, spatial structure and coherence in future have to be made to structure the arrangement of icons and their placement among unchanged landscape features. Besides, information on densities, for example by calculating detailed population and employment distributions when land use is allocated, could improve the quality of the 3D visualisation. This information can be used then to create more variation in icons representing specific land use types.

Maybe more time should have been spent beforehand to the definition of the requirements for such an application. The aim to create geovisualisations on a national or regional scale proved to be ambitious, and it took a lot of time to find a satisfying solution for the technical problems that were met. Hoogerwerf et al. (2005b) discuss the effectiveness of geo-visualisations in

PSP as a result of 1) the broad range of users that need to perform a broad range of tasks with geo-visualisations (Slocum et al. 2001), 2) the broad variety of cognitive issues that are involved when 3D geovisualisations are used as communication tools, including various interaction methods (Bill et al. 1999, Dransch 2000, Fairbairn 2001, Yun et al. 2004). Geovisualisations can support *interaction in* the 3D virtual environment (orientation, movement, navigation, explanation, elaboration and manipulation) and *interaction of* the 3D virtual environment (the user is able to define the settings of the viewer (interface) mode that could influence the way the 3D environment is experienced by the user). Interaction here also stretches itself to the ability of users to give feedback on proposed transitions via text messages, a forum, or voting systems (Lammeren and Hoogerwerf 2003).

An effort to create an overview of interaction methods was initiated by Hoogerwerf et al. (2005b). They introduce a conceptual framework that integrates the theoretical fields of spatial planning, participatory planning and communication to adjust geo-visualisations to a specific planning issue and context in order to communicate spatial information effectively and efficiently to all actors. The framework supports developers by dealing with the divergent prerequisites to geovisualisation of users, planning phases, the level of participation, the communication protocol, and the interface, including the devices.

The development of geo-visualisation as required needs, besides the framework topics to be met, also tools to construct these. Though GIS software is offering a wide variety of 3D transformations and related visualisation techniques, the worldwide breakthrough of virtual flight via Google-Earth illustrates clearly the rapid changes and impact of geo-visualisation. Also in game-technology and internet communities developments are fast. An example are the sceneries for flight simulators made by Dutch flight simulator enthusiasts, these already cover the whole of the Netherlands (http://www.nl-2000.com/nuke/). Adding data to Google Earth is rather easy and applications as Visual Scan might derive more from game-technology.

However, structuring the information needs by definition and preparation of the elements of the framework, as mentioned above, is the main and most tedious job. The framework can be applied as well to coupling of the output of a land use model to a 3D visualisation to assess EES effects. First research on the datasets is needed, especially with respect to their spatial resolution and extent, as these have a direct relation with the minimum element depicted in the 3D scene. A next step is to determine if the 3D scene needs to be observed from a stationary single viewpoint or that the user wants to control the position of the viewer, either by user input or a predefined path to specific items or areas of interest. If the user needs an indication of the orientation and position of the viewer, or the observed objects, the designer must add additional controls to the application such as a map, or a north arrow (Verbree et al. 1999). So far, the construction of a geovisualisation allows for the observation of the 3D scene, but does not incorporate controls to interact with objects in the scene. This requires some sort of selection to specify which of the objects will be part of the interaction. Predetermined types of interaction may involve explanation of the objects using pop-ups, or the object can contain an external link to additional information on an Internet website.

However, the purpose of the geovisualisation may require the implementation of a mechanism of interaction rather than a predefined method. This calls for manipulation, which concerns the change of the objects in the 3D scene, or even their appearance. Manipulation implies that the configuration of the 3D scene is altered. The mechanisms returning the consequences of a manipulation on the object collection must often be defined as an explicit element of the geovisualisation. This feedback is generally related to a specific simulation or model (e.g. to trigger the Land Use Scanner simulation) that takes the alterations made by the user and utilises these to generate a new result. The last step in the framework concerns the level of believability and information intensity. Adding an Avatar reference mode could enhance the feeling of immersion.

Construction of a hybrid extension of VisualScan, as the Geo-Visualisation interface with the Land Use Scanner, could support at least all requirements of the framework. This hybrid extension must combine the Land Use Icon and Landscape Feature approach and contain additional assumptions on future spatial structure and coherence. If the output of the Land Use Scanner model includes information on densities as well, the icons can become more heterogeneous, what reduces their synthetic appearance. Such link between land use models and 3D-visualisation could finally show, after positive usability reviews, that this mission is not impossible.

6. REFERENCES

Al-Kodmany K., 2002, "Visualization Tools and Methods in Community Planning: From Freehand Sketches to Virtual Reality", *Journal of Planning Literature,* **17**: 189-211

Appleton, K. and A. Lovett, 2003, "GIS-based visualisation of rural landscapes: defining 'sufficient' realism for environmental decision-making", *Landscape and Urban Planning,* **65**(3): 117-131.

Appleton, K., and A. Lovett, 2005. "GIS-based visualisation of development proposals: Reactions from planning and related professionals", *Computers, Environment and Urban Systems,* **29**(3): 321-339.

Ball, J., 2002, "Towards a methodology for mapping 'regions for sustainability' using PPGIS", *Progress in Planning,* **58**: 81-140.

Bill R., Dransch D., and Voigt C., 1999, "Multimedia GIS: concepts, cognitive aspects and applications in an urban environment", in: Raper (ed.) *Spatial Multimedia and Virtual Reality,* Taylor & Francis.

Bishop, I.D., 1994, "The role of visual realism in communicating and understanding spatial change and process", in: Hearnshaw and Unwin (eds.), *Visualization in Geographical Information Systems.* Wiley, Chichester.

Bishop, I.D., and E. Lange, 2005, *Visualization in Landscape and Environmental Planning. Technology and Applications.* London and New York, Taylor & Francis.

Bishop I.D., R.B. Hull Iv, and C. Stock, 2005, "Supporting personal world-views in an envisioning system", *Environmental Modelling and Software,* **20**: 1459-1468.

Bloemmen, M., A. Ligtenberg, R. van Lammeren, 2005, *Approaches for the use of geovisualization in participatory spatial planning process.* Wageningen University, Wageningen.

Borsboom-van Beurden, J.A.M., W.T. Boersma, A.A. Bouwman, L.E.M. Crommentuijn, J.E.C. Dekkers, and E. Koomen, 2005, *Ruimtelijke Beelden. Visualisatie van een Veranderd Nederland in 2030,* RIVM rapport 550016003. RIVM, Bilthoven.

Chen, C.,1999, *Information Visualisation and Virtual Environments,* Springer-Verlag, London.

DiBiase, D., 1990, "Visualization in earth sciences", *Bulletin of Earth and Mineral Sciences, Pennsylvania State University,* **59**: 13-18.

Dransch D., 2000, "The use of different media in visualizing spatial data", *Computers and Geosciences,* **26**: 5-9.

Dijkstra, J., J. van Leeuwen and H. Timmermans, 2003, "Evaluating design alternatives using conjoint experiments in virtual reality", *Environment and Planning B: Planning and Design,* **30**: 357-367.

Fairbairn D., G. Andrienko, N. Andrienko, G. Buziek, and J. Dykes, 2001, "Representation and its relationship with cartographic visualization: a research agenda" *CaGIS* 28.

Geertman, S., 2002, "Participatory planning and GIS: a PSS to bridge the gap", *Environment and Planning B-Planning & Design,* **29**(1), p. 21-35.

Groen, J., E. Koomen, M. Piek, J. Ritsema van Eck, A. Tisma, 2004, *Scenario's in kaart. Model- en ontwerpbenaderingen voor toekomstig ruimtegebruik,* NAI Uitgevers/Ruimtelijk Planbureau, Rotterdam/Den Haag.

Hoogerwerf, T., 2003, "The use of virtual reality in spatial planning : a study on realism requirements in the levels of public participation". *Thesis report* GIRS2003-34., 70 p.

Hoogerwerf, T., 2005a, *Geo-visualization approaches.* Wageningen University, Wageningen.

Hoogerwerf, T., and R. van Lammeren, 2005b, "The suitability of visualisations with a different degree of realism for participatory spatial planning". *WU-CGI- Web-article,* 2005a.h

Hudson-Smith, A., S. Evans and M. Batty, 2005, "Building the Virtual City: Public Participation through e-Democracy", *Knowledge, Technology & policy,* **18**(1): 62-85.

Klijn, J., W. Vullings, M. van den Berg, H. van Meijl, R. van Lammeren, T. van Rheenen, A. Veldkamp, P. Verburg, H. Westhoek, B. Eickhout, 2005, "The EUruralis study: technical document". *Alterra-rapport* 1196, ISSN 1566-7197. 215 pp.

Krause, C., 2001, "Our visual landscape managing the landscape under special consideration of visual aspects", *Landscape and Urban Planning,* **54**: 239-254.

Kwan, M.P., and J. Lee, (2003, "Geovisualization of Human Activity Patterns Using 3D GIS: A Time-Geographic Approach", in Goodchild and Janelle (eds*), Spatially Integrated Social Science: Examples in Best Practice*, Oxford University Press, Oxford.

Lammeren, R. van and T. Hoogerwerf, 2003, "Geo-virtual reality and participatory planning: Virtual Landscape Position paper version 2.0", CGI-rapport2003-07. Wageningen University and Research, Wageningen.

Lammeren, R. van, R. Olde Loohuis, A. Momot, and S. Ottens, 2004,, "VisualScan: 3D visualisations of 2D scenarios", *CGI-report* 2004-09, ISSN 1568-1874, Wageningen University/Centre for Geo-Information, Wageningen.

Lange, E., 2001, "The limits of realism: perceptions of virtual landscapes", *Landscape and urban planning*, **54**: 163-182.

Lynch K., 1960, *The Image of the city*, MIT Press, Cambridge.

McEachren, A.M., R. Edsall, D. Haug, R. Baxter, G. Otto, R. Masters, S. Fuhrman and L. Qian, 1999, "Exploring the potential of virtual environments for geographic visualization", Paper presented at *Annual Meeting of the Association of American Geographers*. Honolulu, 23-27 March, http://www.geovista.psu.edu/publications/aag99vr/fullpaper.htm

MNP, 2004, *Quality and the future. Sustainability Outlook*, Netherlands Environmental Assessment Agency, RIVM, Bilthoven.

Momot A., 2004, *Visualization of land use scanner data*, Wageningen University and Research/Centre for Geo-Information/RIVM, Wageningen/Bilthoven

Nijs, A. de, R. Kuiper, and L. Crommentuijn, 2005, *Het landgebruik in 2030. Een projectie van de Nota Ruimte*. Milieu- en Natuurplanbureau, Bilthoven.

Slocum, T., C. Blok, B. Jiang, A. Koussoulakou, D. Montello, S. Fuhrmann, S. and N. Hedley, N., 2001, "Cognitive and Usability Issues in Geovisualisation". *Cartography and Geographic Information Society*, **28**(1): 61-75.

Tijbosch, H., A. Bouwman, M. Hilferink, M. van der Beek, E. Koomen, 2006, *Ruimtescanner 2005: introductie van het discrete allocatiemechanisme*, MNP-rapport 500075001/2006.

Tress, B. and G. Tress, 2003, "Scenario visualisation for participatory landscape planning-a study from Denmark", *Landscape and Urban Planning*, **64**(3): 161-178.

Verbree, E., G. van Maren, R. Germs, F. Jansen and M.J. Kraak, 1999, "Interaction in virtual world views – linking 3D GIS with VR", *International Journal of Information Science*, **13**(4): 285-396.

VROM, EZ, VWS and LNV, 2004, *Nota Ruimte*. Ministerie van Volkshuisvesting, Ruimtelijke Ordening en Milieu, Den Haag.

Yun, L., C. Yufen, et al., 2004, "Cognition theory-based research on adaptive user interface for geo-visualization system". *Proceedings 12th Int. Conf. on Geoinformatics*, University of Gävle, Sweden, 7-9 June 2004.

Wilkens, A., 2005, *GeoVR to interact with spatio-temporal models. The manipulation of spatio-temporal models from a 3D computer environment*, Wageningen University/Centre for Geo-Information, Wageningen.

Zube E.H., D.E. Simcox and C.S. Law, 1987, "Perceptual Landscape Simulations: History and Prospect", *Landscape Journal*, **6**: 62-80.

Multi-Agent Models for Movement Simulation

Crowd Modeling and Simulation

The role of multi-agent simulation in design support systems

Stefania Bandini[1], Sara Manzoni[1], and Giuseppe Vizzari[1,2]
[1]*Laboratory of Artificial Intelligence (L.Int.Ar.)*
Department of Informatics, Systems and Communications, University of Milan-Bicocca, Italy
[2]*NOMADIS Laboratory*
University of Milan-Bicocca, Italy
{bandini,manzoni,vizzari}@disco.unimib.it

Keywords: Artificial intelligence, Agent technology, Simulation

Abstract: The paper presents a Multi Agent Systems (MAS) approach to crowd modelling, based on the Situated Cellular Agents (SCA) model. This is a special class of Multilayered Multi Agent Situated System (MMASS), a model providing an explicit representation of the environment which has a relevant role in supplying agents a context allowing them to act and interact (among themselves and with the environment). The paper will briefly introduce the model and a methodology for the analysis of a crowd scenario and the design of SCA based crowd simulations. The adoption of this kind of system allows evaluating an architectural design with reference to the behaviour of pedestrian that will act in it, given a behavioural specification for these entities. The system is also able to produce a realistic visualization of the simulation, in order to facilitate communication with involved actors (e.g. in case of participatory decisions).

1. INTRODUCTION

The design of different kinds of environmental structures, at different detail levels, from the corridors or emergency exits of a building to the whole transportation system on urban or regional scale, may benefit from an envisioning of how it will perform, given specific assumptions on the usage conditions and the behaviours of the autonomous entities which will populate it. There is thus a growing interest in models and technologies supporting the simulation of this kind of domains.

Jos P. van Leeuwen and Harry J.P. Timmermans (eds.), Innovations in Design & Decision Support Systems in Architecture and Urban Planning, 105-120.

An innovative trend in supporting architects in their activities is represented by virtual environments in which alternative architectural designs can be visualized and compared by involved actors, in a collaborative decision scheme (Djikstra et al. 2003, Batty and Hudson-Smith, 2005). This kind of approach could be improved by the possibility to include in the virtual environments also an envisioning of pedestrian dynamics in the related architectural structures (Bandini et al. 2004b), given the fact that human movement behaviour has deep implications on the design of effective pedestrian facilities (Willis et al. 2004).

Several continuum models for pedestrian dynamics are based on an analytical approach. A relevant example is represented by *social force* models (Helbing, 1991), in which individuals are treated as particles subject to forces. Other analytical models take inspiration from fluid-dynamic (Helbing, 1992) and magnetic forces (Okazaki, 1979) for the representation of pedestrian flows. A different approach to crowd modelling provides the adoption of Cellular Automata (CA) (Wolfram, 1986), with a discrete spatial representation and discrete time-steps. The cellular space includes both a representation of the environment and an indication of its state, in terms of occupancy of the sites it is divided into. Transition rules must be defined in order to specify the evolution of every cell's state; they are based on the concept of *neighbourhood* of a cell, a specific set of cells whose state will be considered in the computation of its transition rule. Local cell interactions may represent the motion of an individual in the space, and the sequential application of this rule to the whole cell space may bring to *emergent* effects and *collective* behaviours, for instance lane formation and evacuation configurations (Schadschneider et al. 2002).

Even if the CA-based approach is generally better understood than analytical models by experts in different application domains, and more easily applied to model related scenarios, both these approaches share the limit of considering individuals as homogenous entities, and generally do not provide elements of flexibility and dynamism, like changes in behaviour of individuals. This may not represent an issue for large scale simulations, in which a certain degree of approximation is unavoidable and often tackled by the adoption of a stochastic approach, but in other situations it could be relevant to take this kind of information into account. For instance, the evaluation of information signs placement depends on different factors related to their effectiveness, and thus to their visibility. The latter is strongly dependant on the behaviour of individuals moving throughout the environment, their goals and destinations, but even their perceptive capabilities. These factors are relevant in the decision of what directions the active entities will take, and to include these concepts in a CA would require an extremely high number of rules, a very large cell state and probably the extension of the concept of neighbourhood to simulate *at-a-distance* interactions (for instance to model the attractiveness of destination sites). All these considerations lead to consider a Multi-Agent System (MAS) (Ferber, 1999) approach to the modelling or this kind of situation. In fact, a MAS consists of a number of possibly heterogeneous agents that act and interact inside an environment, which enables their perception, interaction and action. Accordingly, Multi Agent Based Simulation (MABS) is based on the idea that it is possible to represent the global behaviour of a dynamic system as the result of interactions occurring among an assembly of agents with their own operational autonomy. In particular the Situated Cellular Agents (SCA) model (Bandini et al. 2006) is a situated MAS model, whose spatial structure represents a key factor influencing agents' choices on their actions and in determining their possible interactions. The model has been successfully applied in different contexts, and in particular its focus on the modelling of the environment as well as its inhabiting agents and their interactions, make it particularly suitable for simulation of actual physical systems.

The following section briefly introduces the SCA model, while the adopted modelling methodology is described in section 3. A case study characterized by a complex mixture of competitive and cooperative pedestrian behaviour in underground station crowd dynamics will then be described; preliminary results of the modelling and simulation experience will follow. Section 0 discusses the role of advanced systems for the visualization of simulation dynamics; conclusions and future developments end the paper.

2. THE SITUATED CELLULAR AGENT MODEL

The Situated Cellular Agent model is a specific class of Multilayered Multi-Agent Situated System (MMASS) (Bandini et al. 2002) providing a single layered spatial structure for agents environment and some limitations to the field emission mechanism. A thorough description of the model is out of the scope of this paper, and this aim of Section is to briefly introduce it to give some basic notion of the elements that are necessary to describe the methodology.

A *Situated Cellular Agent* is defined by the triple $< Space, F, A >$ where *Space* models the environment where the set A of agents is situated, acts autonomously and interacts through the propagation of the set F of fields and through reaction operations. Figure 1 shows a diagram of the two interaction mechanisms provided by the model.

Space is defined as a not oriented graph of sites. Every *site* $p \in P$ (where P is the set of sites of the layer) can contain at most one agent and is defined by the 3–tuple $<a_p, F_p, P_p>$ where:

- $a_p \in A \cup \{\perp\}$ is the agent situated in p ($a_p = \perp$ when no agent is situated in p that is, p is empty);
- $F_p \subset F$ is the set of fields active in p ($F_p = \varnothing$ when no field is active in p);
- $P_p \subset P$ is the set of sites adjacent to p.

A SCA agent is defined by the 3–tuple $< s, p, \tau >$ where τ is the *agent type*, $s \in \Sigma_\tau$ denotes the *agent state* and can assume one of the values specified by its type (see below for Σ_τ definition), and $p \in P$ is the site of the *Space* where the agent is situated. As previously stated, agent *type* is a specification of agent state, perceptive capabilities and behaviour. In fact an agent type τ is defined by the 3–tuple $<\Sigma_\tau, Perception_\tau, Action_\tau>$. Σ_τ defines the set of states that agents of type τ can assume. $Perception_\tau : \Sigma_\tau \rightarrow [\mathbf{N} \times W_{f1}] \ldots [\mathbf{N} \times W_{f|F|}]$ is a function associating to each agent state a vector of pairs representing the *receptiveness coefficient* and *sensitivity thresholds* for that kind of field. $Action_\tau$ represents instead the behavioural specification for agents of type τ. Agent behaviour can be specified using a language that defines the following primitives:

- *emit(s,f,p)*: the *emit* primitive allows an agent to *start the diffusion of field f* on p, that is the site it is placed on;
- *react*($s, a_{p1}, a_{p2}, \ldots, a_{pn}, s'$): this kind of primitive allows the specification a *coordinated change of state* among adjacent agents. In order to preserve agents' autonomy, a compatible primitive must be included in the behavioural specification of all the involved agents; moreover when this coordination process takes place, every involved agent may dynamically decide to effectively agree to perform this operation;
- *transport(p,q)*: the *transport* primitive allows defining *agent movement* from site p to site q (that must be adjacent and vacant);
- *trigger(s,s')*: this primitive specifies that an agent must *change its state* when it senses a particular condition in its local context (i.e. its own site and the adjacent ones); this operation has the same effect of a reaction, but does not require a coordination with other agents.

For every primitive included in the behavioural specification of an agent type specific preconditions must be specified; moreover specific parameters must also be given (e.g. the specific field to be emitted in an emit primitive, or the conditions to identify the destination site in a transport) to precisely define the effect of the action, which was previously briefly described in general terms.

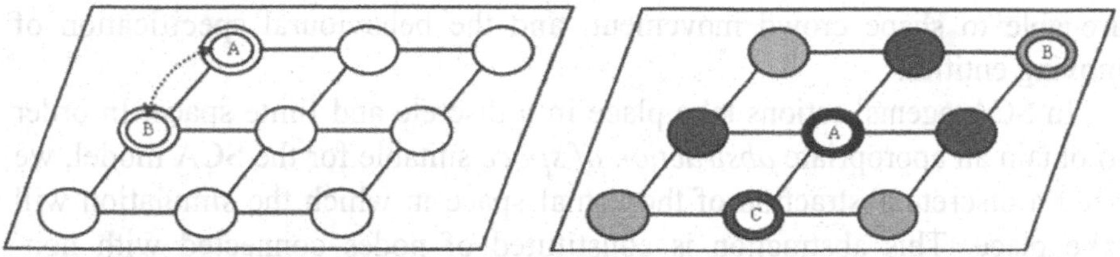

Figure 1. A diagram showing the two interaction mechanisms provided by the SCA model: two reacting agents on the left, and a field emission on the right.

Each SCA agent is thus provided with a set of sensors that allows its interaction with the environment and other agents. At the same time, agents can constitute the source of given fields acting within a SCA space (e.g. noise emitted by a talking agent). Formally, a field type t is defined by $<W_t, Diffusion_t, Compare_t, Compose_t>$ where W_t denotes the set of values that fields of type t can assume; $Diffusion_t: P \times W_t \times P \rightarrow (W_t)^+$ is the diffusion function of the field computing the value of a field on a given space site taking into account in which site (P is the set of sites that constitutes the SCA space) and with which value it has been generated. It must be noted that fields diffuse along the spatial structure of the environment, and more precisely a field diffuses from a source site to the ones that can be reached through arcs as long as its intensity is not voided by the diffusion function. $Compose_t: (W_t)^+ \rightarrow W_t$ expresses how fields of the same type have to be combined (for instance, in order to obtain the unique value of field type t at a site), and $Compare_t$: $W_t \times W_t \rightarrow \{True, False\}$ is the function that compares values of the same field type. This function is used in order to verify whether an agent can perceive a field value by comparing it with the sensitivity threshold after it has been modulated by the receptiveness coefficient.

3. THE ADOPTED METHODOLOGY

The activity of modelling a simulation scenario using a MAS consists in a complex activity, which generally requires the skills of a programmer, a deep acquaintance with the simulation platform, and a good knowledge of the domain that is going to be modelled. Rarely those characteristics can be found in the same person. The methodology proposed in this section provides a guideline directed to experts but also to non-expert users that wish to model a scenario using the MMASS platform as simulation tool. In order to adopt this approach in the crowding context, the modeller must essentially define three elements: the spatial abstraction, in which the simulated entities are situated, the relevant elements of this structure which are able to shape crowd movement, and the behavioural specification of moving entities.

In SCA agents' actions take place in a discrete and finite space. In order to obtain an appropriate *abstraction of space* suitable for the SCA model, we need a discrete abstraction of the actual space in which the simulation will take place. This abstraction is constituted of nodes connected with non-oriented arcs. Nodes represent the positions that can be occupied by single pedestrian once per time. Some of the nodes can be occupied by some agents that constitute part of the environment (in a transit station these are doors, exits, shops etc), and that cannot be occupied by other individuals. SCA space represents thus an abstraction of a walking pavement, but it has to be sufficiently detailed to be considered a good approximation of the real environment surface, and it allows a realistic representation of the movements and paths that individuals would follow.

In this framework we assume that specific *active elements of the environment* can be perceived as reference points influencing (or even determining) the movement of pedestrians. Typically the latter are objects of the environment which constraint agent movement, but also objects that somehow transmit conceptual information (e.g. exit signs or indications). The SCA model provides a simple mean of generating at-a-distance effects that can be exploited to generate attraction or repulsion effects. After space discretization has been defined, active objects in the environment have to be selected, and field types have to be assigned to them. Field type specifies the diffusion and composition function for that type of signal, and to complete the specification of an active element of the environment the emission intensity must also be provided.

The third phase of the methodology is to define the *behaviour of the pedestrians*. The model allows defining heterogeneous agents thanks to the notion of agent type, which comprises the definition of related state, perceptive capabilities and behavioural specification. The behaviour of an agent type, when it is composite and complex, can suitably be segmented in relevant states. Each state represents diverse priorities, and a different attitude of the agent towards movement in the environment. For each state the modeller must be specify the fields emitted by the agent and the sensitivity to fields emitted by other agents.

Moreover, as different fields can be spread over the environment and agents must be able to combine their effects according to a private criterion for action selection. In particular, we chose to define a measure of the utility of the presence of every field type in a given site, to represent its desirability for an agent of a given type in a specific state. Then the definition of conditions for states transition must be modelled. The change of the state of an agent is related to the perception of a specific condition in its current context that determines a change of attitude towards movement.

These three elements define an abstract computational model for a specific situation. The configuration for an experiment in a specific simulation scenario follows these phases, and it requires setting the number, position and initial state of the mobile agents that will populate the simulation. The process ends with a formal statement of what is the goal of the simulation, a precise specification of what has to be observed and how. The variety of possibly monitored parameters, and thus also the number and heterogeneity of distinct monitoring mechanisms, does not allow to define precise guidelines for this phase. Finally, the actual experiments can highlight the need to tune parameters such as field emission intensity and field type characteristics, but also utility values guiding agents' movement.

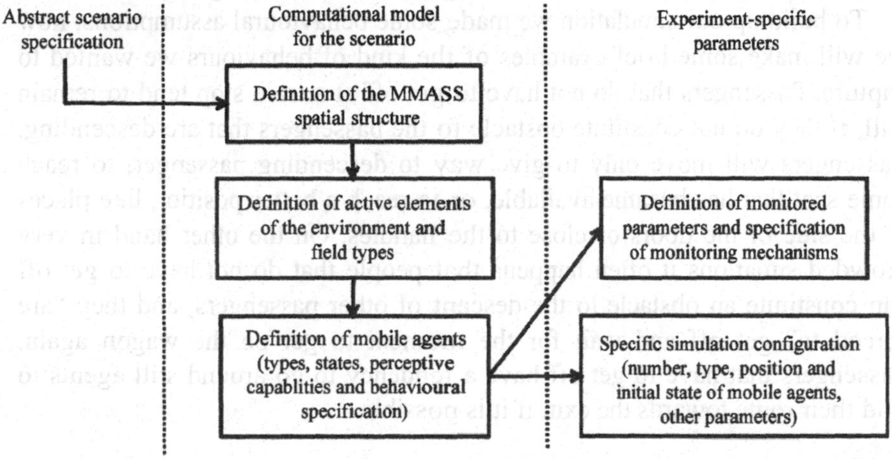

Figure 2. The diagram shows the main phases of the modelling methodology.

4. A CASE STUDY: THE UNDERGROUND STATION

An underground station is an environment where various crowd behaviours take place. Passengers' behaviours are difficult to predict, because crowd dynamics emerges from single interactions between passengers, and between single passengers and parts of the environment, such as signals (e.g. current stop indicator), doors, seats and handles. The behaviour of passengers changes noticeably in relation to the different priorities that characterize each phase of their trips. That means, for instance, that passengers close to each other may display very different behaviours because of their distinct aims in that moment. Passengers on board may have to get off and thus try to reach for the door, while other are instead looking for a seat or standing beside a handle.

Moreover when trains stop and doors open very complex crowd dynamics happen, as people that have to get on the train have to allow the exit of passengers that are getting off. Passengers have to match their own priority with the obstacles of the environment, with the intentions of other passengers, and with implicit behavioural rules that govern the social interaction in those kinds of transit stations, in a mixture of competition and collaboration, to avoid stall situations. Given the complexity of the overall scenario, we decided to focus on a specific portion of this environment in which some of the most complex patterns of interaction take place: the part of platform in presence of a standing wagon from which some passengers are attempting to get off while other waiting travellers are trying to get on.

To build up our simulation we made some behavioural assumptions, now we will make some brief examples of the kind of behaviours we wanted to capture. Passengers that do not have to get off at a train stop tend to remain still, if they do not constitute obstacle to the passengers that are descending. Passengers will move only to give way to descending passenger, to reach some seat that has became available, or to reach a better position like places at the side of the doors or close to the handles. On the other hand in very crowded situations it often happens that people that do not have to get off can constitute an obstacle to the descent of other passengers, and they "are forced to" get off and wait for the moment to get on the wagon again. Passengers that have to get off have a tendency to go around still agents to find their route towards the exit, if it is possible.

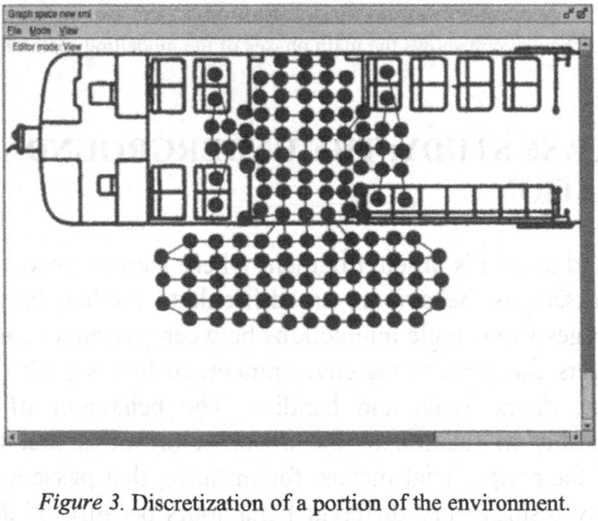

Figure 3. Discretization of a portion of the environment.

Once the train is almost stopped the waiting passengers on the platform identify the entrance that is closer to them, and try to move towards it. If they perceive some passengers bound to get off, they first let them get off and then get on the wagon. In reference to the methodology stated in the previous paragraph, to build an environment suitable for SCA platform, first of all we need to define a discrete structure representing the actual space in which the simulation is set. In our case study we started from an available diagram of an underground wagon. A discrete abstraction of this map was defined, devoting to each node the space generally occupied by one standing person, as shown in *Figure 3*.

In our simulation fields are generated by elements of the environment but also by agents that represent passengers. We identified the following objects as active elements of the environment: *Exits*, *Doors*, *Seats* and *Handles* (see *Figure 4* for their disposition).

Station exits emit fixed fields, constant in intensity and in emission that will be exploited by agents headed towards the exit of the station. Agent-doors emit another field which can guide passengers that have to get off towards the platform, and passengers that are on the platform and are bound to get in the wagon. Seats may have two states: occupied and free. In the second state they emit a field that indicates their presence. An analogous field is emitted by handles, which however are sources of fields characterized by a minor intensity.

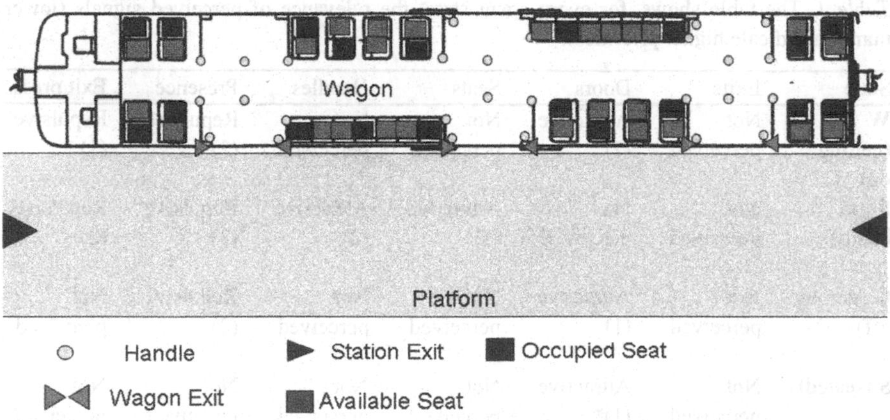

Figure 4. Immobile active elements of the environment.

We have identified the following states for agent of type passengers: *waiting* (*w*), *passenger* (*p*), *get-off* (*g*), *seated* (*s*), *exiting* (*e*). In relation to its state, an agent will be sensitive to some fields, and not to others, and attribute different relevance to the perceived signals. In this way, the changing of state will determine a change of priorities. State *w* is associated to an agent that is waiting to enter in the wagon. In this state, agents perceive the fields generated by the doors as attractive, but they also perceive as repulsive the fields generated by passengers that are getting off, in other words those in state *g*. In state *w* the agent "ignores" (is not sensitive to) the fields generated by other active elements of the environment, such as exits' attractive fields, chairs attractive field and so on. Once inside the wagon, the agent in state *w* changes its state in *p* (passenger), through a trigger function activated by the perception of the maximum intensity of field generated by agent-door type. Agent in state passenger is attracted by fields generated by seats and handles, and repulsed by fields related to passengers that are getting off. In state *g* the agent will instead emit a field warning other agents of its presence, while it is attracted by fields generated by the doors. Once passed through the wagon door the agent in state *g* changes its state to *e* (exiting) and its priority will become to find the exits of the station. *Table 1* summarizes the sensitivity of the passenger to various fields in relation to their state. The table's cells provide also the indication about if the perceived field is attractive or repulsive and the relevance associated to that field type.

All passengers except those in state *g* emit a presence field that generally has a repulsive effect, but a lesser one with respect to the "exit pressure" generated by agents in get-off state.

Table 1. The table shows, for every agent state, the relevance of perceived signals (lower numbers indicate higher priorities).

State	Exits	Doors	Seats	Handles	Presence	Exit press.
W (getting on)	Not perceived	Attractive (2)	Not perceived	Not perceived	Repulsive (3)	Repulsive (1)
P (on board)	Not perceived	Not perceived	Attractive (1)	Attractive (2)	Repulsive (3)	Repulsive (2)
G (getting off)	Not perceived	Attractive (1)	Not perceived	Not perceived	Repulsive (2)	Not perceived
S (seated)	Not perceived	Attractive (1)*	Not perceived	Not perceived	Not perceived	Not perceived
E (exiting)	Attractive (1)	Not perceived	Not perceived	Not perceived	Repulsive (2)	Not perceived

* = *The door signal also conveys the current stop indication.*

5. PRELIMINARY RESULTS

A simulator implementing the previously introduced model was realized exploiting the SCA model: only a subset of the overall introduced model was implemented, and more precisely active objects of the environment and passenger agents in state *w*, *g*, *e*, *p*. *Figure 5* shows a screen-shot of this simulation system, in which waiting agents move to generate room for passenger agents which are going to get off the train. The system is synchronous, meaning that every agent performs one single action per turn; the turn duration is about one third of second of simulated time. The goal of a small experimentation as this one is to qualitatively evaluate the modelling of the scenario and the developed simulator. The execution and analysis of several simulations shows that the overall system dynamics and the behaviour of the agents in the environment is consistent with a realistic scenario, and fits with our expectations. To determine this evaluation, we executed over 100 simulations in the same starting configuration, which provides 6 passengers located on an underground train in state *g* (i.e. willing to get off), and 8 agents that are outside the train in state *w* (i.e. waiting to get on). In all simulations the agents achieved their goals (i.e. get on the train or get out of the station) in a number of turns between 40 and 80, with an average of about 55 turns. Nonetheless we noticed some undesired transient effects, and precisely oscillation, "forth and back" movements and in few simulations static forms providing "groups" facing themselves for a few turns, until the groups dispersed because of the movement of a peripheral element. These phenomena, which represent minor glitches under the described initial conditions, could lead to problems in case of high pedestrian density in the simulated environment.

Instead of modifying the general model, in order to introduce a sort of agent "facing" (not provided by the SCA model), we allowed agents to keep track of their previous position, in order to understand if a certain movement is a step back. The utility of this kind of movement can thus be penalized. Instead, in order to avoid stall situations, the memory of the past position can also be exploited to penalize immobility, lowering the utility of the site currently occupied by the agent whenever it was also its previous position.

These correctives were introduced in the behavioural specification of mobile agents, and a new campaign of tests was performed to evaluate the effect of these modifications in the overall system dynamics. By introducing these correctives, the occurrence of oscillating agent movement was drastically reduced, and the penalization of immobility simplified the solution of stall situations among facing groups. In all simulations the agents were able to achieve their goals, but the number of turns required to complete agents movement is between 28 and 60, with an average of about

35 turns. The analysis and identification of other significant parameters to be monitored, in this specific simulation context and in general for crowding situations, is object of current and future developments.

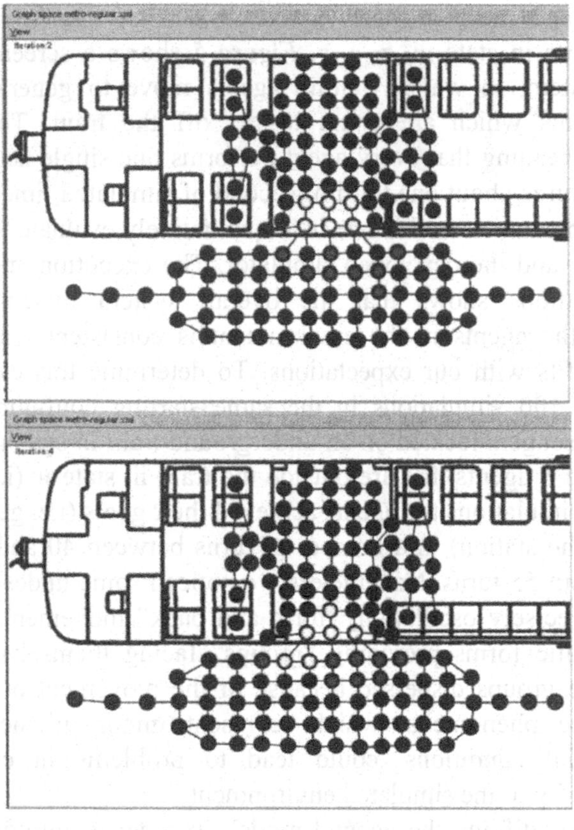

Figure 5. Two screenshots of the underground simulation. On the first one light gray agents are inside the train and going to get off, while dark agents are standing outside and are going to get on. On the second, the latter have made some rooms for the former to get off.

6. THE ROLE OF 3D VISUALIZATION

A relevant part of the project in which this work has been developed provides the generation of effective forms of visualization of simulation dynamics, to simplify its understanding by non experts in the simulated phenomenon. In particular, the developed simulator can be integrated with a 3D modelling and rendering engine (more details on this integration can be found in Bandini et al. 2004b), and a sample screenshot of the animation of the simulation dynamics is shown in *Figure 6.* One of the applications developed to implement SCA based simulations exploits a simulator based on a bidimensional spatial structure representation and an existing commercial 3D modelling instrument (3D Studio MAX[1]). The simulator has been developed as experimentation and exploitation of a long term project for a platform for MMASS based simulations.

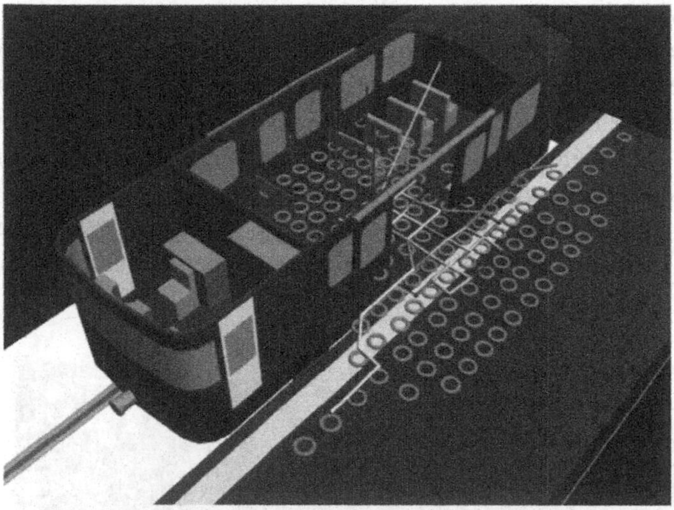

Figure 6. A screenshot of the 3D modelling of the simulation dynamics.

In different situations it can be useful, for sake of communication with non-experts, to obtain a more effective visualization of simulation dynamics. To do so, the bidimensional simulator produces a log-file provided with a fixed-record structure, in which every record is related to a node of the spatial structure or the position of an agent with reference to this structure. Initially, the simulator prints the structure of the environment, then the starting position of each agent. For every iteration of the simulation the new position of every agent is also printed. This file is parsed by a 3D Studio Max script which generates a plane and walls related to the spatial structure, nodes related to sites, and bipeds related to agents. Splines are then generated starting from the discrete positions assumed by various agents, and represent bipeds' movement. This process introduces modifications to trajectories defined by the bidimensional simulator whose sense is to give a more realistic movement to agents' avatars. This application was tested in the

[1] http://www.autodesk.com/3dsmax

discussed case study, and a screenshot of the resulting animation is shown in *Figure 6*.

A current activity provides a different approach, integrating the MAS based simulation system with a real-time 3D engine (Irrlicht[2]). This different architecture supports a dynamic visualization of the simulated system (e.g. it is possible to change the point of view, and even observe the simulation from the actual point of view of a given pedestrian), and it also possible to interact with the simulation while it is running (e.g. to alter specific environmental properties). Moreover, the availability of a 3D model of the environment, in addition to its discretization, supports further enhancements of mechanisms related to agent perception. In particular, it is possible to adopt the SCA field emission, diffusion, perception mechanism as a light way of identifying entities which might perceive a visual signal, but then invoke a more precise but computationally more expensive mechanism exploiting the 3D model to effective evaluate the perception of the signal.

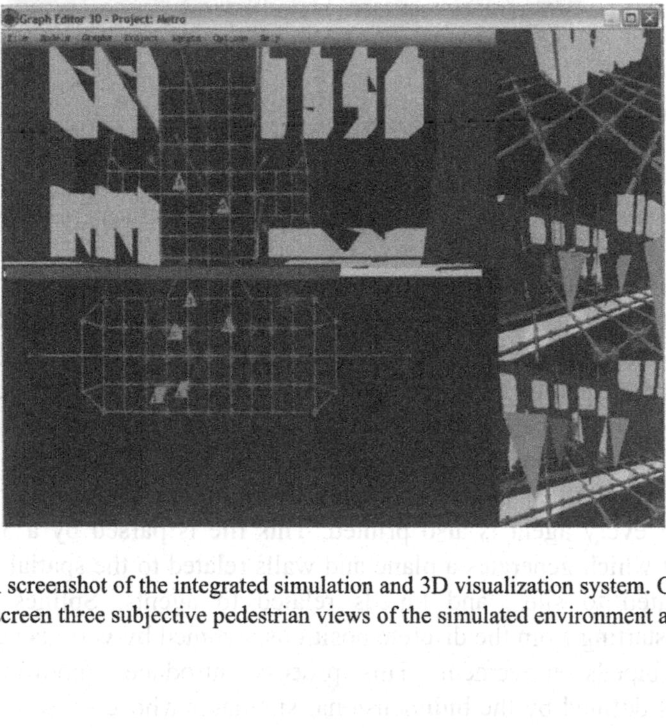

Figure 7. A screenshot of the integrated simulation and 3D visualization system. On the right part of the screen three subjective pedestrian views of the simulated environment are shown.

[2] http://irrlicht.sourceforge.net/

7. CONCLUSIONS AND FUTURE DEVELOPMENTS

The paper presented a research project aimed at defining, designing and implementing novel instruments supporting architectural designers by simulating the dynamics of crowds situated in the designed environments. A methodology for the modelling of crowds with SCAs has been shown, and then a case study related to a complex modelling scenario has been introduced. Preliminary results of this case study have also been illustrated and discussed. Finally, an introduction to an experience in the 3D visualization of simulated dynamics, which can be used for a more effective communication of simulation results, have been described. Future developments are aimed at refining both the methodology and the MMASS platform, in order to better support the modeller/user, in the construction of complex simulation scenarios. In particular the platform provides only preliminary user interfaces and modules aimed at supporting the definition of an active environment, and parameters for specific simulations, and a deeper analysis of current practices of designers must be carried out in order to better set this kind of instrument in their activities.

8. ACKNOWLEDGEMENTS

This work is preliminary result of the Social Mobile Entities in Silico (SMES) project, and was partly funded by the New and Old Mobility Analysis and Design for the Information Society (NOMADIS) laboratory, in the context of the Quality of Life in the Information Society (QUA_SI) multi-disciplinary research programme in Information Society.

9. REFERENCES

Bandini, S., S. Manzoni, C. Simone, 2002, "Dealing with space in multi-agent system: a model for situated MAS," in Castelfranchi, C., Johnson, L. (eds.) *Proceedings of the 1st International Joint Conference on Autonomous Agents and Multi Agent Systems (AAMAS 2002)*, ACM Press,1183-1190.

Bandini, S., S. Manzoni, G. Vizzari, 2004a, "Situated Cellular Agents: a Model to Simulate Crowding Dynamics," *IEICE - Transactions on Information and Systems*, **E87-D**(3): 669-676.

Bandini, S., S. Manzoni, G. Vizzari, 2004b Crowd Modelling and Simulation: Towards 3D Visualization. In: *Recent Advances in Design and Decision Support Systems in Architecture and Urban Planning*, Kluwer Academic Publisher, 161-175.

Bandini, S., G. Mauri, and G. Vizzari, 2006, "Supporting Action-At-A-Distance in Situated Cellular Agents," *Fundamenta Informaticae*, **69**(1-2): 251-271.

Batty, M., A. Hudson-Smith, 2005, "Urban Simulacra: From Real to Virtual Cities, Back and Beyond," *Architectural Design*, **75**(6): 42-47.

Batty, M., J. Desyllas, E. Duxbury, 2003, "The discrete dynamics of small-scale spatial events: agent-based models of mobility in carnivals and street parades," *International Journal of Geographical Information Science*, **17**(7): 673-697.

Dijkstra, J., J.P. van Leeuwen, H. Timmermans, 2003, "Evaluating Design Alternatives Using Conjoint Experiments in Virtual Reality," *Environment and Planning B*, **30**(3): 357-367.

Ferber, J., 1999, *Multi-Agent Systems*. Addison-Wesley.

Helbing, D., 1991, "A Mathematical Model for the Behavior of Pedestrians," *Behavioral Science*, **36**: 298-310.

Helbing, D., 1992, "A Fluid-Dynamic Model for the Movement of Pedestrians," *Complex Systems*, **6**(5): 391-415.

Okazaki, S., 1979, "A Study of Pedestrian Movement in Architectural Space, part 1: Pedestrian Movement by the Application on of Magnetic Models," *Transactions of A.I.J.*, **283**: 111-119.

Schadschneider, A., A. Kirchner, and K. Nishinari, 2002, „CA Approach to Collective Phenomena in Pedestrian Dynamics," in: *5th International Conference on Cellular Automata for Research and Industry, ACRI 2002*, volume 2493 of Lecture Notes in Computer Science, 239-248, Springer-Verlag.

Robert E. Shannon, 1998, "Introduction to the Art and Science of Simulation," in: *Proceedings of the 1998 Winter Simulation Conference*, pages 7–14, IEEE Computer Society.

Willis, A., N. Gjersoe, C. Havard, J. Kerridge, R. Kukla, 2004, "Human movement behaviour in urban spaces: implications for the design and modelling of effective pedestrian environments," *Environment and Planning B*, **31**(6): 805 – 828.

Exploring Heuristics Underlying Pedestrian Shopping Decision Processes

An application of gene expression programming

Wei Zhu and Harry Timmermans
Eindhoven University of Technology

Keywords: GEPAT, Decision process, Bounded rationality, Satisficing heuristic, Utility maximization, Model comparison

Abstract: Most analytical pedestrian behavior researches use utility-maximizing models and have paid less attention to models based on alternative behavioral theories such as bounded rationality. Consequently, there is a lack of deeper explorations into the decision processes of pedestrians. This lack of such alternative models may also be the result of inappropriate methods to estimate such models. For this reason, the paper first introduces a modeling platform GEPAT which has the GEPAT which has the ability to estimate parallel functions using a multi-gene-sectional chromosome structure and to facilitate building models using processors emulating simple decision mechanisms. The going-home decision of pedestrians in Wang Fujing Street is taken as an example to illustrate the use of GEPAT. The most important conclusion from a comparison of the MNL, hard cut-off, soft cut-off and hybrid model is that the satisficing heuristic fits better to the problem structure, at least in this case, than the utility-maximizing rule does. This example also shows the flexibility of GEPAT as a modeling toolbox and the power of estimating complex models.

1. INTRODUCTION

Mathematical modeling represents a way of understanding pedestrian behavior in shopping environments and supporting urban planning by predicting pedestrian behavior under alternative plans. Earlier research used gravity models to explain aggregate pedestrian movement patterns. Examples are Borgers and Timmermans (1985, 1986), Hagishima, Mitsuyoshi, et al. (1987) and Berry, Epstein, et al. (1988). Since the 1990s,

Jos P. van Leeuwen and Harry J.P. Timmermans (eds.), Innovations in Design & Decision Support Systems in Architecture and Urban Planning, 121-136.

the focus of this research tradition has shifted to the microscopic, individual level. The emergence of discrete choice models (DCM) in the 1970s is one of those major reasons that lead to this shift. Many research projects have been carried out using DCM to explore the decision mechanisms of pedestrians by assuming that the pedestrian make choices among alternative shopping places.

Another major reason for the shift in focus to microscopic models is the development of computer technology. The agent-based modeling technique, a hot topic recently, largely inherits the idea of object-oriented programming (OOP). Some agent-based computer simulation systems are available (e.g., Dijkstra and Timmermans, 2000, Haklay and O'Sullivan, 2001, Kerridge, Hine and Wigan, 2001) to mimic pedestrian behavior and support planning.

A good design and decision support system for retail planning relies heavily on reliable models of pedestrian behavior that are embedded in these systems. In turn, the reliability largely depends on the appropriateness of the assumed decision processes, which lead to behavior, and how this is represented by the model. Appropriateness here means the closeness between the decision mechanisms represented by the model and the real decision processes. Although many models have been suggested, the exploration of underlying decision processes is still scant. Simulation techniques are useful to construct very sophisticated decision diagrams, however, they are still unable to identify the behavioral rules underlying behavior. Although the DCM is based on a theory of human behavior and has proven its applicability in a variety of domains, the appropriateness to its behavioral underpinnings in the context of pedestrian movement has largely gone untested. One potential problem is that the assumed decision process may be too simple. According to Bettman (1979), the decision process of consumers should be composed of several inter-related information processing nodes, such as attention, evaluation, decision, learning, just to mention a few key ones. Another potential problem is that the assumed mechanisms may be too complicated, in the sense that pedestrians are assumed to have perfect knowledge of all choice options, and choose the option with the highest utility.

In contrast to such fully rational models, more realistic models might be built based on the bounded rationality theory proposed by H. A. Simon in 1947, who argued that heuristics are crucial in making fast, frugal, and good enough decisions (Todd, 1999, Todd and Gigerenzer, 2003). Different professions have formulated different definitions of heuristics. In the context of this study, we mean simple decision rules such as ignorance-based decision, one-reason decision, elimination heuristics and satisficing heuristics (Todd, 1999). Since the boundary between simplicity and complexity is vague, we only have intuitive belief that mental activities searching for information, using limited information and applying non-compensatory rules are simpler than those getting full information and applying compensatory rules. Such belief is partly supported by Gigerenzer, Todd and ABC Research Group (1999) whose experimental outcomes showed the superiority of simple models over complex models. However, a similar comparison, especially based on real-world behavioral data, has not yet been carried out in the context of pedestrian behavior.

The aim of the present study therefore is to conduct such a comparison. Before the results of this comparison are described in the third section, we will introduce, in the second section, a modeling platform—GEPAT which we developed to construct, calibrate and compare behavioral models. The fourth section will discuss the implications of these results for model building and concludes the paper.

2. GEPAT

2.1 Why GEPAT?

GEPAT is the acronym of Gene Expression Programming as an Adaptive Toolbox. It is a computer program for constructing and calibrating models. The core algorithm is gene expression programming (GEP, Candida Ferreira, 2001), one kind of genetic algorithm (GA). Linear regression models, general linear models, multinomial logit models (MNL), and other classical non-linear models can now be calibrated using different software such as SAS and SPSS. However, when the degrees of non-linearity, discontinuity and non-differentiability of the model function increase, it is difficult or even impossible for these software packages to calibrate the model. Genetic algorithms have been proven to perform satisfactory in solving such problem numerically. Because behavioral models could be very complicated, using GA may have some advantages.

In the case of pedestrian decision research, one usually only observes the behavioral outcome while the decision process remains hidden. It is very well possible that different decision processes could generate the same behavioral outcome (Todd and Gigerenzer, 2003). Thus, comparing different models may be less arbitrary than relying on a single model. GEPAT facilitates such a modeling process with building blocks, representing simple information processing nodes. It becomes much easier to manipulate and reuse these building blocks to specify models than to repeatedly write specific codes for every single model.

2.2 Features of GEPAT

2.2.1 Get Simultaneous Solutions

In GEP, a target function is derived from a code sequence which is composed of numbers, variables and operators. The code is transformed, mated with other codes to give offspring. Usually, the structure of the code is similar to Figure 1. The fundamental element of this structure is the *condon* where operators and operands are randomly generated and stored. Several numbers of codons compose the *gene*, the basic element forming functions after being translated. Both the length (the number of codons) and the number of the genes can be extended to include more information and create more complex functions. Among the genes, a space is designated for the *link function* to work together. All of these elements compose the *chromosome*. After the evolutionary process, the best chromosome with the highest fitness value is preserved and returns the target function.

When a decision process is a system composed of several information processing nodes, one function may well not be enough to represent the whole system. Calibrating model with parallel, inter-related functions is inevitable, which is impossible for the single chromosome structure in GEP. We extended this structure. In GEPAT, an additional element, *gene-section*, is designed (Figure 2). The gene-section is just the chromosome in GEP and the chromosome here is the composition of gene-sections. Each gene-section is an individual function block.

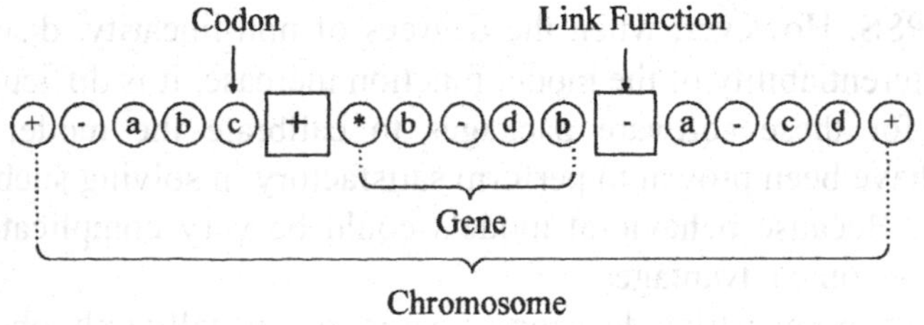

Figure 1. Chromosome structure in GEP.

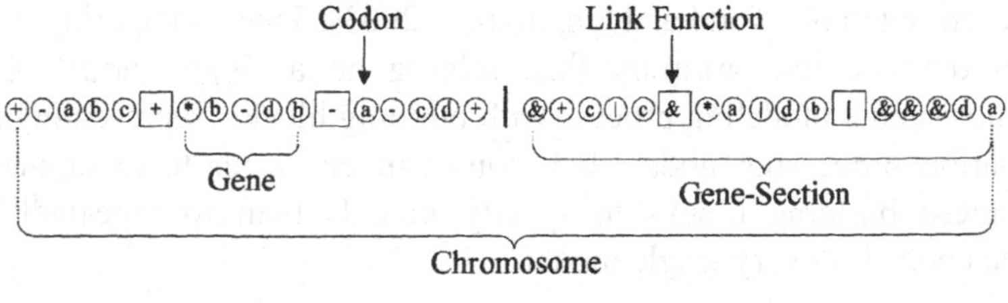

Figure 2. Chromosome structure in GEPAT.

This structure brings more flexibility. First, each gene-section can have its own length according to the complexity of the target function. Second, each gene-section can be assigned a specific type of function (e.g. arithmetic or logical) according to the nature of the task we want it to solve. Third, the gene-sections can communicate with each other with predictions and by-products of the function. This is the most important feature for modeling decision processes where information flows are frequently exchanged among mental operators. With this structure, getting simultaneous solutions for parallel functions becomes viable. In addition, when there is no priori model specification, GEPAT is capable of generating the model automatically. It is also efficient in searching the parameter space when the model is strictly defined.

2.2.2 Test Different Models

By organizing simple building blocks, called *processors*, models with different specifications and complexity can be tested. Each processor has input and output interfaces to receive information from and send information to other processors. For example, *satisficer* represents satisficing heuristics, that is the decision maker acts when the value of an option is over the threshold (cutoff); *maximizer* represents the decision rule choosing the option with maximum (minimum) utility; *memorizer* represents the function of memory and stores information for later use; *updater* cooperates with memorizer to update information when necessary; *collector* collects pieces of information and send them out as a whole when needed; *evaluator* evaluates the chromosome by fitness value and records the best chromosome in the generation; etc..

Figure 3 illustrates an example workflow of GEPAT. It starts from the data source transferring data to the receivers. If the model needs memorized information, the data is first transferred to the memorizer, then to the gene-sections. If the model is run for the first time, an initial generation of chromosomes is randomly generated. Usually, each gene-section corresponds to a processor about whose decision mechanisms we care most. In this example, gene-section1 corresponds to a maximizer so that in the end we can get a function telling how the pedestrian trades off factors and chooses the option. The gene-section2 corresponds to a satisficer. It can extract some judgement thresholds by using logical functions. After the information being transferred and modified in other processors, the evaluator evaluates each chromosome. If this is not the end of the iteration, the chromosomes are modified by gene operations according their fitness values. If the iteration ends, GEPAT outputs the results including the best chromosomes and their corresponding functions. Manipulating GEPAT is no more difficult than drawing such a flowchart. Each building block is represented by a visual element. By linking them and setting parameters properly, we can test many complex schemes of pedestrian shopping decision process.

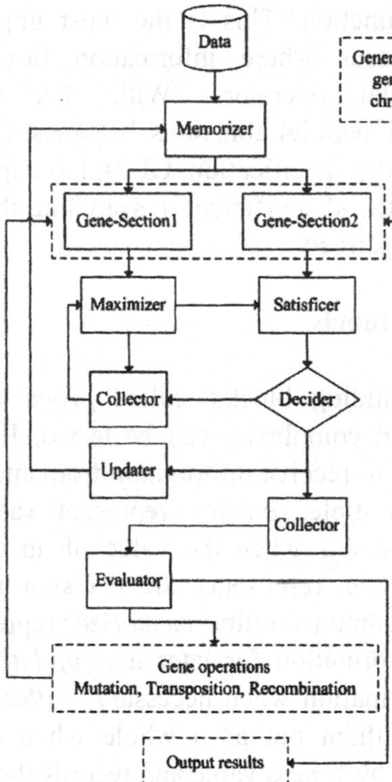

Figure 3. Sample workflow of GEPAT.

3. MODEL COMPARISON

The going-home decision of the pedestrian is taken as an example for the model comparison. The pedestrian shopping decision process can be differentiated into several stages. We assume that before the pedestrian patronizes a store, he/she must decide whether to go home/stop shopping. Building the appropriate model for this decision stage is important, especially for supporting plans based on number of patronages, because it influences the number of pedestrians who continue shopping.

3.1 Data

The data is collected from a survey carried out in Wang Fujing Street, the major shopping street in Beijing, China. The major section of this linear street where most retail stores are located is about 1,200 m long of which 534 m is the pedestrian street (Figure 4). Twenty undergraduate students administrated the survey on May 17th (Monday) and 22nd (Sunday), 2004. On 9 survey spots evenly distributed along the street from 11:00 to 20:00, they randomly asked pedestrians who completed their shopping trip to fill out a questionnaire. The surveyor recorded, based on the respondent's recall, every spatial point (entry/exit point and store) where he/she stopped since he/she entered the survey area. The activity type, expenditure, start and end time of the shopping trip were recorded sequentially in detail. The sample consists of 760 valid respondents of which 275 (36.2%) participated on May 17th and 485 (63.8%) on May 22nd.

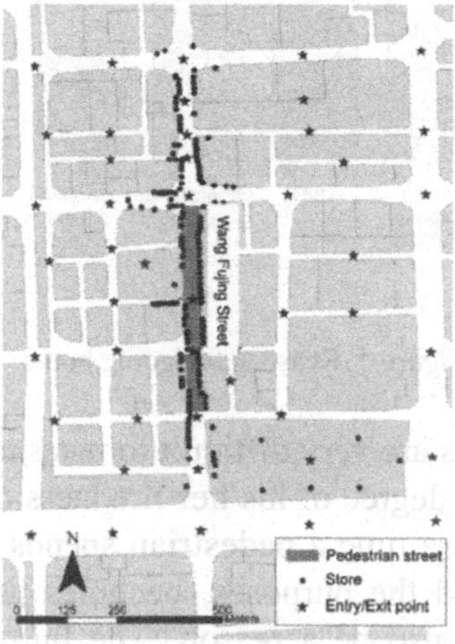

Figure 4. Survey Area.

The percentage of male respondents is 54%, implying that female respondents represent 46%. The sample was categorized into three age groups, young (16-29), middle-aged (30-49) and old (>=50). Their percentages are 53%, 34% and 13% respectively. A total of 689 records with complete multi-stop information were selected for calibrating the model. Each stop in every record is treated as an independent decision process. But we excluded the first stop, that is, the first visit of the pedestrian after entering the shopping street, from the data set, because the decision of not going home is certain. Finally, 2,741 observations were used in the model comparison.

3.2 Model Specifications

The pedestrian may decide to go home for many reasons. The survey also asked the respondents their major reasons for ending the shopping trip (Figure 5). Thirty-five percent of the respondents answered that they had fulfilled their purposes. Tiredness is the second most frequently given answer, representing 30% of the respondents. These two reasons and the fourth largest reason, planned places visited (14%), represent 80% of the respondents. It suggests that a satisficing heuristic could be the major decision mechanism of most pedestrians in Wang Fujing Street in deciding to go home.

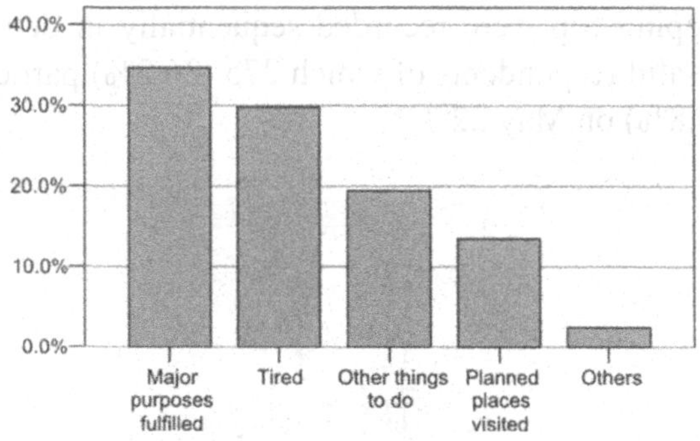

Figure 5. Reasons for going home.

However, these factors are very difficult to measure. We do not know the pedestrian's purpose, the degree of his/her tiredness or visit plans. Time is a substitute factor. The more time a pedestrian spends on shopping, the more likely he/she has fulfilled the purposes, becomes tired, or visited planned places. Time itself may also be a reason. The third largest reason (20%), other things to do, implies more or less that there is a cut-off time after which the pedestrian stops shopping.

We only recorded the start time and the end time of pedestrians' shopping activities because of the difficulty for the respondents to remember the exact time of performing every activity. An estimation technique was used to estimate in-between real time. Based on the known spatial points of the activities, the estimation was carried out using a simple distance/speed equation in the grid space. The shortest path rule was adopted. Two kinds of real time were used in the model: relative time and absolute time, both in minutes. Relative time refers to the time elapsed since the pedestrian started the shopping trip. Absolute time refers to the time difference between the current activity time spot and the 0:00 base.

We compare the rational utility-maximizing model with the bounded rational heuristic models. The former is represented by the classical MNL and the latter is represented by cut-off models adopting the satisficing heuristic.

3.2.1 MNL

In the MNL framework, we assume two choice options for pedestrian's decision making, shopping and going home. Their observable utilities are specified as,

$$\begin{aligned} V_s &= \beta_1 * RT + \beta_2 * AT \\ V_h &= \beta_3 \end{aligned} \tag{1}$$

where V_s is the utility of shopping which is the sum of relative time (*RT*) and absolute time (*AT*) weighted by their parameters, β_1 and β_2, respectively. It is hypothesized that the utility of shopping should decrease as time increases, so these two parameters should be negative. We set the linear combination of factors in this utility function for the convenience of the standard estimation procedure in SAS, although the true function may take any other forms, but this specification is enough to explore the general relationship between time and going-home decision in this study. The utility of going home, V_h, is represented by the single parameter β_3 whose sign is not hypothesized. Then the probability of going home is,

$$P_h = \frac{\exp(V_h)}{\exp(V_s) + \exp(V_h)} \tag{2}$$

MNL is a classical rational model which assumes that the utility is calculated as the weighted sum of all explanatory variables by the decision maker. However, the pedestrian may not necessarily use both *RT* and *AT* for decision. Another unrealistic point is that it adopts the compensatory rule which states that a smaller value in one variable can be (at least partially be) compensated by higher values of one or more other variables, so that, for example, even when the absolute time is very late (e.g., the stores are going to close at 22:00), the pedestrian could still decide to shop if he arrives at the shopping street at exactly this time as long as the utility from *RT* is large enough than that from *AT*.

3.2.2 Hard Cut-off Model

In the hard cut-off model, we use cut-offs for factors and avoid using compensated utilities. We hypothesize two cut-offs, a higher cut-off and a lower cut-off, for *RT* and *AT* (Figure 6). The higher cut-off is set as a physical or psychological limit of the pedestrian: once this cut-off has been passed, a pedestrian must go home. For example, it may be related to tiredness and the pedestrian does not have more energy to shop. In contrast, the lower cut-off is the threshold that must be reached by a pedestrian to continue shopping. It may mean that the pedestrian has not fulfilled the purposes or he/she has more time to spend on shopping when the values of factors are below these lower cut-offs. Unlike the MNL which assumes that all factors are considered simultaneously in the decision, the hard cut-off model includes a three-step information search and judgment procedure.

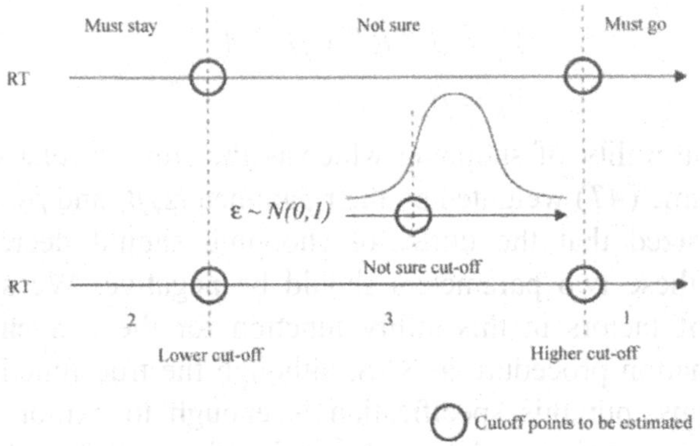

Figure 6. Hard cut-off model.

Step 1: The pedestrian thinks first whether either the value of *RT* or *AT* is above the respective higher cut-off. If it is true, he/she will go home. That is,

$$P_h = 1, \text{ if } RT \geq HC_{RT} \text{ or } AT \geq HC_{AT} \tag{3}$$

where HC_{RT} is the higher cut-off for *RT*, HC_{AT} is the higher cut-off for *AT*.

Step 2: If neither *RT* nor *AT* is above its higher cut-off, the pedestrian checks whether either *RT* or *AT* is below the respective lower cut-offs, RT_{LC} and AT_{LC}. If it is true, he/she will continue shopping. That is,

$$P_h = 0, \text{ if } RT < LC_{RT} \text{ or } AT < LC_{AT} \tag{4}$$

Note that, within the above two steps, the probability of going home either equals 0 or 1, because the cut-offs are hard.

Step 3: If neither *RT* nor *AT* is below its lower cut-off, we assume that the pedestrian is in the *Not-Sure* state. That is, his/her decision cannot be made on these two factors, but on the unobserved factor ε which we assume to be standard normally distributed which can be observed from many other social phenomenon. Another cut-off C_{NS} is hypothesized for ε. If ε is larger than C_{NS}, the pedestrian will go home; otherwise he/she will continue shopping. That is,

$$P_{hNS} = 1, \text{ if } \varepsilon \geq C_{NS} \tag{5}$$

$$P_{hNS} = 0, \text{ if } \varepsilon < C_{NS}$$

Rewriting equation 5 into probabilities, we get

$$P_{hNS} = 1 - F_{NS}(C_{NS}), \text{ if } (RT \geq LC_{RT} \text{ and } RT < HC_{RT}) \text{ and } (AT \geq LC_{AT} \text{ and } AT < HC_{AT}) \quad (6)$$

where $F_{NS}()$ is the cumulative density function of the standard normal distribution.

To sum up, there are five parameters to be estimated in the hard cut-off model, LC_{RT}, HC_{RT}, LC_{AT}, HC_{AT} and C_{NS}. Because C_{NS} is a constant, estimating it is same as estimating the probability of going home when the pedestrian is in the not-sure state, P_{hNS}.

3.2.3 Soft Cut-off Model

The cut-offs in the hard cut-off model are assumed to be constant for every observation of every pedestrian. In reality, this is unrealistic because pedestrians have different habits, purposes, schedules, and/or taste variations. These factors may cause cut-offs to differ among pedestrians or shopping stages of the same pedestrian. Incorporating heterogeneities into the model specification usually makes a model more general and may improve its performance.

We make the hard cut-offs soft by assuming that cut-offs are iid normally distributed across observations. For example, the lower cut-off of RT is assumed to be normally distributed with mean LCM_{RT} and standard deviation $LCSD_{RT}$. The cumulative density function is $F_{LCRT}()_0$, that is,

$LC_{RT} \sim N(LCM_{RT}, LCSD_{RT}), F_{LCRT}()_0$ where the 0 at the lower right corner means that the distribution is left-truncated at 0 because the time cannot be less than 0.

For the same reason,

$HC_{RT} \sim N(HCM_{RT}, HCSD_{RT}), F_{HCRT}()_0$

$LC_{AT} \sim N(LCM_{AT}, LCSD_{AT}), F_{LCAT}()_0$

$HC_{AT} \sim N(HCM_{AT}, HCSD_{AT}), F_{HCAT}()_0$

$C_{NS} \sim N(CM_{NS}, CSD_{NS}), F_{CNS}()$

For RT, the probability of the variable larger than the lower cut-off is,

$$P(RT \geq LC_{RT}) = F_{LCRT}(RT)_0 \quad (7)$$

The pedestrian must go home if RT is larger than the lower cut-off and the higher cut-off simultaneously. The probability is,

$$P_{MGRT} = F_{LCRT}(RT)_0 * F_{HCRT}(RT)_0 \quad (8)$$

The pedestrian must stay if RT is less than the lower cutoff and the higher cutoff simultaneously. The probability is,

$$P_{MSRT} = (1 - F_{LCRT}(RT)_0) * (1 - F_{HCRT}(RT)_0) \quad (9)$$

The pedestrian is not sure to go or stay if *RT* is larger than the lower cutoff and less than the higher cutoff. The probability is,

$$P_{NSRT} = 1 - P_{MGRT} - P_{MSRT} \quad (10)$$

For the same reason, probabilities for *AT* are,

$$P_{MGAT} = F_{LCAT}(AT)_0 * F_{HCAT}(AT)_0 \quad (11)$$

$$P_{MSAT} = (1 - F_{LCAT}(AT)_0) * (1 - F_{HCAT}(AT)_0) \quad (12)$$

$$P_{NSAT} = 1 - P_{MGAT} - P_{MSAT} \quad (13)$$

We still assume $\varepsilon \sim N(0,1)$ when the pedestrian is in the not-sure state. This time because the C_{NS} is not constant, the probability of going home depends on the joint distribution of the two variables, assuming their independence. That is,

$$P_{hNS} = P(\varepsilon \geq C_{NS}) = P(C_{NS} - \varepsilon \leq 0) = F_{NS}(0) \quad (14)$$

where $F_{NS}()$ is the cumulative density function of the normal distribution $N(CM_{NS} - 0, \sqrt{CSD_{NS}^2 + 1})$.

The logic of the soft cut-off model is the same as the hard cut-off model. With the above probabilities, the probability of a pedestrian going home is,

$$P_h = P_{MGRT} + P_{MGAT} - P_{MGRT} * P_{MGAT} + P_{NSRT} * P_{NSAT} * P_{hNS} \quad (15)$$

For the same reason as the hard cut-off model, the estimation of CM_{NS} and CSD_{NS} can be substituted by the estimation of the single probability P_{hNS}. Thus, in the soft cut-off model, a total of 9 parameters are to be estimated.

3.2.4 Hybrid Model

The pedestrian might also decide in such a way that simple rules are used first; if simple rules cannot help to decide, more complex rules will be applied. Similar arguments have been made by Bettman (1979) and Grether and Wilde (1984). The hybrid model incorporates both the cut-off model and the utility-maximizing model. It is specified as the soft cut-off model with the only difference in the not-sure probability, P_{hNS}. In the hybrid model, P_{hNS} is not constant but derived from,

$$\beta_3 - \beta_1 * RT - \beta_2 * AT + \varepsilon \geq 0 \quad (16)$$

where β_1, β_2, and β_3 are parameters with the same meanings as in MNL. ε is a random factor assumed to be standard normally distributed, so that,

$$P_{hNS} = 1 - F_\varepsilon(\beta_1 * RT + \beta_2 * AT - \beta_3) \quad (17)$$

The probability now is dependent on the trade-off between the utility of going home and the utility of shopping. In total, eleven parameters are to be estimated in the hybrid model.

3.3 Results and Comparison

The MNL model is calibrated with SAS and cut-off models are calibrated with GEPAT, as the parameter finder, using a hill-climbing algorithm as the local search method to complement GEP as the global search method. The results are shown in Table 1. Three statistics are used for model comparison. All the models are calibrated based on the maximum likelihood (ML). Akaike's Information Criteria (AIC) is calculated as the comprehensive index of goodness-of-fit against model complexity. Each model is simulated 20 times, drawing random numbers when necessary, to give the average percentage of correct hits as another statistic for the comparison.

Table 1. Results of model calibrations.

MNL		Hard Cut-off		Soft Cut-off		Hybrid	
P	Value	P	Value	P	Value	P	Value
β_1	-0.007	LC_{RT}	29.797	LCM_{RT}	132.048	LCM_{RT}	0.000
β_2	-0.008	HC_{RT}	674.966	$LCSD_{RT}$	83.976	$LCSD_{RT}$	327.290
β_3	-10.501	LC_{AT}	809.840	HCM_{RT}	676.000	HCM_{RT}	676.992
-	-	HC_{AT}	1313.169	$HCSD_{RT}$	0.010	$HCSD_{RT}$	0.010
-	-	P_{hNS}	0.308	LCM_{AT}	927.851	LCM_{AT}	916.544
-	-	-	-	$LCSD_{AT}$	87.422	$LCSD_{AT}$	85.820
-	-	-	-	HCM_{AT}	1305.591	HCM_{AT}	1377.659
-	-	-	-	$HCSD_{AT}$	104.161	$HCSD_{AT}$	230.719
-	-	-	-	P_{hNS}	0.752	β_1	-0.047
-	-	-	-	-	-	β_2	0.000
-	-	-	-	-	-	β_3	-3.502
ML	-1121.200	-1381.830		-1070.599		-1077.843	
AIC	2248.400	2773.660		2159.199		2177.687	
Sim	0.546	0.656		0.743		0.744	

The parameters of MNL are estimated as expected. Its ML is only higher than that of the hard cut-off model. The parameters of the hard cut-off model could be interpreted as follows. The lower cut-off for RT is about 30 min, suggesting that all the pedestrians shop for at least half an hour. However, this cut-off could be too small for the whole sample because the model is too clear-cut to allow any mis-prediction which will lead to a very low ML. The same problem could also apply to the lower cut-off for AT, which is about 13:30. For this reason, only 19.1% of the sample can be explained by the lower cut-offs, although completely right, while the shopping behavior actually represents 74.7% of the sample. The effectiveness of the higher cut-offs is weak for that only 0.6% of the sample can be predicted as going home. The remaining 80.3% of the sample is explained by the random variable where the going-home probability under the not-sure state is 0.308. The hardness of the specification of the hard cut-off model makes its ML the least among all models.

The introduction of heterogeneity into the soft cut-off model clearly raises the ML, which is the highest. The means of the lower cut-offs increase (about 132 min for RT and 15:20 for AT), compared to the ones in the hard cut-off model, and their standard deviations are also reasonable. This directly leads to the increase in shopping behavior within the sample and the increasing probability of going-home in the not-sure state. However, the means of the lower cut-offs are similar to those in the hard cut-off model and their effects remain weak. Because the higher cut-off for AT is still large, about 22:00, its standard deviation can only enclose limited number of observations.

The ML of the hybrid model is slightly lower than that of the soft cut-off model, but the mean of the lower cut-off drastically changes to 0 with a large standard deviation, others remain similar. This means that, on average, the utility-maximizing part substitutes some of the predictability of *RT*. However, this substitution effect is more competitive rather than complementary so that it detriments the overall ML a little bit.

The same conclusions can be drawn based on the AIC-statistic. Although the soft cut-off model has more complexity (the number of parameters), its AIC is still the best (lowest). In contrast, the more complexity in the hybrid model brings a lower ML and AIC, probably implying that the utility-maximizing part is just mis-specified. As for the simulation statistic, an interesting question exists between the MNL model and the hard cut-off model, where the latter simulates behavior much better than the former, inversing the relationship represented by ML and AIC. This raises the question on how to judge the goodness of a model with conflicting goodness-of-fit statistics. A deeper exploration into the properties of the model is required and we only give a first explanation here. The MNL model may be performing better in terms of the ML-statistic because of its better averaging ability to enclose more variance, just like using a big round shape (the model) to wrap an irregular shape (the data). But because the hardness of the hard cut-off model (just like a slim ellipse shape) limits its flexibility, it tries to wrap the most significant feature of the irregular shape (may be the most frequently happening event in the data) while ignores other parts. The more variability of the MNL implies more uncertainty during simulation while the hard cut-off model hits the most important data effectively by predicting more certainly with modest variability. In this sense, we argue that the hard cut-off model fits better to the problem structure of the going-home decision.

4. CONCLUSION

Most analytical pedestrian behavior research has used utility-maximizing models and paid less attention to models of other behavioral theories such as bounded rationality, which may lead to a deficiency in deeper explorations into the decision processes of pedestrians. Such deficiency could also be a result of inappropriate methods to estimate the model. For this reason, the paper first introduces the modeling platform GEPAT with the ability to estimate parallel functions using a multi-gene-sectional chromosome structure and to facilitate building models using processors emulating simple decision mechanisms.

The going-home decision of the pedestrians in Wang Fujing Street is taken as an example to illustrate the use of GEPAT. The major conclusion derived from the model comparison is: (1) the satisficing heuristic fits better to the going-home decision than the utility-maximizing rule, suggesting the bounded rational behavior of pedestrians; (2) pedestrian behavior is heterogeneous; introducing heterogeneity into the model specification is appropriate and effective; (3) lower cut-offs, as the baseline of decision, are much more important than high cut-offs, reinforcing the hypothesis that pedestrians are satisficers.

This example has also shown the flexibility of GEPAT as a modeling toolbox and the power of estimating complex models. Because that GEPAT uses numerical estimation algorithm, it causes a common problem of unawareness of the user whether the estimator reaches the global optimum. In this case, although we tried the estimation procedure for each model several times and got similar results, but usually not the same, it still cannot guarantee the global optimum. Such procedure costs intensive computation resource and time. Improving the estimation efficiency of GEPAT will be the major task of our next step, with other additional functionalities such as a parameter sensitivity test.

5. REFERENCES

Berry, B. J. L., B. J. Epstein, A. Ghosh, et al. 1988, *Market Centers and Retail Location: Theory and Applications*, Prentice-Hall, London.

Bettman, J. R., 1979, *An Information Processing Theory of Consumer Choice*, Addison-Wesley Publish Company, Reading.

Borgers, A. and H. Timmermans, 1985, "A Model of Pedestrian Route Choice and Demand for Retail Facilities within Inner-City Shopping Areas", *Geographical Analysis*, **18**(2): 115-128.

Borgers, A., and H. J. P. Timmermans, 1986, "City Centre Entry Points, Store Location Patterns and Pedestrian Route Choice Behavior: a Microlevel Simulation Model" *Socio-Economic Planning Science*, **20**(1): 25-31.

Dijkstra, J. and H. Timmermans, 2002, "Towards a Multi-agent Model for Visualizing Simulated User Behavior to Support the Assessment of Design Performance", *Automation in Construction*, **11**: 135-145.

Ferreira, C., 2001, "Gene Expression Programming: A New Adaptive Algorithm for Solving Problems", *Complex Systems*, **13**(2): 87-129.

Gigerenzer, G., P. M. Todd and ABC Research Group, 1999, *Simple Heuristics that Make us Smart*, Oxford University Press, New York.

Grether, D., and L. Wilde, 1984, "An Analysis of Conjunctive Choice: Theory and Experiments", *Journal of Consumer Research*, **1984**(10): 373-385.

Hagishima, S., K. Mitsuyoshi, and S. Kurose, 1987, "Estimation of Pedestrian Shopping Trips in a Neighborhood by using a Spatial Interaction Model", *Environment and Planning A*, **19**(9): 1139-1152.

Haklay, M., and D. O'Sullivan, 2001, "'So go downtown': Simulating Pedestrian Movement in Town Centers", *Environment and Planning B*, **28**(3): 343-359.

Kerridge, J., J. Hine, and M. Wigan, 2001, "Agent-Based Modeling of Pedestrian Movements: The Questions that Need to be Asked and Answered" *Environment and Planning B*, **28**(3): 327-341.

Todd, P. M., 1999, "Simple Inference Heuristics versus Complex Decision Machines", *Minds and Machines*, **9**(4): 461-477.

Todd, P. M., and G. Gigerenzer, 2003, "Bounding Rationality to the World," *Journal of Economic Psychology*, **24**(2): 143-165.

SCALE

A street case library for environmental design with agent interfaces

Chiung-Hui Chen[1] and Mao-Lin Chiu[2]

[1]*Department of Information and Design, Asia University*

[2]*Department of Architecture, National Cheng Kung University*

Keywords: Agent interface, Behaviour, Simulator, Street design

Abstract: Urban space provides a context for human interaction. Recently, urban planning has largely placed the user at the street as the centre of infrastructural design, with significant implications for the perceived attractiveness of user environments. However, visual observation is often difficult for verifying planning goals. The simulation of pedestrian behaviour is important for physical planning, but such research is scarce. In this study, we adopt an empirical approach for generating reactive path following. Further, we implement scenarios as computer scripts with agent-based interfaces to identify navigational patterns. Moreover, we built a hierarchy of individual behavioral models and define a behavior production system to control the agent. Key attributes of streets such as rest space, utilities, landmarks, and buildings have space tags as identifiers to associate streets with related activities.

1. INTRODUCTION

As an important element of urban form, streets function as social spaces, commercial spaces, cultural spaces, as channels of movement and as symbolic representations of local tradition and culture. Since street spaces are not only compromised of physical elements but also of people who are moving and using them, informal street activities emerge as an integral part of street life.

Urban designers faced with the task of designing such spaces, needs a tool that will allow different designs to be compared in terms of their

Jos P. van Leeuwen and Harry J.P. Timmermans (eds.), Innovations in Design & Decision Support Systems in Architecture and Urban Planning, 137-150.

attractiveness and effectiveness. Therefore, this paper develops an agent interface approach for creating a street simulator of user behaviours in urban street environments. We implemented the agent interface as an individual-based simulation as part of a proposed project called "SCALE" (A Street Case Library for Environmental design). The project is demonstrated to examine differences between the simulation and the real environment, and illustrate the methodology, which includes observation, analysis, prototype system and evaluation. We also present possible applications and discuss social impacts of these applications.

2. CULTURAL-BASED INFORMAL STREET ACTIVITIES

Urban streets are one of the important physical elements of cities. The street is a place where human activities are concentrated. It mainly functions as a channel of movement that connects one place to the other. The multi-functions of a street have been recognized by various scholars including Jacobs (1961), Rykwert (1986), Czaenowsky (1986), Moughtin (1992), Rapoport (1987) and Jacobs (1993). The roles of streets in urban life can be summarized as follows: a street is a channel of movement, a communication space, a place of social and commercial encounter and exchange, a place to do business, a political space and also a symbolic and ceremonial space in the city.

Streets of Asian cities are significant in the context of urban public life. Asian streets have culturally and traditionally served the city as a public space, a place where people come together for commerce, to eat and to socialize. Cultural-based informal street activities refer to street events, art performances, traditional foods, culture-based goods such as crafts etc that form the life of a street. According to Rapoport (1987), activity in any given setting is primarily culturally based in that it is the result of unwritten rules, customs, traditions, habits, and the prevailing lifestyle and definition of activities appropriate to that setting. Rapoport (1987) further argued that cultural variables are primary for any activity, including walking and others, occurring in streets. It is culture that structures behavior and helps explain the use or non-use of streets and other urban spaces – or of other settings. Thus, the use of streets by pedestrians is primarily culturally based as a physical environment. Given this culturally based predisposition towards obeying unwritten rules of proper street use, people can also be influenced by physical variables.

3. SYSTEM DESCRIPTION

In virtual environments, the physicality is different. People would consider navigating virtual spaces to be quite different from navigation in physical spaces as there is no body to be moved. However, we treat them as essentially similar activities. For example, we provide the user with a steering agent interface without the full range of sensory inputs of a physical body. In addition, "Guide Map" can be used to provide navigational information and can be supplemented with additional detail about the tags in the environment. In other words, "maps are social tags", they are there to give information and help people explore, understand and find spaces of their interest. Even when we are not directly looking for information we use a wide range of cues, both from features of the environment and from the behaviors of other people to manage our activities.

3.1 Define the Scenarios, user Behaviors and Agent Simulator

SCALE proposes that a digital city should provide a universal location identifying system for Space-tags and other location-aware information. SCALE is a platform to support digital data and is a media suitable for advertising and city guide information. Space-tags are virtual objects that can be accessed only within a limited area and limited time period. Location-restricted digital objects are stored in a database on a server which is required to manage Space-tags objects.

In addition, by using avatar agents, users can walk around and experience the virtual environment on a "human scale", and can see 3D scenes from the viewpoint of a walking avatar. Thus, avatars can access Space-tags with portable devices such as mobile phones. They walk around in a street and find Space-tags that can be only found at that location, and also can put Space-tags at the location where he or she is, which can be found by other avatars nearby. This function also enables local communication applications. In order to clarify the relations between the problem domains in this research, we provide a conceptual framework in terms of five perspectives: Agent interface, User behaviors and Date analysis mode, Simulation environment, Street condition, and the real application of street design. The relations are shown in *Figure 1*.

3.2 The SCALE and the Agent-Interface

The virtual environment is defined by a plane(x, y) of a street space, limited by a grid frame. In the area defined by this frame different Space-tags can be created. These tags represent the following external stimuli, characterized into six groups: (1) seats, (2) trees, greenery space, (3) signs, (4) shops, (5) parking, (6) public services (toilets, etc). The avatar agent perceives these stimuli and acts upon them, as shown in *Figure 2.*

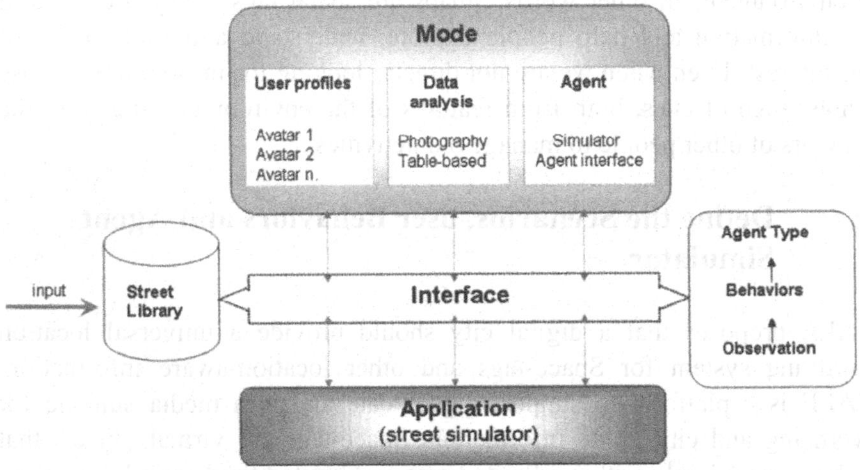

Figure 1. A conceptual architecture of the SCALE.

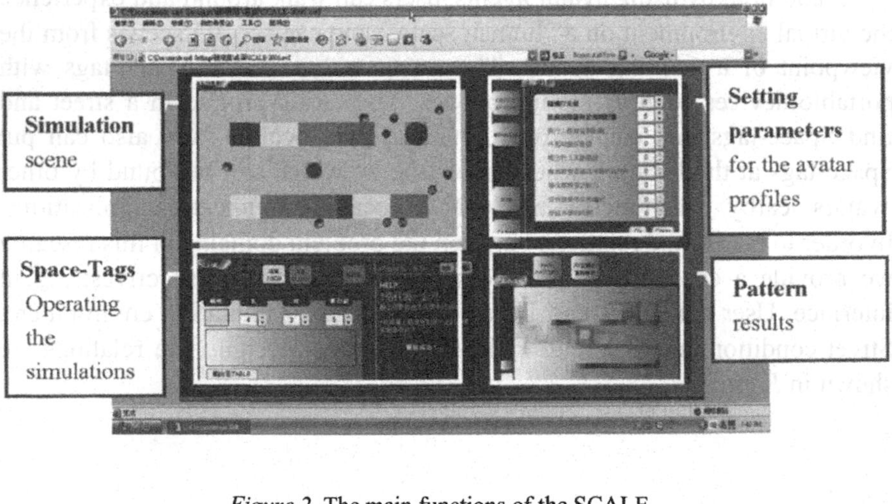

Figure 2. The main functions of the SCALE.

The interface provides the following functions:

- List all the Space-Tags that can be accessed;
- Creation of new Space-Tags;
- Detect these Space-Tags.

A user is shown as an avatar (male, female or elder). Space-tags are shown with icons around the user. The avatar agent instance proceeds as follows:

(1) Read the state information for the avatar including position, direction, velocity, and viewpoint.
(2) If the state has changed, use the data of the previous step and go to (4). Otherwise, calculate the conditions and locomotion parameters for the next step and go to (3).
(3) Calculate one step and store the data in memory.
(4) Display the one-step.
(5) Go back to (1).

4. METHODOLOGY

Data collection methods of pedestrian behaviour can be broadly classified into two categories: "tracking", and "fixed-point observation". Tracking involves following pedestrians who pass through any given place. Fixed-point observation implies counting the number of pedestrians passing through a specific place in a given time interval. Tracking is effective for obtaining data of pedestrian route choice behaviour, while fixed-point observation is effective in obtaining data for a particular place.

4.1 Observation Study

Past studies on the relationship between people's activities and spatial information given by the environment suggest that such studies can provide general guidance for urban design. However, it is not at all easy to collect enough data for such studies; observing people's behavior in a physical environment requires time and it is also difficult to identify the factors that influence user behavior. By using agent technology, we investigated the characteristics of activities in Taichung City, Taiwan. By limiting the subject of the study to observing the relationship between visual stimuli and the behavior of people in the study area, we started testing the applicability of the developed system as a tool for observing people's behavior in window shopping and strolling activities.The procedure and findings of this study will be described in the next sections.

4.2 Data acquisition

The data of noticeable features in a street is classified into two major categories i.e. the physical and non-physical features. Physical features refer to all tangible features in the street, especially architectural features, which can be classified into four major categories, landmarks, places, shops and malls. Landmarks refer to all physical features in the street which have monumental or symbolic meanings to people. Non-physical features refer to all the intangible, non-permanent and mobile features in the street which can be classified as general ambience and specific ambience. General ambience represents the nuances of the street, scenery, and also people's activities which can be perceived during experiencing the street. The best way to perceive ambience is by walking through the street and experiencing the nuances of the street space. Specific ambience represents all features mentioned by the perceiver, such as street vendors, foods, culture, etc. The category of noticeable features in a street shows the diversity of elements in the street as perceived by the respondents.

4.2.1 Participants: Twenty Pedestrians

Twenty individuals participated in the study, including ten males and ten females. The age of the participants varied between 18 and 29, with the average being 23. The occupational backgrounds of the participants varied. Sixteen were students, four were housewife. None of the participants knew each other. Seventeen of twenty participants had never visited the shopping street before. Respondents were asked to exchange remarks on what they saw and what they thought, just as they would if they were walking in the street. In addition, remarks they made during the real walk were collected. *Figure 3* illustrates the sequence of images with position and location detection during walking in a street.

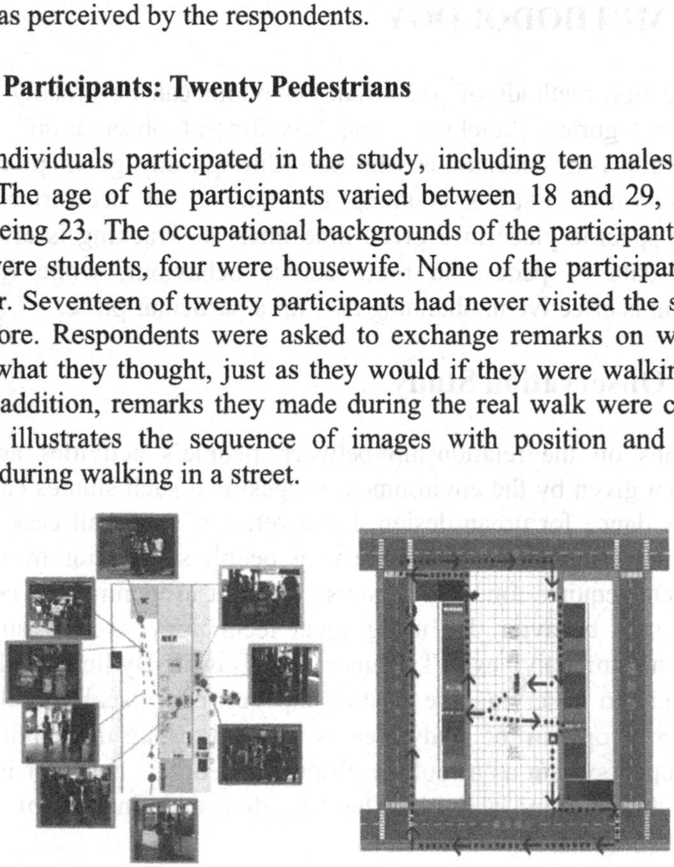

Figure 3. Sequence of images with position and location detection during walking in a street.

4.3 Attempted Evaluation of Simulate the Pedestrian Patterns

Our previous studies (Lin and Chiu, 2003) found that a digital city is a social information infrastructure for urban life, while users or participants within the digital city are often foreign to the environment without navigational aids. A scenario is defined as a script with roles, scenes, events, and time sequence (Chen, Lin and Chiu, 2004). In other words, the scenarios of role-play are designed according to the background and the requirement of potential users, such as age, gender, occupation, and visiting situation (individual or group). Therefore, users can define their roles in the play and can select an established scenario, change the scenes or time sequence of the established scenario to have a personal experience.

4.3.1 Experiment 1: Agent as Navigational Aids

Anders (1998) argues that "The memory palace could resurface as a model for future collective memory allowing users to navigate stored information in an intuitive spatial manner" and that "…cyberspace will evolve to be an important extension of our mental processes" allowing us to"…create interactive mnemonic structures to be shared and passed from one generation to the next."

In other words, the proposed "Memory Palaces" could serve as an ever-evolving repository of people's knowledge that users can discuss, learn from, contribute to, and collectively build upon. As shown in *Figures 4* and *5*, a sample scenario simulates and describes what users experience in the street environment.

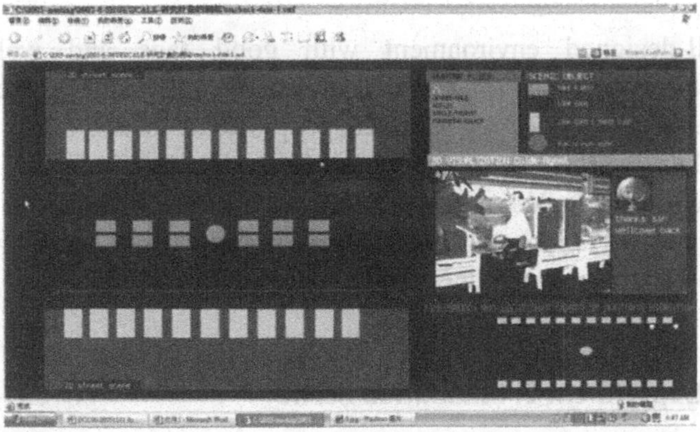

Figure 4. A window snapshot of the scenario that user's explore experience in the street environment and navigation with the agent-interface.

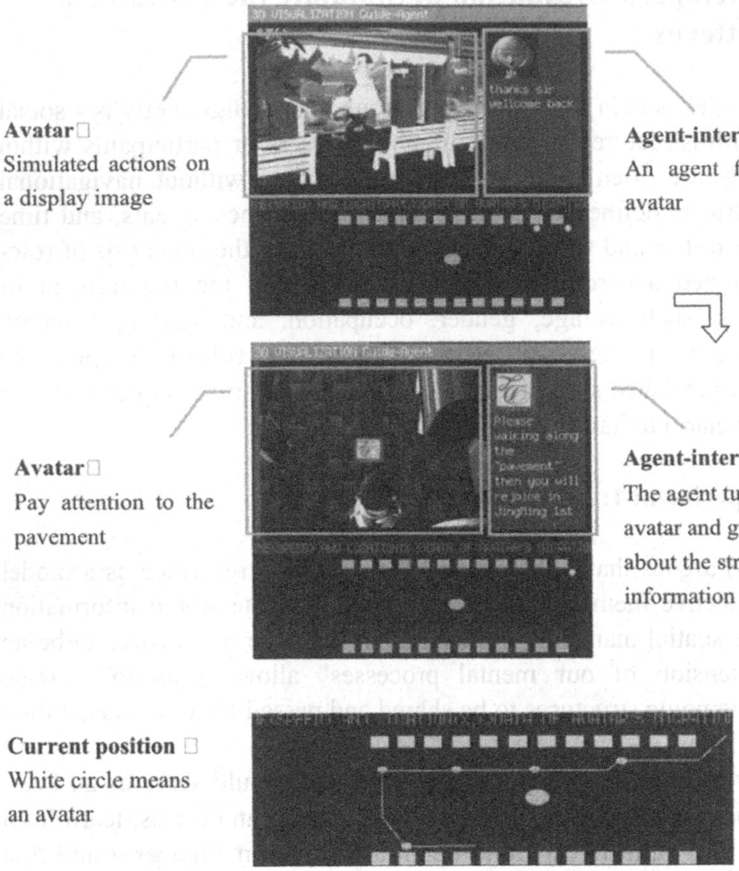

Figure 5. The user has his own route and sequence to visit the scenes at the street environment.

A well-designed environment with good tags and well-designed navigational aids such as a guide agent will be conducive to good navigation. When navigating a street, people tend to ask other people for advice rather than study maps, and information from people is usually personalized and adapted to suit the individual's needs. Meanwhile, people explore streets by talking to or following the trails of others, or walk into a sunny street to see something. These experiences all have an impact on navigation.

4.3.2 Experiment 2: Simulation Scene

The experiment presents a series of simple simulation runs to convey the idea that a change of street environment can have different effects on movement patterns and the use of space. We create five simulation scenarios, differentiated by the Space-tags and varying the character of the composed elements. In this series we mix three types of pedestrians together in one scene. Each scene has the same number of pedestrians and the same proportion. In addition, in each simulation the run period is ten minutes. The resulting movement patterns of scenes are presented in *Figures 6 -10.*

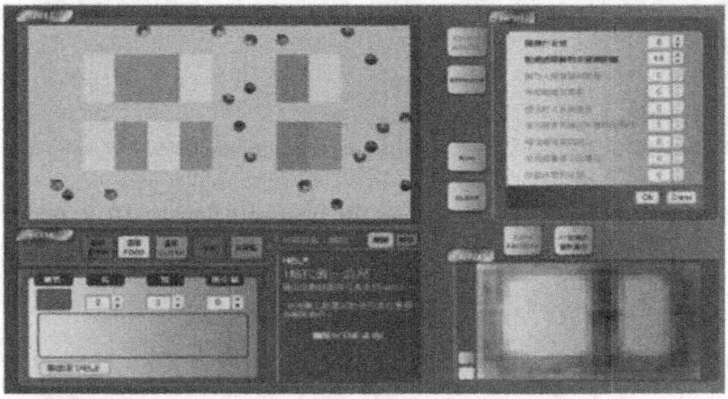

Figure 6. Scenario-1 : Three types of space-tags infill, male=10, female=10, Random walking=0, detect distance=10, run time=10minutes.

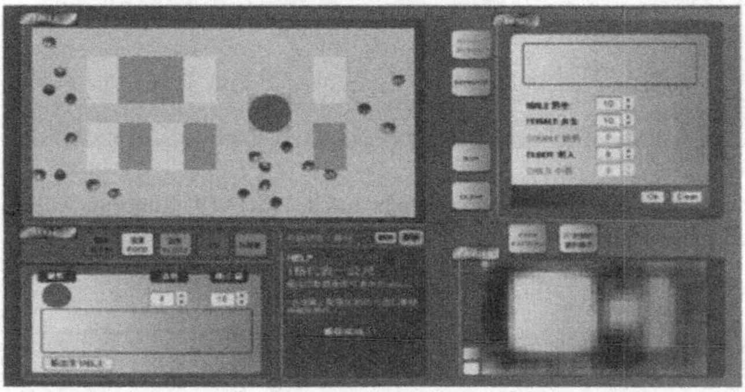

Figure 7. Scenario-2: Four types of space-tags infill, male=10, female=10, Random walking=10, detect distance=10, run time=10minutes.

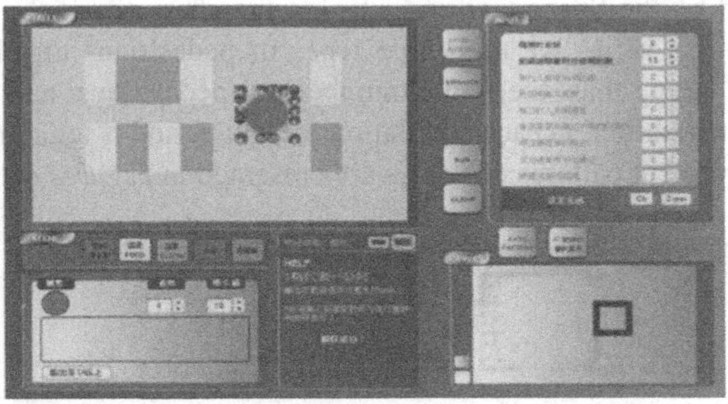

Figure 8. Scenario-3: Four types of space-tags infill, male=10, female=10, Random walking=0, detect distance=10, run time=10minutes.

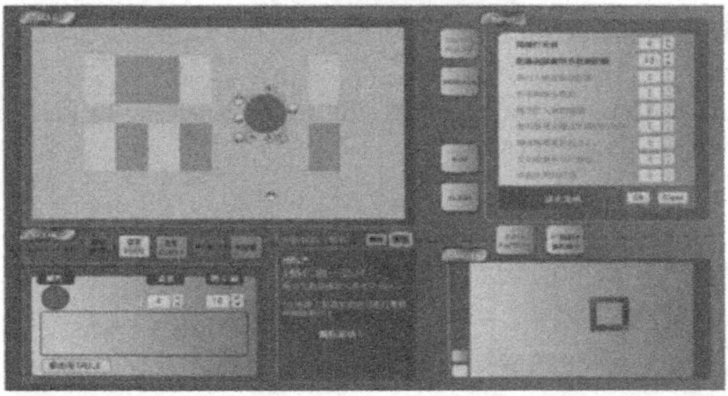

Figure 9. Scenario-4: Four types of space-tags infill, elder=10, Random walking=0, detect distance=10, run time=10minutes.

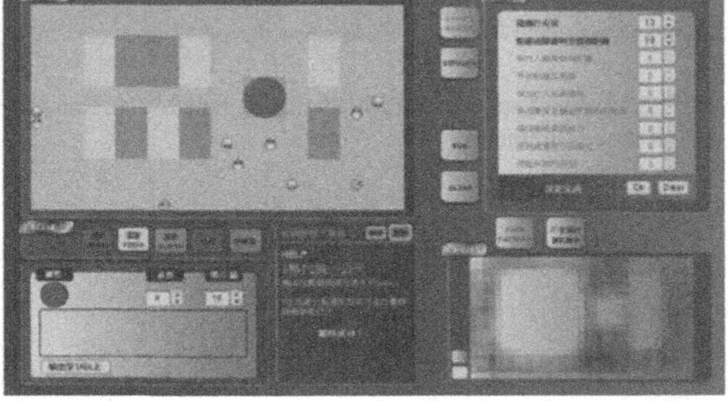

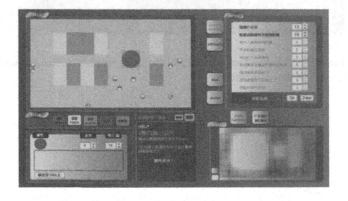

Figure 10. Scenario-5: Four types of space-tags infill, elder=10, Random walking=10, detect distance=10, run time=10minutes.

4.3.3 Study Results

Comparing these five patterns, we clearly find that differences in space configuration and the location of the attractions affect the distribution of spatial movement patterns. We also learn that movement patterns are influenced by time of day, different numbers and proportion of wandering and purposive behaviors. All the differing effects are based on the interaction between each avatar and the elements of the environment. The patterns generated from this experiment do not give much information for space analysis and evaluation, but the trace of movement pattern is a useful technique for graphically recording the dynamic quality of behavior and environment relationships. Thus, the urban designer can test the effects of any design changes through the SCALE simulators on pedestrian behaviours before their implementation, allowing to explore how the layout of urban street space affects human behaviours within it, and what kinds of factors induce a more enjoyable walking experience.

5. DISCUSSION

The above observation suggests that quality of visual information has an important influence on guiding people. A visual indication that shops were located further, such as a continuous street surface design, integrated design of street signs or street furniture, was an important factor for guiding pedestrians.

5.1 Street and Human Activity

Through studies in the street we found that:

(1) Even during walking, pedestrians' conversation related mostly to topics prompted by visual stimuli.

(2) Pedestrians tended to walk within the sections of streets featuring large shops, an arcade, and integrated street surface designs.

(3) Pedestrians carefully looked at the changes found along the streets as well as shops or other spatial objects related to the subjects of their interest, but they often chatted while moving and paid little attention to the street scene.

(4) Pedestrians mainly changed course at an intersection at which they could easily identify their location, in order to avoid becoming lost. In other words, they would seldom enter streets that they could not see down or that seemed to contain nothing of interest.

5.2 The Cognition Map and Human Memory

The preliminary results of the study indicate that people tend to think in activities, events and needs. Which general routes do users take? What are the most interested places? How do people move and place themselves in an urban street? Are there well-traveled routes that may indicate a particular problem solving strategy? Which places are multi-way branching places; pass through places or destination areas? In our study, the Space-Tags and avatar agent provides navigation support for pedestrians of a virtual street environment and helps designers with the organization and the layout of street environment content and the tuning of interaction possibilities.

However, first-time visitors acquire information about the virtual environment that is then augmented and modified on return visits. The memory of places we have visited may guide us back to those locales. The classic model of spatial learning proposes that individuals first acquire route knowledge, then begin to remember the existence of landmarks, trees and signs, although not their location. In addition, SCALE can serve as a tool to research people's social activities in streets. With Space-Tags, we can implement attractions and entertainment scenes at events or activities, like festivals. If a city offers many attractions by Space-Tags, many people will visit there, explore only those they need, like sightseeing information, and enjoy them.

5.3 Navigational Patterns

It is important for street designers to know the created places where people want to stay. The analysis of navigational patterns includes the directional targets and moving paths in streets. The key aspects of two different but related factors: (1) object identification represented by space tags which is concerned with understanding and classifying the activities in the virtual environment; (2) intentions of exploration which is concerned with finding out about a local environment and how that environment relates to other environments.

In other words, object identification is concerned with finding interesting configurations and information of objects, and exploration focuses on understanding what exists in an environment and how the tags are related. Space Tags in the virtual environment have different meanings for different people.

6. CONCLUSION

The system described in this chapter is based on a set of space tags associated with street attributes such as utilities, landmarks, and buildings. These tags are associated with attributes, which allows representing user patterns that can be triggered by the avatar agent.

An important research issue concerns the degree of changeability that is desirable or acceptable for users. As indicated, as agents may become more "adaptive", the virtual environment itself tends to become more complex (e.g. more colors, maps, icons), and consequently multiple agents as needed may be present at the same time. The question is how users will respond to the abundance of stimuli that is present when several additions are combined.

A further step in this research project will be the integration of multi-agents in the virtual environment, which means that agents will have to be able to communicate and cooperate with one another while assisting users.

7. ACKNOWLEDGEMENTS

This project is supported by the Taiwan National Science Council, grant-NSC 94-2211-E-468-001.

8. REFERENCES

Alexander, C., Sara, I., & Murray, S., 1977, *A Pattern Language*. New York: Oxford University Press.

Arthur, P. and R. Passini, 1992, *Way-finding. People, Signs, and Architecture*. McGraw-Hill.

Chen, C.H., H.T. Lin and M.L. Chiu, 2004, "A scenario-based agent system for digital city interaction," *proceedings of CAADRIA 2004*, Korea, p. 693-705.

Czarnowsky, Thomas V., 1986, "The Streets as Communications Artifact." In Stanford Anderson (ed.) *On Streets*, Cambridge, Massachusetts and London: The MIT Press.

Helbing, D., P. Molnár, I.J. Farkas, and K. Bolay, 2001, "Self-organizing pedestrian movement." *Environment and Planning B: Planning & Design* **28**(3): 361-83.

Hillier, B., A. Penn, J. Hanson, T. Grajewski, and J. Xu, 1992, "Natural movement or configuration and attraction in urban pedestrian movement." *Environment and Planning B: Planning and Design* **19**.

Hook, K., Benyon D.R. and Munro, A., 2003, *Designing Information Spaces: The Social Navigation Approach*. Springer-Verlag, London.

Ishida, T., Nakanishi, H., 2003, "Designing scenarios for social agents." In: Zhong, N., Liu, J., Yao, Y. (Eds.), *Web Intelligence*. Springer, Berlin, p. 59-76.

Jacobs, Allan B., 1993, *Great Streets*, Massachusetts Institute of Technology, USA.

Jiang, B., 1999, "SimPed: Simulating Pedestrian Flows in a Virtual Urban Environment." *Journal of Geographic Information and Decision Analysis (GIDA)*, **3**(1): 21-30.

Lin, H.T., Chiu, M.L., 2003, "From Urban Landscape to Information Landscape-Digital Tainan as an Example," *Automation in Construction* **12**: 473-480.

Schelhorn, T., O'Sullivan, D., Haklay, M. and Thurstain-Goodwin, M., 1999, "STREETS: An Agent-Based Pedestrian Model," *Centre for Advanced Spatial Analysis Working Paper Series*, Paper 9, UCL.

Turner, A., and A. Penn, 2002, "Encoding natural movement as an agent-based system: an investigation into human pedestrian behavior in the built environment." *Environment and Planning B: Planning & Design* **29**(4): 473-490.

Whyte, W.H., 1980, *The Social life of Small Urban Spaces*, Washington D.C.: The Conservation Foundation.

Morozumi, M., T. Uchiyama, M. Tanae, and Y. Sueshige, 2002, "Linked QTVR System for Simulating Citizen's Strolling Activities," *Proceedings of ACADIA*, Pomona p. 219-229.

Therakomen, P., 2001, MouseHaus.class : *The Experiments for Exploring Dynamic. Behaviors in Urban Places*, Master Thesis. Department of Architecture, University of Washington, USA http://depts.washington.edu/maushaus/newMH/mousehaus/mousehaus/mousehaus.html

Moughtin, Cliff, 1992, *Urban Design: Street and Square, Butterworth Architecture*, Great Britain.

Rapoport, Amos, 1977, *Human Aspects of Urban Form: Towards a Man-Environment Approach to Urban Form and Design*, Pergamon Press, New York, USA.

Rapoport, Amos, 1987, "Pedestrian Street Use: Culture and Perception." In Moudon, A.V, *Public Streets for Public Use*, Van Nostrand Reinhold Company Inc., New York.

Rykwert, J., 1986, "The Street: The Use of Its History." In Stanford Anderson (ed.) *On Streets*, Cambridge, Massachusetts and London: The MIT Press.

Approach to Design Behavioural Models for Traffic Network Users

Choice of transport mode

Jean-Marie Boussier, P. Estraillier, D. Sarramia, and M. Augeraud
L3i of La Rochelle University

Keywords: Urban traffic simulator, Virtual city, Multi-agent system, Behavioural model, Transport mean, Taguchi's method, Belief theory

Abstract: Our research work concerns the development of a multimodal urban traffic simulator designed to be a tool of decision-making aid similar to a game wherein the user-player can test different scenarios by immersion in a 3D virtual city. Our approach is based on the activity-based model and the multi-agent technology. The implemented result is a hybrid simulator connecting numerical simulation and behavioural aspects coming from real data. This paper is focused on two points: firstly, we introduce how a final user (the traffic regulator) instantiates and assembles components so as to model a city and its urban traffic network; secondly, we present the use of Dempster-Shafer theory in the context of discrete choice modelling. Our approach manipulates input variables in order to test realistic representations of behaviours of agent categories in a decision-making process. The traffic modelling is based on a questionnaire elaborated from standard arrays of Taguchi. The significant variables and interactions are determined with the analysis of variance which suggests a reduced model describing the behaviour of a particular social category. The belief theory is used to take into account the doubt of some respondents as well as for the preferences redistribution if the number of alternatives changes. The effects of external traffic conditions are also quantified to choose a 'robust' alternative and to use the agents' memory.

1. INTRODUCTION

The search for realistic representations of behaviour in travel demand modelling is explained by the need to assess impacts of strategies of transportation planners and engineers. Over the past few years, simulation

Jos P. van Leeuwen and Harry J.P. Timmermans (eds.), Innovations in Design & Decision Support Systems in Architecture and Urban Planning, 151-166.

has assumed a predominant role within the framework of an activity-based approach because of its flexibility, its realism and strong behavioural foundations. But it is well-known that, in order to capture intelligence of human groups, it is necessary to have interactions between mental simulations and real world. In this context of travellers' behaviour study, researchers use many approaches where individual or traveller groups are the main unit. One good method to achieve this goal is the use of multi-agent simulation; there are few operational tools, e.g. ALBATROSS (Arentze, Hofman, et al. 2000), TRANSIMS (Raney and Nagel, 2004) and FAMOS (Pendyala, Kitamura, et al., 2005). The chain of activities included in a diary is the result of an individual decision process and in some cases it integrates a negotiation step with a partner. Discrete choice utility-maximizing models often used do not always succeed in reflecting true behavioural mechanisms underlying travel decisions (Pozsgay and Bhat, 2001). The second approach emphasizes the need for rule-based computational process models. Some of these models use a set of decision trees. Recent works show that Bayesian networks are particularly valuable to capture the multidimensional nature of complex decisions (Verhoeven, Arentze, et al. 2005).

This paper is structured as follows. Firstly, an overview of particularities of our multimodal urban traffic simulator is given as well as the multi-agent architecture and the principle of our adaptive ground model in order to be able to build a three-Dimensional city. Secondly we propose an approach for decision-making process linking the belief theory and particular questionnaire for testing quantitative or qualitative input variables; our approach is shown in the case of behaviours of driver agents choosing a transport mode. The last section presents the interest of our approach to design scenarios used by our simulator for various categories of people who may change preferences according to different external conditions.

2. PARTICULARITIES OF OUR SIMULATOR

Somebody in charge of traffic regulation must use our simulator for two purposes: as a simulation decision aid-making tool or as an electronic game. For that, he must have the possibility to manipulate a great number of input variables (socio-demographic, economic, urban, transport...) in order to test scenarios in the short or long term.

We want to give a spatial representation which reflects the real environment of the user. The basic idea is to consider that an urban traffic network can be modelled using square blocks. Each block having predefined properties, infrastructure, and functioning rules represents a part of the studied system. The user can instantiate and assemble components in order to model a city and its urban traffic network, see *Figure 1*.

For a realistic representation of the travel demand, the basic idea is to develop a framework to test easily many input variables effects as well as to predict the behaviour of travellers for a given scenario. For that, a great number of categories of traveller agents must be defined in order to propose a 'mean' behaviour based on a similar perception of effects of variable changes. The classification is based on socio-demographic characteristics as age, gender, employment status, presence of a physical handicap.

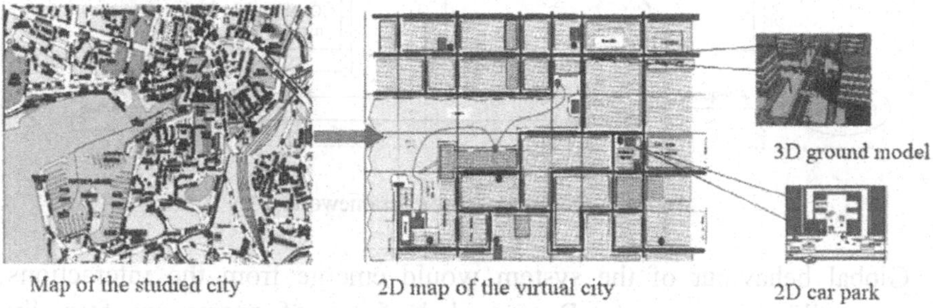

Figure 1. Global view of spatial representation of a city.

Another aspect is that behaviour models must be able to illustrate the evolution of agent preferences or of transportation system characteristics. In the short time, the memory effect of agents must be taken into account (e.g. the predicted result for the trip duration was modified by external traffic conditions). In the long term, new alternatives must be tested for a realistic representation of the decision-making process; that means the classification of choices per agent must be done by taking into account effects of a new alternative independently of the others.

2.1 Simulator Architecture

We provide a framework to build easily the model of a system existing or future (Augeraud, Boussier, et al. 2005). This framework is depicted in *Figure 2*. The multi-agent system (MAS) is divided in three subsystems. The first one concerns the urban traffic simulation. Agents participate to the activity of the city (vehicles, bicycles, pedestrians…). The second one concerns information system service behaviours. In this part, agents model employees and computing system of the information system itself. Actions of agents belonging to this subsystem are the result of interactions between agents of the simulation of the Urban Traffic System. These interactions may generate new vehicles, new pedestrians… into the system. The third one is dedicated to decision aid making objective of our tool. The system produces a synthesis of the whole analysis taking into account the statistics wanted by the user and by collecting simulation data from the other subsystems.

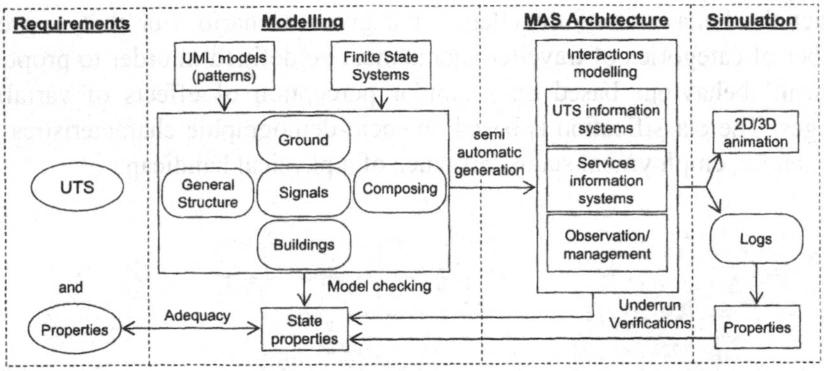

Figure 2. Our proposal for a framework.

Global behaviour of the system would emerge from the interactions between all various agents. Because behaviours of agents are formally expressed using finite states machines, structure properties and some functional properties may be verified by using model-checking techniques. UML and Coloured Petri Net formalisms are used to represent states and interrelations between agents.

2.2 Design of the Adaptive Ground Model

The used structure is based on graphs using a top-down approach and an analysis of the domain. We use a particular type of nodes called connection node to connect blocks. When blocks are assembled, connection nodes disappear because they represent nothing for the urban traffic network. This scheme is applied to nodes of all networks without type differentiation (pedestrian, vehicle…). Blocks are built to be linked in a seamless way.

Figure 3-a shows that graphs are placed on a square area. Graphs are built by taking into account the infrastructure and the various entities.

There are two types of areas: the area which can be used by agents, and the area only displaying scenery. So cars and pedestrians can only move on their dedicated network of the system. A space is reserved for buildings. They are placed to model real social or economical attractive locations in the city. The associated 3D model is presented in *Figure 3-b*.

In our tool, we can change the buildings and we can decide if a node is usable or not.

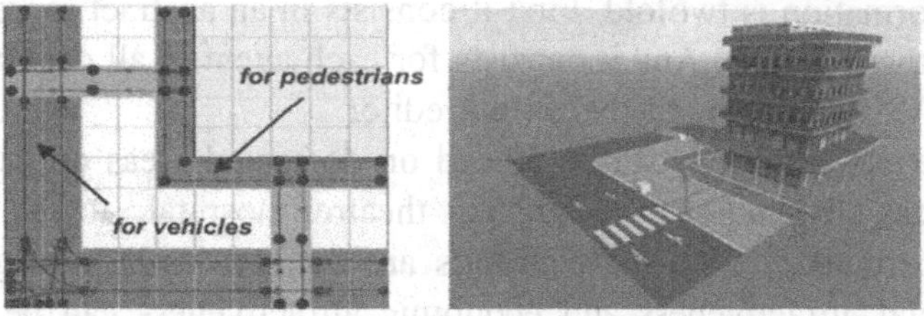

Figure 3. a) A part of the cross graph model. b) Three-Dimensional representation.

Figure 4 shows the model of common crossing which is composed of four blocks. We can see the corresponding 3D representation.

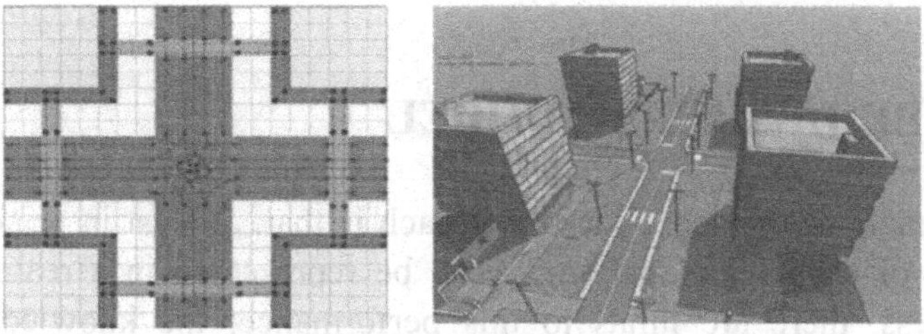

Figure 4. The cross graph model and the 3D representation.

This realism is exploited in the short run within the framework of a video game and will lead to the development of a decision-making tool aid out of urban matter. We are designing a technique of optimization allowing the breaking up of the graphic elements into many independent sub elements. With this opening, we will be able to set up a system of procedural construction of the buildings, to adapt precisely their forms and their functions to space available (Parish and Müller, 2001). With current techniques, it is not possible to describe easily buildings on great surfaces. We will base our works on techniques of image processing to define one description language of such buildings. From the viewpoint of buildings and infrastructures, one key point is the estimation of their cost and their environmental impact on urban traffic.

The design of a system includes the following ordered steps:

- Assembling two blocks, by means of the graphical editor, creates a new bigger block. At the same time an abstract view of the system is built by means of pedestrian and vehicle networks; the result is a directed graph. A data structure is bounded to each vertex. It carries all the information required by a vehicle about its environment.
- The signs and vertical signals, such as traffic light or road panels are edited. The graphical editor produces 3D graphical representation from templates and also produces information for the MAS. This

information is twofold. First it consists in an abstract representation of the roadmap. And it consists for each agent in all data needed by MAS and also by the behaviour editor.

- Scenery elements can be placed on dedicated areas on each block. Such elements can be bank, theatre, hospital, town hall, and supermarket… As the centroïds are defined in the 4-stage model, social attractiveness and economic attractiveness can be given to those elements.
- The MAS is designed and, in input, uses data provided during the second step. Then, agents can be set either as common agents or as agents' generators. They can be dragged and dropped on the previously edited ground.

3. BEHAVIOURAL MODEL

The advantage of the agent-based approach is that, at least in principle, the behaviour of each individual entity can be represented in a realistic way. Nevertheless, there are limits to this performance: the knowledge about human behaviour must be performed and the necessary input data available.

3.1 Approach to Design the Behavioural Model

There are several problems to solve: how to introduce a great set of input variables without decreasing simulator performances? How to quantify the effect of qualitative variables? For a studied city, how to take into account the effects of its particularities on the perception of some variables (accessibility, congestion …)? Interesting works (Bos, van der Heijden, et al. 2005) also added qualitative variables to describe some aspects of the travel demand. There are based on results of specific questionnaire using orthogonal tables (Louviere, 1987) which generally give interesting results for the classification of main variables in a decision making process. However, how to take into account uncertain responses given by some respondents? What are agent behaviours in a dynamic framework: travel conditions change, an alternative (choice) appears or disappears? How is the memory effect used?

By coupling two methods, we answer to the above questions. Taguchi's method (called Design Of Experiments) (Taguchi, 1987) is typically used in the optimization of industrial process for understanding the relationship between process parameters and the desired performance characteristic. Dempster-Shafer theory (called Belief theory) (Shafer, 1976) provides an approach efficient to take into account doubt or ignorance during a decision-making process.

The reliability of this approach is based on the possibility offered to regulator to define finely social categories of citizens. Because of their homogeneity, we consider that a 'mean' behaviour can be modelled for all decision-making process. The study is done for one particular category (male, about 22 years old, students without physical handicap). The approach is illustrated for analysis of transport mode choice (on foot, bicycle, car and bus).

Many models of transport mode choice have been developed in which the mode choice is typically conceptualized as a function of characteristics of alternative travel modes and a set of individual and household characteristics; see (Hess, Polak, et al., 2005) for a good overview. Recently, techniques of Artificial Intelligence were also used for the decision making process: see (Postorino and Versaci, 2002) about Fuzzy Logic and (Verhoeven, Arentze, et al., 2005) about Bayesian Decision Networks.

In our approach, the Taguchi's method and the Belief theory are conjointly used like *Figure 5* shows it:

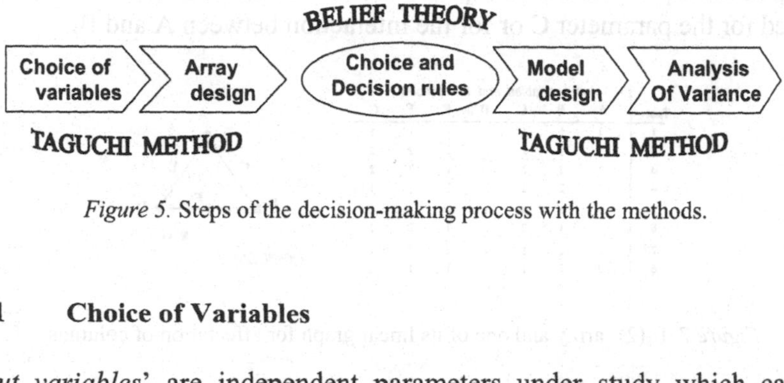

Figure 5. Steps of the decision-making process with the methods.

3.1.1 Choice of Variables

'*Input variables*' are independent parameters under study which can be controlled. The '*level*' of input variable refers to the number of values (states) of the variable to be analysed. The paper aim being the description of our methodology and not the proposal of a model for modal choice, we did not focus our reflection about selection of input variables (5 variables with 2 levels). The characteristic 'quality' to be optimized is the output or the '*response*' variable to be observed. After preliminary tests, we concluded a set with more than 6 linguistic variables is laborious to be manipulated by the respondents of the questionnaire. Input variables, corresponding levels, responses and corresponding evaluations are presented in *Figure 6*.

Input variables	level 1	level 2
Weather	favourable	not favourable
Risk of incidents	low	great
Distance	short	high
Conditions to travel	with parcel	without parcel
Timing	to be in hurry	not to be in …

possible Responses	
Linguistic evaluation	**Crisp value**
very rarely	1
sometimes	2
frequently	3
very frequently	4

Figure 6. Input variables and responses.

3.1.2 Design of Questionnaire

The Taguchi's method uses fractional factorial designs which are a subset of full factorial designs (using fewer runs, generally lower than 20). The selected array is the $L_8(2)^7$ array (matrix). Varying several design variables simultaneously may have interactive effects on the studied response; when the effect of one variable depends on the level of another one, an interaction exists. The linear graph is used to affect columns (see *Figure 7*). For example, the input variable A is put in the first column; the third column can be used for the parameter C or for the interaction between A and B.

Test	*Input variables and interactions* A	B	C	D	E	F	G
1	1	1	1	1	1	1	1
2	1	1	1	2	2	2	2
3	1	2	2	1	1	2	2
4	1	2	2	2	2	1	1
5	2	1	2	1	2	1	2
6	2	1	2	2	1	2	1
7	2	2	1	1	2	2	1
8	2	2	1	2	1	1	2

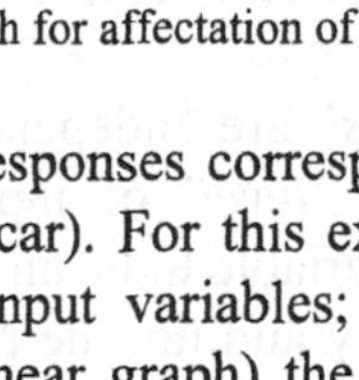

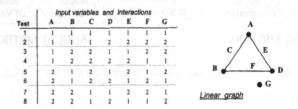

Figure 7. $L_8(2)^7$ array and one of its linear graph for affectation of columns.

Figure 8 presents two scenarios and its responses corresponding to two alternatives for transport mode (bicycle and car). For this example, only 5 columns (A, B, C, D, G) are affected to input variables; the two other columns can be used to compute (see the linear graph) the effects of two interactions: 'Weather'- 'Conditions to travel' (column E) and 'Conditions to travel'- 'Risk of incidents' (column F). People must give a response for each transport mode (linguistic evaluation) and for all scenarios.

For a dynamic analysis (number of choices which changes), deliberately no classification of transport modes is required to respondents for a given scenario. We proposed only 4 responses for each mode. Somebody can give the same evaluation for different modes. For our analysis 1,472 responses were treated.

Scenario	Input variables: Weather (A)	Risk of incidents (B)	Distance (C)	Conditions to travel (D)	Timing (G)	Answer of a respondent: Bicycle =V	Car =C
1	favourable	low	near	with parcel	to be in hurry	sometimes	sometimes
:	....	....	....	....	....	....	....
6	not favourable	low	far	without parcel	to be in hurry	very rarely	frequently
:	....	....	....	....	....	....	....

Figure 8. Input variables and responses for two alternatives (bicycle and car).

3.1.3 Classification of Given Answers about Choices

All scenarios are concerned by a non-negligible percentage of doubt or ignorance. The classification is done with the Dempster-Shafer theory.

Let us take $\Theta = \{ H_1, H_2, H_3, H_4 \}$ to be the set of hypothesis which make up the frame of discernment with H_i potential alternatives.

In our case, the frame of discernment is: $\Theta = \{\text{on foot, bicycle, bus, car}\}$.

The assigned probability $m_\Theta(A)$ measures the belief exactly assigned to A and represents how strongly the evidence supports A. A basic probability assignment is a function which is called a mass function and satisfies:

$$2^\Theta = \{\Phi, H_1, \ldots H_1 \cup H_2, \ldots \Theta\};\ m_\Theta : 2^\Theta \to [0,1];\ m_\Theta(\Phi) = 0;\ \sum_{A \subseteq \Theta} m_\Theta(A) = 1$$

where 2^Θ is the power set of Θ; Φ is the null set; A is any subset of Θ, $m_\Theta(\Theta)$ is the degree of ignorance.

For each respondent and for each scenario, the normality assumption can be recovered by dividing each mass by a normalisation coefficient. For a respondent who answered 'frequently for on foot', 'sometimes for bicycle', 'very rarely for bus' and 'frequently for car', the belief masses would be:

m(on foot, car) = 0.66 ; m(bus) = 0.11 ; m(bicycle) = 0.22.

The discounting operation consists in taking into account the reliability of source (respondent) R_j (α^{R_j}); the discounted belief function is:

$$\forall A \neq \Theta,\ m_\Theta^{R_j\,\alpha}(A) = \alpha^{R_j} \cdot m_\Theta^{R_j}(A);\quad m_\Theta^{R_j\,\alpha}(\Theta) = \left(1 - \alpha^{R_j}\right) + \alpha^{R_j} \cdot m_\Theta^{R_j}(\Theta)$$

For us, all beliefs are reduced by the factor α computed with a linear function: if the same evaluation is given for the 4 alternatives, α=0.25; for 3 alternatives, α=0.5; for 2 alternatives, α=0.75 and α=1 when no redundancy.

After the successive application of the orthogonal addition (Janez, 1996), a single belief function is obtained for all focal elements and subsets of the discernment framework.

After that, for each scenario, the mass of the subsets which are not singletons is redistributed by the pignistic transformation

$$\forall H_i \in \Theta, \quad P_\Theta(H_i) = \sum_{\substack{A \in 2^\Theta \\ H_i \subset A}} \frac{1}{|A|} m_\Theta(A) \tag{1}$$

where $P_\Theta(H_i)$ is pignistic probability for H_i; $|A|$ is the cardinality of A.

Figure 9 presents an example for a scenario and the results after pignistic transformation. In order to give a global evaluation for each scenario j and each mode k, a score is defined according to *Equation 2*:

$$S_j^k = \lambda \cdot P_\Theta{}^j(H_k) \tag{2}$$

where λ is an arbitrary crisp value and $P_\Theta^j(H_k)$ is the pignistic probability of k mode belonging to the scenario j.

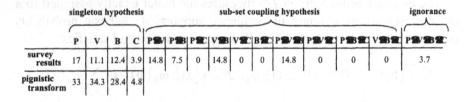

	singleton hypothesis				sub-set coupling hypothesis										ignorance
	P	V	B	C	P∪V	P∪B	P∪C	V∪B	V∪C	B∪C	P∪V∪B	P∪V∪C	P∪B∪C	V∪B∪C	P∪V∪B∪C
survey results	17	11.1	12.4	3.9	14.8	7.5	0	14.8	0	0	14.8	0	0	0	3.7
pignistic transform	33	34.3	28.4	4.8											

Figure 9. Mass distribution and pignistic probabilities for transport alternatives (in percentages).

For the studied scenarios it is possible to give a classification of transport modes. But the scores are only computed for 8 configurations corresponding to the $L_8(2)^7$ array. From here, we are going to see the how to assign the scores to the 24 other scenarios.

3.1.4 Model Design

One of our hypothesis being the homogeneity of a category, we search a 'mean' behaviour which is representative for the studied category. The basic idea is the superposition principle; we adopted an additive model based on Analysis of Means (ANOM) proposed by Vigier (Vigier, 1988).

$$S = S_{mean} + (a_1 \quad a_2)A + (b_1 \quad b_2)B + ... + \begin{pmatrix} a_1b_1 & a_1b_2 \\ a_2b_1 & a_2b_2 \end{pmatrix} AB + ... + \varepsilon \tag{3}$$

where S_{mean} is the mean value of all responses; a_i are matrix elements representing the mean effect of variable A at level i and a_ib_j are matrix elements for the mean effect of the interaction AB when A is at level i and B at level j:

$$a_i = S_{mean}(A_i) - S_{mean} \;;\quad a_ib_j = S_{mean}(A_i, B_j) - S_{mean} - a_i - b_j \tag{4}$$

with $S_{mean}(A_i)$ is the mean value of responses of the scenarios where A is at level i; $S_{mean}(A_i, B_j)$ is the mean value of responses when A is at level i and B at level j.

For example, the variable A is at level *1*, the variable B is at level *2* and …, then the score S is computed as follows:

$$S = S_{mean} + a_1 + b_2 + a_1b_2 + ...$$

A numerical application is applied from *Equation 3* to produce the determining part of the model giving the score associated to the bicycle use:

$$S_b = 2.3 + (0.081 - 0.081)A + (0.022 - 0.022)B + (0.345 - 0.345)C + (-0.085 \; 0.085)D + (0.059 - 0.059)G + \begin{pmatrix} -0.02 & 0.02 \\ 0.02 & -0.02 \end{pmatrix} AD + \begin{pmatrix} 0.04 & -0.04 \\ -0.04 & 0.04 \end{pmatrix} BD$$

3.1.5 ANalysis Of VAriance (ANOVA)

Significant variables and interactions are classified by comparing the variance of the mean effects of input variables V_A and interactions V_{AB} with the residual variance V_r.

for an input variable: $$V_A = \frac{m}{n_A(n_A - 1)} \sum_i a_i^2 \tag{5}$$

for an interaction: $$V_{AB} = \frac{m}{n_A n_B (n_A - 1)(n_B - 1)} \sum_i \sum_j (a_i b_j)^2 \tag{6}$$

for the residual: $$V_r = \frac{1}{\gamma_r} \sum_i \sum_j \left(S_{ij_{resp}} - S_{ij_{th}}\right)^2 \tag{7}$$

with m: number of tests; n_A, n_B: number of levels for the variables; a_i^2, $(a_ib_j)^2$: mean effects of variables and interactions; γ_r: residual degree of liberty; S_{ijresp}, S_{ijth} : the measured and theoretical value of the response.

The Fischer-Snedecor test allows to obtain a reduced model and to assess the score. In the case of bicycle use, the significant input variables are the distance, the travel conditions and the weather:

$$S_b = 2.3 + (0.081 \quad -0.081)A + (0.345 \quad -0.345)C + (-0.085 \quad 0.085)D$$

Because of the homogeneity of the category hypothesis, we claim that the same model can be used to estimated 2^5 scenarios. Similar reduced models are obtained for the three other alternatives (bus, car, on foot). E.g. for the car, the significant input variables are: distance, risk of incidents, weather.

4. INTEREST OF OUR APPROACH TO DESIGN SCENARIOS FOR THE SIMULATOR

The same approach can be used for the choice of the destination or the car park. According to the perception that a category of users has for the traffic characteristics of their city, it is possible to obtain different behavioural models for different cities. For the same city, different categories can have a different behaviour. See our work about car park choice done by two individual categories (Boussier, Estraillier, et al., 2005).

4.1 Behavioural model per defined categories

The regulator defines only characteristics of agent categories (age, gender, employment status, presence of a physical handicap) and the input variables to be tested for a step of traffic module (car park choice, destination or mode choice). The Taguchi's tables and their corresponding linear graphs are stored in a library; an orthogonal array is suggested according to the number of variables to be studied. Interactions are also suggested to refine the model. After the collection of questionnaire results, the regulator distributes the answers according to each category.

Scores are computed and the number of information is reduced with the analysis of variance for each category. That means that even if initially the number of input variables was great, the simulator performance will not be affected by the data number needed by agents for the decision-making processes. If an agent must do a choice between transport modes, it is possible to compute scores and to classify alternatives according to the highest score among all scenarios. Initially, agents of the same category have the same behavioural model used to define their behaviour for all scenarios; the differences between agents' behaviours result from individual diaries and constraints (married or not, departure time, traffic conditions, and memory effect).

4.2 Redistribution of Preferences

The inconvenient of classical models is the difficulty to reclassify alternatives in a dynamic context. In our approach, the classification takes into account added or eliminated alternatives.

If one of alternatives is not possible (e.g. the student has not a private car), we can use the conditioning law of Dempster which consists in no taking into account the mass of the focal elements (singletons) which become impossible. The mass is transferred on the hypothesis which could be truths (Janez, 1996). After renormalisation, another mass distribution and a new set of pignistic probabilities can be computed. For example, let us consider in *Figure 9* the case where the bus is an impossible alternative. The redistribution of 'preferences' is given in *Table 1* after the conditioning.

Table 1. New mass distribution using conditioning law of Dempster and new pignistic probability without the 'bus' alternative.

	F	V	C	F∪V	F∪C	V∪C	F∪V∪C
Initial survey results	28.0	29.6	4.4	33.8	0.0	0.0	4.2
New pignistic transform	46.3	47.9	5.8	-	-	-	-

We have also tested the possibility offered to a regulator to redistribute preferences using the conditioning law of resemblances. In our example, one chosen criterion could be the pollutant mode (car, bus) or not (on foot, bicycle). In the absence of the 'bus' alternative, all preferences (singletons and subsets including bus) are distributed on the 'car' alternative.

It is also possible to add a new alternative (e.g. tramway); the respondents complete the questionnaire concerning this new hypothesis. The new discernment framework is: $\Theta' = \{on\ foot, bicycle, bus, car, tramway\}$. A new mass distribution with corresponding scores is established for each scenario and each mode after the pignistic transformation.

4.3 Effects of External Conditions

It is accepted that the result of choice previously done can be affected by 'parasites' or external conditions as accessibility, safety, traffic information which are independent of input variables. A jointly study of the effects of input variables and external conditions must be done. For that, we used a combination between two tables; see the example in *Figure 10*. In this table, the score 1 corresponds to all scenarios evaluated for a good accessibility, a good safety during the travel and with real time information. The scores are assessed as previously shown (mass distribution, pignistic transformation, model elaboration steps).

For a bicycle, a coupled model symbolically can be presented as follows:

$$S_b = S_{bmean} + \underbrace{A + B + C + D + G + AD + BD}_{array\ L_8} + \underbrace{H + I + J}_{array\ L_4}$$

where S_{bmean} is the general average of all 32 responses. The matrix elements of input variables and interactions are computed with *Equation 3* and *Equation 4* where the mean value of a score for a scenario is the average between the scores (computed on lines). For the elements of *H, I, J,* the average value is computed in columns.

$L_4(2)^3$ — *External conditions*

$L_8(2)^7$ — *Input variables*

							1	2	3	4
						Information (J)	yes	no	no	yes
						Road safety (I)	good	bad	good	bad
						Accessibility (H)	good	good	bad	bad
Scenario	Weather (A)	Risk of incidents (B)	Distance (C)	Conditions to travel (D)	Timing (G)	Score	1	2	3	4
1	favourable	low	short	with parcel	in hurry		2.8	2.1	2.2	1.9
:	....	....	....	....	....		....	....	....	....
6	not favourable	low	great	without parcel	in hurry		1.6	1.4	1.2	1.3
:	....	....	....	....	....		....	....	....	....

Figure 10. Double arrays for the bicycle evaluation.

Then, for a homogeneous category of agents it is possible to compute a score for the 256 (=32*8) scenarios for each transport mode. For greater configurations, questionnaires would be laborious.

There are two interesting uses for this coupling.

Firstly, let us suppose that an agent wants to perform a scenario where the mean score associated to the bicycle is nearly the same as for the car (a limit value may be established by the regulator). We claim that the agent will choose the most robust (cautious) solution, i.e. the transport mode for which the travel is not strongly affected by external conditions. The most robust solution is obtained by computing the ratio signal/noise *(S/N)* for the *j* scenario and the *k* mode (Taguchi, 1987).

$$\left(\frac{S}{N}\right)_j^k = 20 log\left(\frac{S_{mean\,j}^{\;k}}{\sigma_j^k}\right) \tag{8}$$

where $S_{meanj}^{\ k}$: average of scores for scenario *j* and for the mode k_j; σ^k: square of variance of the scores attributed for the scenario *j* and the mode *k*.

The chosen mode for the scenario *j* corresponds to the maximum of *S/N*. Of course, only the part of the model corresponding to L_8 array is taken into account. If the scenario is not included in the array $L_8(2)^7$, it is possible to design models of *S/N* for each transport mode (*Equation 3* and *Equation 4*).

Secondly, let us imagine that, for a scenario with input variables defined beforehand, the choice of transport mode is done. The travel duration is computed for the scenario corresponding to the best external conditions in order to minimize the trip duration. This case is the first vertical column of the L_4 array, i.e. a good accessibility, a good safety during the travel and a system giving information. While the agent travels, traffic conditions change (e.g. congestion, real time information ...); so, the consequence is the increase of trip duration. An agent will use this knowledge as follows. The next time for this scenario, the taken score will be that of the configuration of external conditions corresponding to the last trip of this scenario. Another classification of modes could be suggested.

5. CONCLUSIONS AND PERSPECTIVES

The conceptual framework underlying a forthcoming multimodal simulator based on multi-agent technology has just been presented. Our aim is to provide a framework in which people in charge of traffic regulation will only have to manipulate existing network components in order to model their system. To help decision maker to build his system, graphical interfaces are offered in our framework to design a block and to assemble blocks together. The realistic representation of travellers' behaviour is modelled using results from questionnaires made with the Taguchi's method. An example illustrates the choice of transport means done by students, where effects of quantitative and qualitative variables are tested. The belief theory is used for a preliminary distribution of preferences for transport modes; the answers of all respondents are taken into account even if there is doubt or ignorance in their decision process. Using the Analysis of Variance, significant variables are defined and, in hypothesis of a homogenous category, a reduced model illustrating a 'mean' behaviour is designed for all scenarios. We showed how the belief theory is conjointly used to redistribute preferences if the number of alternatives changes. The effect of external variables as the 'accessibility', 'safety', 'real-time information' is also discussed during the Taguchi's approach use. So, we proposed a way to use the memory effect for an agent in order to modify its initial choice.

A future work will be to replace the crisp values, used for the score computation, by fuzzy variables in order to better quantify the uncertain character of some answers to the questionnaire. The choice of path will be differently solved because the number of alternatives is too great. Our approach becomes relatively 'greedy' in computing times. For that, we will use a classification technique based on combined supervised learning for system diagnosis using Dempster-Shafer theory.

6. REFERENCES

Arentze, T.A., F. Hofman, H. Van Mourik, and H.J.P. Timmermans, 2000, "ALBATROSS: a multi-agent rule –based model of activity pattern decisions", *Journal of the Transportation Research Board*, **1706**: 136-144.

Augeraud, M., J.M. Boussier, F. Collé, P. Estraillier, and D. Sarramia, 2005, "Simulation approach for urban traffic system: a multi-agent approach", *Proceedings of International Conference on Industrial Engineering and Systems Management, Marrakech, Morocco.*

Bos, I., R. van der Heijden, E. Molin, and H.J.P. Timmermans, 2005, "Traveler preference for park&ride facilities: empirical evidence of generalizability", *Proceedings of the 84th Annual Meeting of the Transportation Research Board, Washington DC, USA.*

Boussier, J.M., P. Estraillier, M. Augeraud, and D. Sarramia, 2005, "Agenda elaboration of driver agents in a virtual city", *Proceedings of Multi-agents for modelling Complex Systems, workshop of ECCS, Paris, France.*

Hess, S., J.W. Polak, A. Daly, and G. Hyman, 2005, "Flexible substitution patterns in models of mode and time of day choice: new evidence from the UK and the Netherlands", *Proceedings of the 84th Annual Meeting of the Transportation Board, Washington DC, USA.*

Janez, F., 1996, *Fusion de sources d'information définies sur des référentiels non exhaustifs différents: solutions proposées sous le formalisme de la théorie de l'évidence*, Thèse de doctorat ISTIA, Nantes.

Louviere, J., 1987, "An experimental design approach to the development of conjoint based simulation systems with an application to forecasting future retirement choices", in: Golledge and Timmermans (eds.) *Behavioral Modeling in Geography and Planning*, Croom Helm Publisher, London.

Parish, Y. and P. Müller, 2001, "Procedural modeling of cities", in: SIGGRAPH (eds.) *SIGGRAPH 2001: Proceedings of the 28th annual conference on Computer Graphics and Interactive Techniques, Los Angeles, USA*, ACM Press, New-York, p. 301-308.

Pendyala, R.M., R. Kitamura, A. Kikuchi, T. Yamamoto, and S. Fujii, 2005, "FAMOS: the Florida activity mobility simulator", *Proceedings of the 84th Annual Meeting of the Transportation Research Board, Washington DC, USA.*

Postorino, M.N. and M. Versaci, 2002, "A fuzzy approach to simulate the user mode choice behaviour", *Proceedings of the 13th Mini-EURO Conference, Bari, Italy.*

Pozsgay, M.A. and C.R. Bhat, 2001, "Modeling attraction-end choice for urban recreational trips: implications for transportation, air quality and land-use planning", *Research report, University of Texas at Austin.*

Raney, B. and K. Nagel, 2004, "An improved framework for large scale multi-agent simulation of travel behavior", *Proceedings of the 4th Swiss Transport Research Conference, Conference, Monte Verita, Swiss.*

Shafer, G., 1976, *A mathematical theory of evidence*, Princeton University press, Princeton.

Taguchi, G., 1987, *System of Experimental Design: vol. 1-2*, Don Clausing UniPub/Kraus International Publications, Dearbon.

Verhoeven, M., T.A. Arentze, H.J.P. Timmermans, and P.J.H.J. van der Waerden, 2005, "Modeling the impact of key events on long-term transport mode choice decisions: a decision network approach using event history data", *Proceedings of the 84th Annual Meeting of the Transportation Research Board, Washington DC, USA.*

Vigier, M., 1988, *Pratique des plans d'expériences: méthodologie Taguchi*, Les Editions d'Organisation, Paris.

Shape Morphing of Intersection Layouts Using Curb Side Oriented Driver Simulation

Michael Balmer and Kai Nagel[1]
Swiss Federal Institute of Technology Zurich
[1]Technische Universitaet Berlin

Keywords: Agent simulation method, Intersection layout, Evolutionary algorithm

Abstract: In a traffic network, capacities of parts of the network restrict the amount of transport that can be handled by this network. The capacity of a given traffic network element is not fixed, but influenced by parameters such as number of lanes, maximum speed, weather, view horizon, and so on. These parameters also define the maximum capacity of complicated intersections. Special shapes of intersections, particularly in urban regions, may further increase or decrease their capacity.

This paper investigates an evolutionary algorithm to automatically improve the geometrical layout of parts of an urban network according to externally specified criteria. The paper consists of two main parts. In the first part, a simulation model is described which is able to produce realistically behaving vehicles only by using information about the curb side locations of the roads. This avoids the need to use lane connectivity, signal plans, etc. - which are details that would change during a change of the intersection layout. In the second part of the paper, the simulation changes the road and intersection layouts based on the behaviour of the vehicles. Using a feedback loop allows one to optimize the capacity of the modelled road system while its spatial extents are minimized.

As a case study, a special roundabout is examined: "Central" in downtown Zurich, Switzerland. The particularity of this roundabout is that it partially behaves like a roundabout but also contains two uncontrolled intersections. Due to its central position in the city, the roundabout is very busy with both individual cars and public transport vehicles.

Jos P. van Leeuwen and Harry J.P. Timmermans (eds.), Innovations in Design & Decision Support Systems in Architecture and Urban Planning, 167-183.

1. MOTIVATION

Traffic simulation methods are widely accepted in transportation research. They are used to answer a large variety of questions like traffic demand (e.g. VISEM, www.ptv.de), capacity and breakdown estimation (Bernard, 2005), large-scale traffic analysis based on microscopic demand (e.g. MATSIM, www.matsim.org), and so on. On the scope of a single intersection, simulation methods are used to optimize signalization (Shelby, 2004), to estimate intersection capacities using agent based intersection simulation (Manikonda, Levy, et al. 2001), etc. On the other hand, the question about the shape of an intersection is typically an issue of constructors and designers. Capacity reduction caused by the shape of an intersection is normally not considered. This paper shows how to simulate the interaction between the capacity and the shape of a single roundabout.

Typically, the area of a roundabout is limited by existing buildings, necessary pedestrian ways, governmental rules, and other additional constraints. To find a shape for a roundabout using as little space as possible while providing the required traffic capacity is therefore an optimization problem (Campbell, Cagan, et al. 1999, Dijkstra and Timmermans, 2002).

In this paper we demonstrate an agent-based approach (Ferber, 1999) to solve this optimization problem. Section 2 describes how a roundabout is defined. It also gives a brief description about the roundabout chosen for this study. Section 3 describes how an agent based car driver simulation is being developed and applied in order to calculate traffic indicators like congestion or capacity. In contrast to the large variety for microscopic traffic simulation models (CORSIM, 2005, MITSIM, 2005, VISSIM, 2005), the introduced simulation is only based on information of the shape (the curbs) of the scenario. With the knowledge of the calculated indicators of Section 3, a method is shown in Section 4 that uses those to change the shape of the roundabout such that the calculated indicators will be optimized. Section 5 describes some first results of the case study of the "Central" roundabout of Zurich. The paper finishes with a summary and an outlook on further works.

2. MODELLING THE ROUNDABOUT

We define a roundabout by directed street segments. A street segment consists of one or more driving lanes with the same driving direction and does not have any kind of junctions. An incoming street segment of the roundabout additionally holds a defined area inside the street segment shape where car driving agents are allowed to enter the roundabout. Each chosen roundabout holds a defined number of incoming and outgoing street segments. For an outgoing street segment, a similar area is defined where agents are allowed to leave the system. For simplicity we define such an area by a circle C with centre and radius. Entering circles are denoted by C_i^{start} and leaving circles by C_j^{end}. The model of the roundabout provides the following information:

- Description of the curbs. The curbs are represented as geometric primitives like geometric nodes and links.
- Description of the driving routes through the roundabout. The driving routes are necessary for simulating the driver agents. Each car driving agent holds information about the streets where he is going to enter and to leave the roundabout. Each possible combination of entering and leaving street segments has to be associated with a route through the roundabout.

Since we do not want to include any information but the curbs into the model, also the driving routes have to be defined by a set of curb segments.

2.1 Modelling the Curbs

Modelling of the curbs is a straight forward process. Instead of using the real shape of the roundabout, we simplify the curbs as line segments. This generic approach has the advantage of allowing the description of any kind of roundabout including any intersections and also larger networks without the need to handle different kinds of geometric primitive types. The following shows the curb modelling process:

1. Given a shape and the driving direction of a roundabout and an arbitrary origin of a Euclidian coordinate system
2. Define a set N of nodes along all curbs of the given shape
3. Define a set L^{vis} of directed links along the curbs, each connected by a start and an end node from the set N. The direction of the link must follow the driving direction in this street segment.

Note that a link vector has a defined location, direction and length in the coordinate system given by its start and end node.

2.2 Modelling the Driving Routes ("Tunnels")

In order to let the agents travel through the roundabout, a route through it needs to be assigned to each agent. Since we only want to use the curbs as given information about the roundabout, those driving routes (now also called "tunnels") must be described as a set of curb segments (see Helbing, Schweitzer, et al. 1997, for other ways of describing routes).

A tunnel T_{ij} describes the area in which an agent is allowed to drive. The area is described by a set of links. Since a tunnel has exactly one entrance area C_i^{start} and one leaving area C_j^{end}, there are no "holes" in the tunnel, meaning that a tunnel is defined by exactly two curb sides, a left one and a right one. To provide this, we need to add additional "invisible" links to differentiate street segments which are part of a tunnel from those which are not. The following describes the modelling process for all possible tunnels of the given roundabout:

1. Given a set of links L^{vis} as described in Section 2.1.
2. Add a set of "invisible" links L^{invis} for each intersection area, such that each street segment is separated from the intersection area. The total set of links is therefore $L = L^{invis} \cup L^{vis}$.
3. For each pair of incoming and outgoing street segments of the roundabout, define one tunnel described by a set of visible and invisible links such that each tunnel holds exactly two curb sides, a left and a right one, starting from the incoming street segment and ending at the outgoing one.

2.3 "Central" Roundabout of Downtown Zurich

This paper uses the "Central" roundabout of downtown Zurich as a case study. There are several reasons for choosing this roundabout: (a) It is one of the bottlenecks of the Zurich street network. (b) Even if its shape correlates to a roundabout; it still holds two intersections. (c) The number of lanes varies inside the roundabout. Therefore, the amount of space used by a street segment also varies within the roundabout. (d) A major reconstruction was done during summer 2004. This study was done before the reconstruction of the Central roundabout. So, part of the analysis of Section 5 can be done by comparing the results with the situation after the reconstruction.

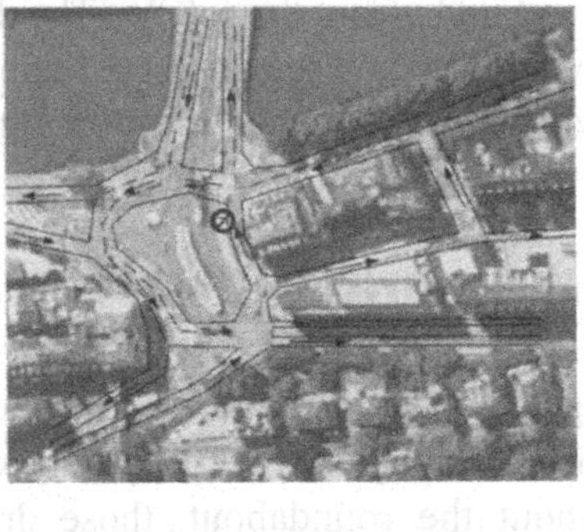

Figure 1. Top view of the "Central" roundabout of downtown Zurich (Source: http://www.sanday.ch, accessed February 2005).

Figure 1 shows the special shape of the "Central". Note that the middle road (where left turns are not allowed) is a "short-cut" for leaving the

"Central" towards the bridge over the river. Therefore the "Central" is not a "real" roundabout anymore. It holds five incoming and five outgoing street segments. This leads us to 25 different tunnels as described in Section 2.2.

3. CAR DRIVER SIMULATION

Given a model of a roundabout as described in Section 2 we build an agent based car driver simulation. The guiding design principle for that simulation is to use a modelling approach that uses as little input data as possible. The model described in the following works without lanes, signals, turn priorities, etc. It is clear that such a model cannot reach the same levels of realism as, say, CORSIM, MITSIM, or VISSIM, but it will turn out that the results are useful enough for the purposes of overall intersection layout. In addition, the approach does not only reduce the amount of data collection, but it also makes it unnecessary to potentially change intersection prioritization as a result of the intersection geometry adaptation process described in Section 4. The car driver simulation is designed according to the following specifications:

1. Every agent is assigned to a specific tunnel and is not allowed to change into another tunnel during the simulation.
2. Given a tunnel, an agent must not drive over the tunnel's curb sides.
3. An agent must start at the incoming street segment and end at the outgoing one of the given tunnel.
4. An agent is not allowed to drive "unrealistically" through the given tunnel. He must not drive backwards and he must not drive extremely apart from the driving direction given by the tunnel (i.e. right-angled or driving in opposite direction).
5. An agent must respect the physical rules of acceleration.
6. An agent tries to drive with a "desired driving speed".
7. An agent can not steer more than a given "maximum steering" constant (otherwise cars could change directions right in place).
8. An agent must respect other agents in the system. He has to decelerate or overtake if a slower agent drives in front of him.

The main idea of this agent based car driver simulation follows the principle of particle simulations with discrete time steps Δt used in various topics in computational science. Assume each tunnel defines a current which flows in the direction of the given directed curbs and assume that a car driving agent is one particle of fluid in that current, then the constraints 1, 2, 3, 5 and 8 of the above list are fulfilled (for laminar flow).

By adding more, partially overlapping currents representing the other tunnels, and by fulfilling constraint number 4, 6 and 7, those particles become agents (Ferber, 1999). This idea will be formalized in the following subsections.

3.1 Defining a Car Driving Agent

As mentioned above, each car driving agent a_k is assigned to a tunnel T_{ij} of the given roundabout. T_{ij} defines the path of a_k through the system. For each point in time t each agent holds a certain amount of information about his current state: position $\vec{a}_k(t)$, driving speed $s_k(t)$, and driving direction $\varphi_k(t)$. Each agent also holds some predefined constant parameters: desired driving speed s_k^{des}, maximum acceleration $acc_k^{\max}$, a shape (for simplicity the shape is defined as a circle with radius r_k), a maximum steering limit, $\rho_k^{\max}$, and finally a maximum allowed angle with respect to a given flow force vector, called $\theta_k^{\max}$. The flow force will be described in detail in Section 3.2. With the two scalars $s_k(t)$ and $\varphi_k(t)$, the agent's velocity vector $\vec{v}_k(t)$ is defined.

To follow the idea of a particle simulation each agent in the system reacts to an external force field (similar approaches in Gloor, Cavens, et al. 2003). For each point in time t during the simulation and on each agent's position $\vec{a}_k(t)$ a force $\vec{F}_{tot}(t)$ needs to be calculated which influences the agent. This force consists of three components, a "flow force" $\vec{F}_{flow}$, a "curb repulsion force" $\vec{F}_{curb}$ and a "neighbour agent repulsion force" $\vec{F}_{neigbour}$:

$$\vec{F}_{tot} = \vec{F}_{flow} + \vec{F}_{curb} + \vec{F}_{neigbour}.$$

The following sections give us a brief overview about the meaning of the three forces (more details in Balmer, Vogel, et al. 2005).

3.2 Flow Force

The flow force field defines the flow of a tunnel T_{ij}. It pushes an agent in the right driving direction through his given tunnel. It is also responsible for letting an agent drive with its desired speed. The flow force is defined by

$$\vec{F}_{flow}\left(\vec{a}_k^{(T_{ij})}(t)\right) = \alpha \cdot \left(s_k^{des} \cdot \hat{v}_{flow}^{(T_{ij})}(\vec{a}_k) - \vec{v}_k(t)\right)$$

This is a standard way to describe (exponential) adaptation to the desired velocity. The desired velocity is given by the flow velocity vector $\vec{v}_{flow}^{(T_{ij})}(\vec{a}_k)$ which defines the flow direction of tunnel T_{ij} at position $\vec{a}_k$, multiplied with the desired speed, s_k^{des}. Throughout this paper, $\hat{v}$ will describe the unit vector of vector $\vec{v}$, i.e. $\hat{v} = \vec{v} / |\vec{v}|$. The parameter α describes the speed of the adaptation.

If the flow force were the only one influencing an agent, then it would be possible for an agent to drive over the curbs of his tunnel. The agent would also not respect other agents. The following two forces prevent those effects.

3.3 Curb Repulsion Force

The curb repulsion force field pushes agents away from the curb sides of a tunnel in order to prevent the agents from driving across them (see Stucki, 2003 for a similar approach). The closer an agent gets to a curb, the stronger he will get pushed away from it. The curb repulsion force is defined by

$$\vec{F}_{curb}\left(\vec{a}_k^{(T_{ij})}\right) = \sum_{\vec{l}_{ij}} \vec{F}_{curb}\left(\vec{a}_k^{(T_{ij})}, \vec{l}_{ij}\right)$$

$$\vec{F}_{curb}\left(\vec{a}_k^{(T_{ij})}, \vec{l}_{ij}\right) = s_k^{des} \cdot \left(\left|\vec{d}\left(\vec{l}_{ij}, \vec{a}_k\right)\right| - r_k\right)^{-\gamma} \cdot \hat{d}\left(\vec{l}_{ij}, \vec{a}_k\right)$$

The distance vector $\vec{d}\left(l_{ij}, \vec{a}_k\right)$ describes the distance between a link l_{ij} and a position $\vec{a}_k$ as shown in *Figure 2*. With parameter $\gamma > 0$ the repulsion force of a link near to an agent is larger than the one of a link far away.

Adding this force to the flow force described in Section 3.2 will guarantee that an agent is driving through his tunnel without crossing the tunnel's curb side.

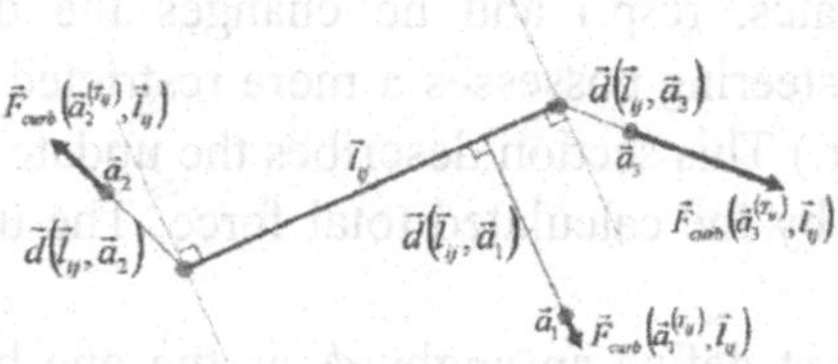

Figure 2. Curb repulsion force on an agent by a given link.

3.4 Neighbour Agent Repulsion Force

By adding the neighbour agent repulsion force field to the total force of an agent in the simulation system, the agents respect their counterparts (Helbing, Farkas, et al. 2000). The neighbour agent repulsion force is defined by

$$\vec{F}_{neigbour}\left(\vec{a}_k(t)\right) = \sum_{\vec{a}_m(t) \in A_{front}(\vec{a}_k(t)),\, a_m \neq a_k} \vec{F}_{neigbour}\left(\vec{a}_k(t), \vec{a}_m(t)\right),$$

where A_{front} is the area in front of agent a_k. *Figure 3* shows when an agent $\vec{a}_m$ is part of the area A_{front}. The repulsion force given by an agent a_m on an agent a_k is

$$\vec{F}_{neigbour}\left(\vec{a}_k(t), \vec{a}_m(t)\right) = s_k^{des} \cdot \left(\left|\vec{a}_k(t) - \vec{a}_m(t)\right| - r_k - r_m\right)^{-\delta} \cdot \frac{\vec{a}_k(t) - \vec{a}_m(t)}{\left|\vec{a}_k(t) - \vec{a}_m(t)\right|}.$$

With parameter $\delta \geq 0$ the neighbour agent repulsion force of an agent a_k near to agent a_m is higher then the one of another agent far away.

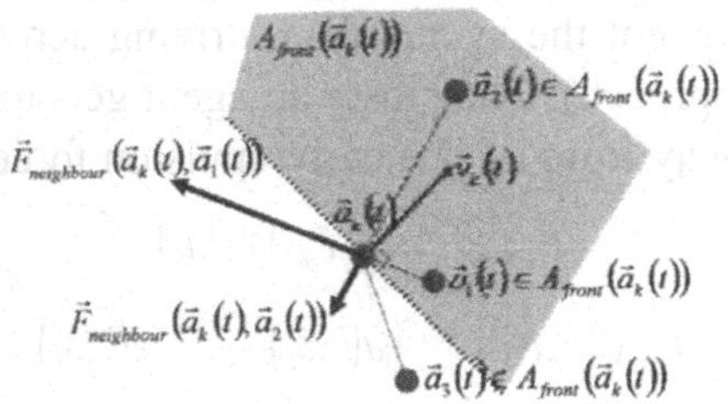

Figure 3. Neighbour agent repulsion force on an agent.

3.5 Acceleration and Steering

As already mentioned at the beginning of Section 3.1 an agent at position $\vec{a}_k$ at time t reacts in two ways to a given force $\vec{F}_{tot}$. In each time step Δt, he accelerates (decelerates, resp.) and he changes the driving direction by steering. (Note that steering possesses a more restricted dynamics than, say, pedestrian behaviour.) This section describes the update rules for the agent's speed and direction by the calculated total force. The update is done in the following two steps.

Step I: Let us first define an angle ϕ as the one between the velocity vector and the total force of an agent at time t. According to the physical rules of motion, an agent at position $\vec{a}_k(t)$ with speed $s_k(t)$, direction $\varphi_k(t)$ and force $\vec{F}_{tot}\left(\vec{a}_k^{(T_{ij})}(t)\right)$ reacts like the following

$$s_k^I(t) = s_k(t) + acc_k(t) \cdot \Delta t$$
$$\varphi_k^I(t) = \varphi_k(t) + \rho_k(t) \cdot s_k(t) \cdot \Delta t$$

The acceleration $acc_k(t)$ of the agent induced by the total force is therefore the abscissa of the coordinate system by the velocity vector. Similar to that, we can define the steering $\rho_k(t)$ as the ordinate value of the total force.

Figure 4 shows the graphical interpretation of these two scalars. To ensure that the calculated values are inside the defined range given by a maximum acceleration and a maximum steering, they have to be reduced to those limits in case that they are out of range.

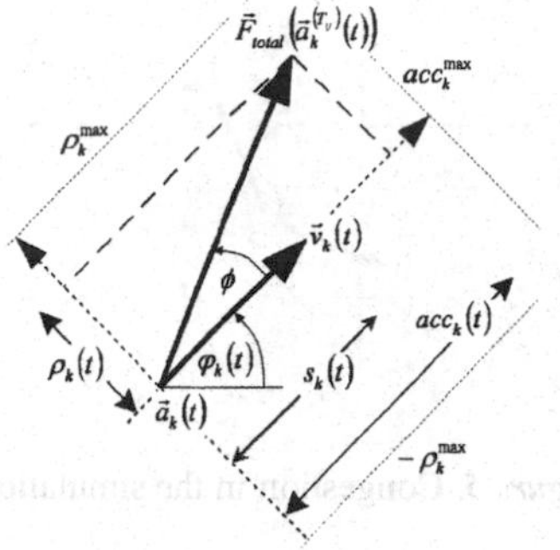

Figure 4. Calculation of acceleration and steering of an agent.

Step II: Let us first define an angle θ as the one between the velocity vector and the flow force of an agent at time t. Since an agent is still allowed to drive backwards (negative speed) and to drive against the given flow direction of his tunnel, we need to correct the velocity vector of *step I* such that it respects these constraints. The correction is defined as

$$s_k(t+\Delta t) = s_k^{II}(t) = \begin{cases} 0, & \text{if } s_k^I(t) < 0 \\ s_k^I(t), & \text{else} \end{cases}.$$

$$\varphi_k(t+\Delta t) = \varphi_k^{II}(t) = \begin{cases} \varphi_{flow}(\vec{a}_k^{(T_{ij})}(t)) + \theta_k^{max}, & \text{if } \theta > \theta_k^{max} \\ \varphi_{flow}(\vec{a}_k^{(T_{ij})}(t)) - \theta_k^{max}, & \text{if } \theta < \theta_k^{max} \\ \varphi_k^I(t), & \text{else} \end{cases}$$

The speed is therefore reset to zero if it is negative and the direction of the velocity vector is turned towards the flow force vector if the angle between those two vectors is too large. The final updated velocity vector is therefore

$$\vec{v}_k(t+\Delta t) = \vec{v}_k^{II}(t) = \vec{v}_k^{II}(s_k^{II}(t), \varphi_k^{II}(t))_{s,\varphi}$$

We can now calculate the position of the agent at time $t+\Delta t$:

$$\vec{a}_k(t+\Delta t) = \vec{a}_k(t) + \vec{v}_k(t+\Delta t).$$

3.6 Congestion

The above described simulation model produces car driving agents who "realistically" drive through a roundabout inside a defined tunnel. They can overtake or follow other agents in the system. Especially slow driving agents can produce tailbacks. But this does not mean that the simulation produces congestion in terms of capacity constraints of a street network. Typically entering lanes and crossroads are the cause of congestion. The simulation developed in this paper is able to reproduce that.

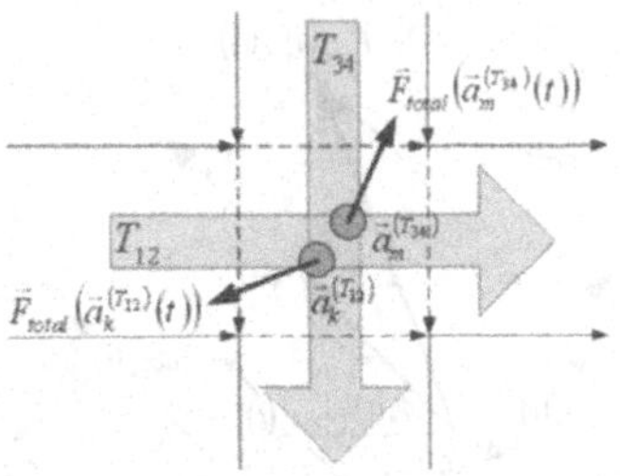

Figure 5. Congestion in the simulation.

Figure 5 gives an example of a congested situation in the simulation. In this example we define a crossroad with two tunnels. One agent uses tunnel T_{12}, the other tunnel T_{34}. Because of the short distance between the two agents, the neighbour agent repulsion force has the largest contribution to the total force of each agent. Therefore the total force is directed more or less opposite to the desired directions of the agents, which means that they have to decelerate and finally stop driving (speed equals zero).

This is a typical "Deadlock" situation, which has to be prevented. Fortunately, it also indicates "difficult" intersection topologies and therefore we can use that information for changing the topology (details in Section 4). Nevertheless, we need to resolve this deadlock situation, which is done in quite a simple way. If an agent's velocity is zero, he starts counting the number of time steps he doesn't go on driving. The higher this number is, the more probable it becomes that he will just drive on in the next time step. That means that the agents then do not respect the other agents anymore for the next time step. Therefore they will just drive across each other.

Of course this is not realistic, but on the other hand, in uncontrolled intersections (and sometimes also in controlled intersections) similar situations occur. Cars are getting stuck similar to a deadlock situation, and then they try to find a gap and "squeeze" themselves trough it without respecting driving rules. The way the simulation handles this is just a simplification. The probability of driving on in a deadlock situation is calculated as followed:

$$p(drive\ on) = \begin{cases} \frac{\#StepsBeingStuck}{100}, & if \quad \#StepsBeingStuck < 100 \\ 1, & else \end{cases}$$

With this simple approach deadlocks can be resolved.

4. SHAPE MORPHING OF THE ROUNDABOUT

The main idea of morphing the shape of a roundabout can be described by the following statement: *Congestion occurs because the street segment (or intersection segment) is too narrow.* This simple statement gives us the idea how we could morph the shape of a roundabout. Everywhere where congestion happens, the roadside corners (nodes) should move away. As we already described in Section 3.6 an agent is in a congested area when he has stopped because of a deadlock situation and eventually drives on without respecting the other agents. In other words, a car driver simulation as described in Section 3 can produce "drive-on" events. In the following, we will use the position of the agent that produces such an event (denoted as the agent's position $\vec{a}_k$) to modify the shape of the roundabout.

Morphing of the shape is done iteratively. This iteration process is presented in the following section which is based on iterative learning processes (Raney and Nagel, 2004).

4.1 Iteration Process

The iteration process is done by the following steps:

1. Given an initial shape of the roundabout (modelled as in Section 2).
2. Run the agent based car driver simulation for a defined time period with simulation step Δt and keep track of all the "drive-on" events.
3. Change the shape of the roundabout by using the information of the "drive-on" events from the previous simulation.
4. Rerun the simulation with the changed roundabout, and so on.

The advantage of this process is that the morphing algorithm can be easily replaced by another. The following section describes one possible algorithm to morph the shape in iteration step 3.

4.2 Morphing Model

The model should provide the following feature: In congested areas the streets should get wider while in non congested areas the streets should shrink. For that we need to move the roadside corners (nodes $\vec{n}$), since the roadside links $\vec{l}$ are defined by their start and end nodes. But to move the nodes we need to know the movement direction. A simple but robust way is to calculate the centroid of each disjoint "non-street" area $A_i^{non-street}$ (defined by the set of its border nodes) and to move the nodes toward or away from these centroids. *Figure 6* gives an example of those centres. It is calculated as the arithmetic average of the border nodes of this area. The normalized moving direction for each node of area $A_i^{non-street}$ is therefore $\vec{m}(\vec{n} \in N_i) = (\vec{c}_i - \vec{n}) / |\vec{c}_i - \vec{n}|$.

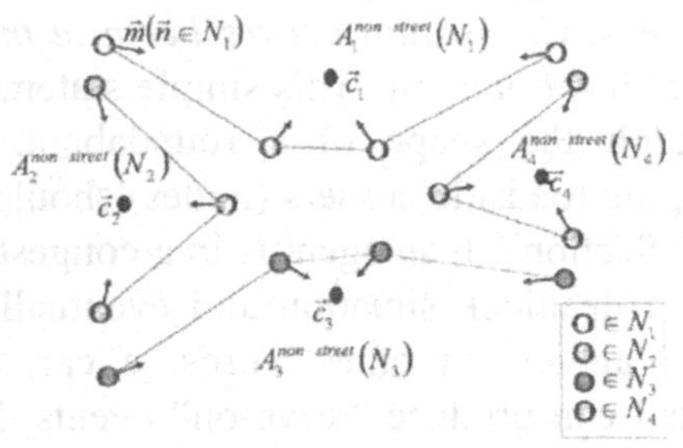

Figure 6. Example of four independent non-street areas and their centres.

Now, we need to calculate an influence parameter $\kappa(\vec{n}, \vec{a}_k)$ of a node by a given "drive-on" event $\vec{a}_k$. The calculation is done inversely proportional to the distance between the node and the event:

$$\kappa(\vec{n}, \vec{a}_k) = \begin{cases} 0, & if \quad |\vec{n} - \vec{a}_k| > r_\kappa^{\max} \\ +1 - |\vec{n} - \vec{a}_k| / r_\kappa^{\max}, & else \end{cases}, \quad \kappa(\vec{n}, \vec{a}_k) = [0,1]$$

The parameter $r_\kappa^{\max}$ defines the maximum radius of influence of an event. If a node is located outside of the influence area of an event, $\kappa(\vec{n}, \vec{a}_k)$ is zero. Since there can be more than just one event during the agent based simulation, we sum up the calculated influences for each node: $\kappa(\vec{n}) = \Sigma_{\vec{a}_k} \kappa(\vec{n}, \vec{a}_k)$

Since we cannot control the absolute number of events which occur during a simulation, the range of $\kappa(\vec{n})$ can vary a lot from between iterations. By normalizing it, we calculate an appropriate value which describes the offset by which a node should be moved towards its corresponding centroid $\vec{c}_i$. Additionally we also want to allow that streets can shrink, which means, that the nodes with minimal influence should move away from their centre. Therefore the "moving length" $l(\vec{n})$ of a node of area $A_i^{non-street}$ with centre $\vec{c}_i$ can be calculated as

$$l(\vec{n}) = \frac{l^{\max} - l^{\min}}{\max_{\vec{n}}(\kappa(\vec{n})) - \min_{\vec{n}}(\kappa(\vec{n}))} \cdot \left(\kappa(\vec{n}) - \max_{\vec{n}}(\kappa(\vec{n}))\right) + l^{\max}.$$

The parameter $l^{\max} \geq 0$ defines the maximal length a node is allowed to move towards its centre, while $l^{\min} \leq 0$ defines the maximal length a node is allowed to move away from its centre. If $l^{\max} = 0$ the street segments do not grow, while with $l^{\min} = 0$, they do not shrink.

5. SETUP AND FIRST RESULTS

Error! Not a valid bookmark self-reference. defines the setup for our simulation. In general it is not that easy to verify the results using the above described model. On the other hand, qualitative comparisons can give us some good indications about the benefits of such a model. The major reconstruction work on the "Central" roundabout, which ended in autumn 2004, enables us to evaluate the results of the shape morphing against some real world experiences.

Table 1. Setup for the "Central" roundabout.

Roundabout Model		**Agents**	
# Nodes	108	$r_k = r$	1.3 m
# visible Links	96	$\rho_k^{max} = \rho^{max}$	$\pi/4$
# invisible Links	26	$acc_k^{max} = acc^{max}$	$+\infty$ m/s^2
# incoming street segments	5	$\theta_k^{max} = \theta^{max}$	$\pi/12$
# outgoing street segments	5	s_k^{des}	U[20,50] km/h *
Radius start/end circles	30 m	**Forces**	* Uniform distribution
# tunnels	25	α	5
# non-street areas	12	β	3
Morphing model		γ	3
r_k^{max}	10 m	δ	3
l^{max}	+1.0 m	**Time Step & Simulation Time**	
l^{min}	−0.2 m	Δt	0.05 sec
		# time steps	4800 (240 sec)

5.1 Results

Car Driver Simulation: As the outcome of the case study shows, the car driver simulation produces the expected results. Agents find their ways through the system inside their tunnels. They do not drive over a curb side and they respect other agents in the system. They also overtake or follow slower agents depending on the width of the street segment. Congested situations occur and dissolve depending on the number of agents in the system. The "drive-on" rule resolves deadlock situations. An important fact is that "drive-on" events occur only in congested areas, which is important for the morphing model.

Morphing Model: Also the morphing model shows the expected behaviour. Congested street areas are getting larger while free flow areas are shrinking. But the shape of the roundabout is getting more and more unrealistic. This happens because of the extremely simple morphing rules. From an engineering point of view, one could say that this result is not usable. Nevertheless the created shape of the roundabout gives us very good indication about areas where the system has too much capacity and vice versa.

Figure 7 shows us the result of the morphing process. The car driver simulation produces two main congestion areas (shown in iteration 0 of *Figure 7*). Those areas are expanding (A_1^{grow} and A_2^{grow} , shown in iteration 10 of *Figure 7*). On the other hand there are several street segments that shrink. Interestingly, almost the whole left part of the roundabout is shrinking (A_4^{shrink} and A_5^{shrink}), even though there are junctions. The streets on that area were built with two or three lanes (see also the schematic drawing in *Figure 1*). This could lead us to a conclusion that at least one lane can be closed. Another interesting shrinking area is A_1^{shrink} . That street segment almost shrinks to the size of an agent (2.6 meter width). It also looks as if this street could be the cause of the two congested areas. Since agents who want to leave the roundabout at the outgoing street segment on the top could also drive along the right loop. So, it is possible to completely close this street segment.

A_1^{stable} shows that there are also some street segments which have more or less the proper capacity (equal to a two lane street segment). Other stable areas can be found at the incoming and outgoing street segments.

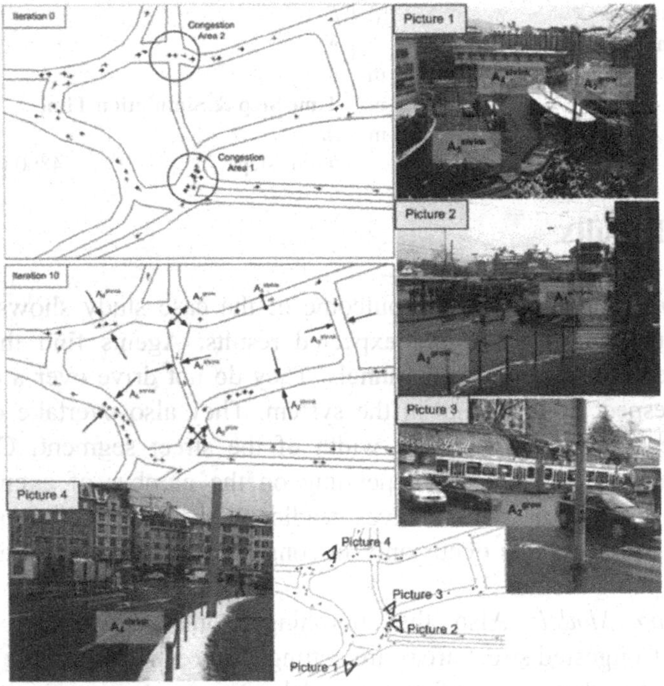

Figure 7. Growing and shrinking areas during the iteration process with comparison to the "Central" after the reconstruction.

5.2 Qualitative Comparison to Reality

Until spring 2004 the shape of the "Central" roundabout was looking like the *Figure 1*. During the peak hours, police were used to control the traffic at the two major congestion areas shown in Iteration 0 of *Figure 7*. During the reconstruction process the street segment at area A_1^{shrink} (see *Figure 7*) was closed for a long time and the drivers were redirected along the right loop. The pictures of *Figure 7* show the shape of the "Central" after the reconstruction was finished. The shrinking and growing areas of *Figure 7* are labelled for better orientation. As we can see in the two areas A_4^{shrink} and A_5^{shrink}, there is a reduction from three to two driving lanes and the centre street segment is reopened again. It is quite fascinating that the simulation shows the same changes. Even more interesting is that the simulation indicates that the middle street segment is causing too much problems and therefore should be closed in this special case. It would be of interest to measure the behaviour of the reconstructed "Central" if we closed this street again for a longer time period.

6. FUTURE WORK

The car driver simulation: At the moment each agent calculates his present total force completely "from scratch" for each link of his tunnel and for all other agents in the system at every time step. This wastes a lot of the available computational performance. There are several issues where it would make sense to pre-compute forces at the beginning of each car driver simulation run (e.g. discretization of space; see also Nishinari, Kirchner, et al. 2001 and Schadschneider, 2001). With a more appropriate data-structure (e.g. Quad-Trees, i.e. the "tree method" in molecular dynamics), neighbour agents could also be found much faster.

It is also of interest to add other traffic participants, like trams and pedestrians. Especially for the "Central" the pedestrians influence the capacity of the roundabout a lot, because the direct way from the Zurich main station to the University goes through the "Central". With this, during the morning and the evening peak the place is "flooded" with pedestrians.

The morphing model: The morphing model is a quite simple. The nodes change their position only along a given line. The model also does not respect geographical constraints like already existing buildings, pedestrian roads, etc. Last but not least the resulting shape of the scenario does not look like streets anymore.

7. SUMMARY

This paper shows two approaches: First, a "realistic" agent based car driver simulation using only the curb side information of the scenario as an input and second, a morphing model for changing the shape of the given roundabout. With a simple iteration process it is shown that good indications can be found for optimizing the shape of the scenario. The iteration process would allow us to replace the given morphing model by a more enhanced one.

Apart from the above, using iteration processes for optimization problems has at least one other advantage: It allows us to separate the problem into pieces such that they are easier to understand, monitor and analyse.

8. REFERENCES

Balmer M., A. Vogel, and K. Nagel, 2005, "Shape morphing of intersections using curb side oriented driver simulation", in: *Proceedings of 5th Swiss Transport Research Conference (STRC)*, Ascona, Switzerland, see http://www.strc.ch (accessed Feb. 2006).

Bernard, M., 2005, "New Design Concepts for Transport Infastructures". in: *Proceedings of 5th Swiss Transport Research Conference (STRC)*, Ascona, Switzerland, see http://www.strc.ch (accessed Feb. 2006).

Campbell, M.I., J. Cagan, and K. Kotovsky, 1999, "A-design: An agent-based approach to conceptual design in a dynamic environment", in: *Research in Engineering Design*, 11(3), Springer-Verlag London Limited, p. 172-192.

CORSIM, Corridor Simulation, 2005, see http://www.ops.fhwa.dot.gov/trafficanalysistools/corsim.htm (accessed Feb. 2006).

Dijkstra, J. and H. Timmermans, 2002, "Towards a multi-agent model for visualizing simulated user behaviour to support the assessment of design performance", in: *Automation in Construction*, **11**: 135-145.

Ferber, J., 1999, *Multi-agent systems. An introduction to distributed artificial intelligence.* Addison Wesley.

Gloor, C., D. Cavens, E. Lange, K. Nagel, and W. Schmid, 2003, "A pedestrian simulation for very large scale applications", in: Koch and Mandl(eds.) *Multi-Agenten-Systeme in der Geographie* 23, Klagenfurter Geographische Schriften, Institut fuer Geographie und Regionalforschung der Universität Klagenfurt, p. 167-188.

Gloor, C., L. Mauron, and K. Nagel, 2003, "A pedestrian simulation for hiking in the Alps" in: *Proceedings of 3th Swiss Transport Research Conference (STRC)*, Ascona, Switzerland, see http://www.strc.ch (accessed Feb. 2006).

Helbing, D., I. Farkas, and T. Vicsek, 2000, "Simulating dynamic features of escape panic", in: *Nature*, **407**: 487-490.

Helbing, D., F. Schweitzer, J. Keltsch and P. Molnar, 1997, "Active walker model for the formation of human and animal trail systems" in: *Physical Review E*, **56**(3): 2527-2539.

Manikonda, V., R. Levy, G. Satapathy, D.J. Lovell, P.C. Chang, and A. Teittinen, 2001, "Autonomous Agents for Traffic Simulation and Control", in: *Transportation Research Record*, 1774, Washington, D.C., p. 1-10.

MITSIM, Microscopic Traffic Simulator, 2005, see http://mit.edu/its/mitsimlab.html accessed Feb. 2005.

Nishinari, K., A. Kirchner, A. Nazami, and A. Schadschneider, 2001, "Extended floor field CA model for evacuation dynamics", in: *Special Issue on Cellular Automata of IEICE Transactions on Information and Systems*, volume E84-D.

Raney, B. and K. Nagel, 2004, "An improved framework for large-scale multi-agent simulations of travel behaviour", in: *Proceedings of 4th Swiss Transport Research Conference (STRC)*, Ascona, Switzerland, see http://www.strc.ch (accessed Feb. 2006).

Schadschneider, A., 2001, "Cellular automaton approach to pedestrian dynamics – theory", in: Schreckenberg and Shama (eds.) *Pedestrian and Evacuation Dynamics*, Springer, p. 75-85.

Shelby, S.G., 2004, "Single-Intersection Evaluation of Real-Time Adaptive Traffic Signal Control Algorithms", in: *Transportation Research Record*, **1867**: 183-192.

Stucki, P., 2003, "Obstacles in Pedestrian Simulations", Diploma thesis, Swiss Federal Institute of Technology, ETH Zurich, Switzerland.

VISSIM, Planung Transport und Verkehr, PTV, 2005, see http://www.ptv.de/cgi-bin/traffic/traf_vissim.pl (accessed Feb. 2006).

Multi-Agent Models for Urban Development

Gentrification Waves in the Inner-City of Milan

A multi agent/cellular automata model based on Smith's Rent Gap theory

Lidia Diappi and Paola Bolchi
Dept. of Architecture and Planning, Politecnico di Milano Technical University

Keywords: Multi agent systems, Housing market, Gentrification, Emergent systems

Abstract: The aim of this paper is to investigate the gentrification process by applying an urban spatial model of gentrification, based on Smith's (1979; 1987; 1996) Rent Gap theory. The rich sociological literature on the topic mainly assumes gentrification to be a cultural phenomenon, namely the result of a demand pressure of the suburban middle and upper class, willing to return to the city (Ley, 1980; Lipton, 1977, May, 1996). Little attempt has been made to investigate and build a sound economic explanation on the causes of the process. The Rent Gap theory (RGT) of Neil Smith still represents an important contribution in this direction. At the heart of Smith's argument there is the assumption that gentrification takes place because capitals return to the inner city, creating opportunities for residential relocation and profit. This paper illustrates a dynamic model of Smith's theory through a multi-agent/ cellular automata system approach (Batty, 2005) developed on a Netlogo platform. A set of behavioural rules for each agent involved (homeowner, landlord, tenant and developer, and the passive fidwellingfl agent with their rent and level of decay) are formalised. The simulations show the surge of neighbouring degradation or renovation and population turn over, starting with different initial states of decay and estate rent values. Consistent with a Self Organized Criticality approach, the model shows that non linear interactions at local level may produce different configurations of the system at macro level. This paper represents a further development of a previous version of the model (Diappi, Bolchi, 2005). The model proposed here includes some more realistic factors inspired by the features of housing market dynamics in the city of Milan. It includes the shape of the potential rent according to city form and functions, the subdivision in areal submarkets according to the current rents, and their maintenance levels. The model has a more realistic visualisation of the city and its form, and is able to show the different dynamics of the emergent neighbourhoods in the last ten years in Milan.

Jos P. van Leeuwen and Harry J.P. Timmermans (eds.), Innovations in Design & Decision Support Systems in Architecture and Urban Planning, 187-201.

1. INTRODUCTION

The massive regeneration processes and significant changes in central neighbourhoods of European cities brings to the fore the gentrification issue. Far from being an unified and standardized process, gentrification occurs in various ways in different cities neighbourhoods, implying a variety of actors. The relevance of the phenomenon has earned many attempts to analyse, measure and compare different urban trends. Recently a special issue of Urban Studies (2003) has presented a wide survey of contributions on this topic.

The complex nature of gentrification was recognised in the mid 1980s by Rose (1984) and Beauregard (1986) which define the process as a "*chaotic concept connoting many diverse if interrelated events and processes [that] have been aggregated under a single (ideological) label and have been assumed to require a single causal explanation*" (Beauregard, 1986, p. 40).

Nevertheless, most of the literature still deals with the stage model drawn in the '70 by the Marxist economist Neil Smith (1979). According to this model, gentrification is a supply driven process, where investors and their capitals intervene in run-down neighbourhoods in order to capitalise on the "rent gap" or on potential increase in value in these neighbourhoods. Investors cause this by buying dwellings, renovating and reselling them to more affluent members of the "middle-high" class.

The Smith's theory act as a counterpoint to the demand side, consumer's sovereignty approach of David Ley (1992) and others, which views in gentrification the rise of a new middle class with new tastes and new utility functions in housing preferences.

The Smith's idea still represents a benchmark in gentrification theories and approaches; this paper develops a Multi Agent model which is based on its stage process.

The process is particularly suitable to be modelled into the Multi-Agent framework. The interest of the theory lies in the interplay among actors involved in the process, each of them being influenced by their knowledge: for property owners, landlords and tenants the knowledge is local and based on observations on the neighbourhood; they decide what to do with their properties on the basis of local information and an interactive imitative behaviour with neighbours.

Developers, big investors and, more recently, new actors coming on the urban scene, like cultural pioneers in search of alternative spaces, invest and locate on the basis of global knowledge of the whole urban real estate market.

It should be noted that, in this theoretical framework, the gentrification modelling fits both with the MAS approach, given the plurality of behavioural rules interacting each other, as well as with the Cellular Automata agents logic, since the proximity in the neighbourhood plays a crucial role. This paper represents a further development of a previous version of the model (Diappi, Bolchi, 2005), which was mainly devoted to explore, through simulations, the key parameter values for the evolution of the system. The simulations have shown an oscillatory, periodic behaviour of the system trends for a wide range of parameters.

2. THE SMITH'S STAGE MODEL

According to Smith's Rent Gap theory, gentrification is the last of successive stages affecting the housing stock cycle. Before briefly illustrating the theory it is worth recalling the Smith's definition of some key variables.

Smith defines *capitalised rent* (CR) the value of the building plus the ground rent, while *potential rent* (PR) is the rent value evaluated if a site were developed for its "*highest and best use*".

During the first cycle of use the ground rent is likely to increase as urban development continues outward, and the house value will begin to decline very slowly. Depreciation will induce a price decrease, but the extent of this decrease will depend on how much the ground rent has also changed in the same period.

In areas which experience initial depreciation, landlords and homeowners, who are aware of the imminent decline, are likely to sell out and seek newer homes where their investments will be safer.

At this stage, there is a tendency of the neighbourhood to convert to rental tenancy unless repairs are undertaken. This transition, typical of declining housing market, is expected to bring some degree of under maintenance which will yield surplus capitals to be invested elsewhere. The outflow of capitals increases the local degradation and individual efforts to counteract this decline are overwhelmed by the "neighbourhood". Even an individual landlord or owner which looked after his property would be forced to charge higher than average rent in the neighbourhood putting his property out of the market. The collective process of "undermaintenance" leads to progressive depreciation of real estate. Those tenants which are able to pay a suitable rent for a good housing quality and are aware of the increasing physical and social decay of the neighbourhood, look for a new dwelling and move away. Their flat, probably put on the market at lower rent, will be occupied by a more disadvantaged household. A filtering down process is taking place.

The physical deterioration and economic depreciation of the inner city neighbourhoods is a strictly logical and rational outcome of housing market. This depreciation creates the objective economic conditions for a rationale response of the market: capital revaluation. Developers decide to invest on the basis of the rent gap between the actual capitalised rent and the potential rent. When the rent gap is sufficiently large, the gentrification process starts. Investments become profitable, capitals flow into the market and determine the physical regeneration of the area, and the household turnover upwards.

Some different actors may lead to gentrification. In many European cities they include few, young and acculturated, single yuppies or households (pioneers), in search of urban niches facilitating alternative urban life styles.

Until now, gentrification has been presented as a spontaneous process. Actually the same process can arise as an undesired effect of public regeneration policies or projects. This usually starts by a single regeneration project, which may further trigger redevelopment in the area and eventually reduce the Gap between the potential and capitalized rent. The subsequent stages involve wealthier middle class households and real estate developers. The final stage is marked by the consolidation of the new upper class nature of these neighbourhoods.

3. THE CASE STUDY: GENTRIFICATION IN MILAN

In the last few years the housing market in Milan has shown one of the higher and sudden increases in prices of the last decades. Over the same period the city has lost approximately ten per cent of his one million and three hundred thousand inhabitants. Although most European and US cities have experienced suburbanisation or disurbanisation, the massive loss of population in Milan has been mainly driven by the speculative nature of real estate market and the rise a new affluent creative class (Florida, 2001) in the design and fashion industry - which characterizes the worldwide known image of Milan.

In the past, the city has experienced many gentrification waves. The first one, experienced in the sixties, affected mainly the core of the city within the Canals (Navigli) ring. The second one, which took place in the seventies, deeply transformed the more popular and dense urban neighbourhoods in the area between the two rings of Canals and of the Spanish walls, and affected also some peripheral areas with a strong environmental character like the radial axes along the canals towards the south east of the city. In the nineties, gentrification expanded in some peripheral areas where the existing old, generally multi-store, industrial buildings were particularly suitable to the modern spaces for the emerging creative activities. However, in the second half of the nineties one further wave, more significant than previous ones, arose in connection with the real estate "bubble" which invested extensively most of the more successful cities in Europe. In this respect, it should be noted that the gentrification wave not only expands in time, but affects also the rings already gentrified in the past, producing further population turnover.

Currently, the centre of Milan is assisting to the transformation of already gentrified prosperous and solid upper-middle class neighbourhoods into much more exclusive and expensive enclaves. This intensified gentrification is happening in few selected areas that have become the focus of intense and conspicuous consumption by a new generation of very rich "financiers". As Neil Smith pointed out "*the hallmark of this latest phase of gentrification is the reach of global capital down to local neighbourhood scale*".

4. THE AGENT BASED MODEL SET UP

The Smith's approach offers the opportunity to investigate the process through a Multi Agent/Cellular Automata model.

It is well known (Batty, 2005) that in all MAS model a clear distinction should be made between the set of objects called agents $\{A\}$ and sets of cells or layers that define different landscapes $\{L\}$. There are M agents, each one indexed by k and defined as existing in the location i and at time t, A_i^k (t), k = 1,2, ….,M. There are W different landscapes and in each landscape λ there are N cells where each location at time t is defined by L_i^{λ} (t), $\lambda = 1,2,\ldots,W$; $i = 1,2,\ldots N$.

The agents react purposively to their environment which is encoded in the landscape and that action and interaction between agents and landscapes can always be defined in these terms.

In our model we consider four types of agents: landlords, tenants, homeowners and developers. The cells are the property units (PU*i*) or real estate units, each of them being characterised by a location *i*, a potential rent PR_i *(t)* and a capitalised rent varying over time, $CR_i(t)$, a maintenance status MS_i *(t)*, ranging from zero (very bad) to one (new).

In this model $CR_i(t)$ coincides with the real estate value or transaction value exchanged on the market in *i* "*given the present land use*" (Smith, 1979, p. 543) at time *t*.

The Potential Rent $PR_i(t)$ is "*the amount of rent) that could be capitalised under the land's highest and best use*". PR_i is then correlated to the demand of real estate investments and to the attractiveness of the city as a whole. This variable is strongly influenced by global factors such as the relative position of the city among national and international urban networks and the relative convenience to invest in real estate instead of in movable goods (the recent real estate speculative bubble is a clear evidence of this). Other local factors such as the long term loan interest rates, the policies on housing market, the tax treatment of real estate are also crucial. Given all these conditions let's assume, for potential rent, a conic form, according to Hoyt (1933) and Wingo (1961) and Alonso (1965), which consider land rent explained in terms of accessibility, being the latter perfectly complementary to transportation costs. Indeed, empirical data on real estate values in Milan confirm the conical form of urban rent (Fig. 1). Nevertheless, potential rent may also represent the expected value after renewal or rehabilitation. In a dynamic perspective the city evolves and some neighbourhoods emerge and gentrify while others decline. It may be assumed that, being equal the distance from the centre, the potential rent at *d* coincides with the neighbourhood's highest real estate rent among all the neighbourhoods included in the ring with *d* distance from the centre, since this area has been able to capture *the highest and best use*. The interpolation of the top values for each radius gives the conic surface of potential rent for the city (Fig. 2).

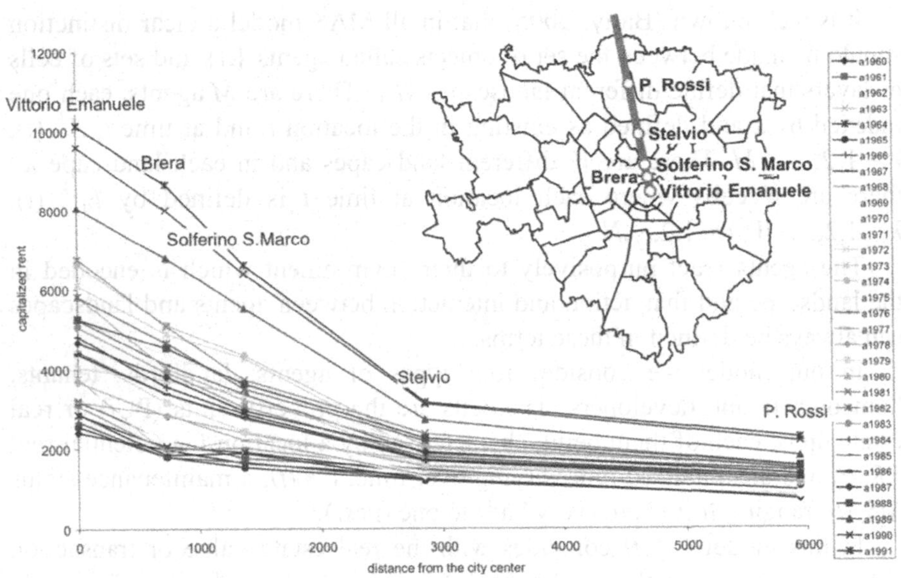

Figure 1. Evolution of the real estate rents in Milan from the center northwards over the period 1961-1988.

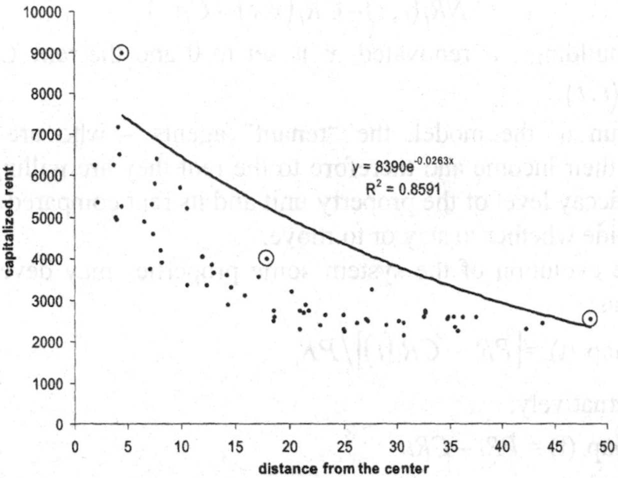

Figure 2. Potential rent is obtained through the interpolation of the rent top values.

In its process dynamics the model identifies the following agents and decision rules:

The initial state of simulations is generated by randomly attributing an age x_i to each property unit i. Its conservation state depends on age according to the function:

$$D_i(x,t) = e^{-\lambda x_i} \tag{1}$$

where λ is a constant annual decay rate. Each time the property unit undergoes rehabilitation, its age restarts from zero.

The capitalised rent is the market value of the property at time t. At the new construction time t_0 his capitalised rent $CR_i(x,t_0)$ coincides with the potential rent PR_i (t_0). Then, with the increasing ageing of the property, $CR_i(x,t_0)$ declines:

$$CR_i(x,t_0) = PR_i(t_0)\,e^{-\lambda x} \tag{2}$$

At each cycle, the owner of the property unit evaluates the advantage in rehabilitating it, knowing that after rehabilitation, the capitalised rent of his building will equal the average rent of property units in the neighbourhood. The expected rent $NR_i(r,t)$ is then calculated as the average of capitalised rent of the neighbourhood, defined over the J PUs within a radius r:

$$NR_i(r,t) = \frac{\sum_{j=1}^{J} CR_j(x,t)}{J}, \tag{3}$$

The cost of rehabilitation $C_i(x)$ increases with the age of buildings. Assuming C_0 as the initial construction cost, set constant for all the PUs:

$$C_i(x) = C_0\left(1 - e^{-\lambda x}\right), \tag{4}$$

If:

$$NR_i(r,t) - CR_i(x,t) > C_i(x) \tag{5}$$

then the building i is renovated, x_i is set to 0 and the rent $CR_i(x,t)$ is equal to $NR_i(r,t)$.

At each run of the model, the "tenant" agents – who are classified according to their income and therefore to the rent they are willing to pay – evaluate the decay level of the property unit and its rent compared to the rent paid, and decide whether to stay or to move.

During the evolution of the system some properties may develop a rent gap, defined as:

$$\text{Rent Gap}_i\ (\text{t}) = [PR_i - CR_i(t)]/PR_i$$

Or alternatively:

$$\text{Rent Gap}_i\ (\text{t}) = PR_i - CR_i$$

At this stage, investors may intervene. At each cycle they have a certain amount of capital available *B(t)* for rehabilitating property. Having global information they can identify for which building the difference between potential rent and capitalised rent is greater than a certain threshold δ:

$$\text{Rent Gap}(i) > \delta \tag{6}$$

Investors begin to invest in those properties having the higher rent gap among those where (6) applies. Its rehabilitation cost is calculated using (4) and subtracted from the available capital. The age of the building x_i is set equal to 0 and the capitalised rent is recalculated. Unlike the situation when rehabilitation is carried out by owners, it is equal to the potential rent:

$$CR_i(t+1) = PR_i(t) \tag{7}$$

The random selection of a PU with adequate rent gap is repeated and the procedure reiterated until the available capital is exhausted.

The whole procedure is repeated for a number of cycles which mainly depends on the parameter values selected. Simulations can be regarded as useful when results are stable over a large number of cycles.

In a previous paper (Diappi and Bolchi, 2005) the authors have illustrated the results obtained through simulations on an abstract city on which one by one the parameters have been implemented within a wide range of values.

5. IMPLEMENTATION OF THE MODEL

The model considers the evolution of the real estate market in Milan from 1993 to 2003. At this stage, our aim is to produce a simplified model of the spatial dynamics of the phenomenon able to show the leading factors which affect real estate dynamics.

The model runs on a Netlogo Platform. Giving the abstract nature of the model, some simplifying assumptions have been made. This implies a geometrical approximation from the irregular network of blocks to a regular grid of cells. Each cell represents a property unit which changes maintenance level and capitalised rent over time. The visual representation of Netlogo allows distinguishing the local dynamics of cells through different colour and shades. The model has been set up on the basis of the following conditions:

The study area has been divided in 67 zones according to the real estate submarkets identified by Borsino Immobiliare (Real estate stock exchange) of the Chamber of Commerce of Milan, an institution which provides every six months for each zone the real estate quotations:

- for sale and for rental
- for different land uses: dwellings, offices, shops and industrial buildings and
- for age of the property (new, recent (5-30 years) and old (> 30 years)) .

The average values for residential use has been assumed in the model as capitalised rent. The potential rent has been estimated as a conic function which interpolates the highest values for each ring. In each zone the share of buildings for each maintenance status (very good, good, bad and very bad), has been derived from 1991 and 2001 National Census Data. At the beginning in each zone the maintenance status is allocated to each property through a Monte Carlo procedure. Other parameters are:

Annual Decay rate $\lambda = 0,02$
Available capital = 100.000
Construction cost for each building = 1000
Neighbourhood radius =10 cells
Capitalised rent ranges from 1500 to 7000

The measure of the rent gap threshold turned out to be crucial in our experiments. Indeed, there are two possible options: the first one measures, for each cell, the rent difference between PR and CR; the second one considers the ratio of CR over PR. In both cases investors begin to invest in those properties having the higher rent gap. We assumed:

Rent Gap = (PR_i-CR_i) or $\left[PR_i - CR_i(t)\right]/PR_i$

These first experiments do not include the landlord/tenant actors and do not consider explicitly the residential mobility. It should be said that rented dwellings, excluding social housing stock, in Milan represent only 13% of the total housing stock on free market.

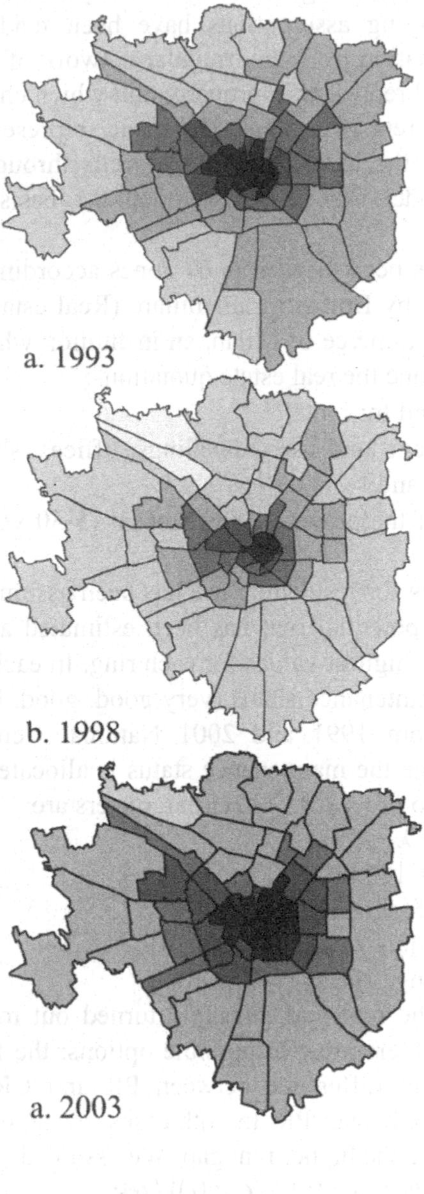

Figure 3. The evolution of the average real estate values in Milan in the period 1993-2003 (Source: Borsino Immobiliare Chamber of Commerce of Milan).

6. FIRST RESULTS

The Data on the real estate values in the area is represented in Fig. 3 a,b,c. All the three figures show that the rents are declining from the central topical values to the periphery. Nevertheless, some relevant differences affect the tree images. First of all the peak values, and consequently all the other values, drop in the middle of the period considered, namely 1998. As it is well known in real estate economics, this market is usually undergoing a cyclical trend of ten years. In the model this typical trend can be included by modifying the potential rent function.

Second, when the new wave of demand emerges, the centre of the city is the first to be involved; here the prices increase at higher rate than in the rest of the urban area.

The first simulation A assumes the above mentioned parameter values and the rent gap measured in absolute terms. Given the conic form of potential rent, high profits turn out to advantage big investors which are able to put massive investments in the centre.

When the rent gap is measured as ratio $\left[PR_i - CR_i(t)\right] / PR_i$ (simulation B), the range of investors involved widens, while also in areas with relative low PR, it should be convenient, for agents with limited budget, to invest, taking advantage of the relative cheap prices of the properties.

This is shown in simulations A (Fig. 4) and B (Fig. 5). The simulations A, in which investments are taking place according to the highest rent gap, show continuing investments in the central and semi-central areas, while peripheries are abandoned to a constant decline; whereas simulation B, number of cycles being equal (100), shows a more complex structure. It is worth noting here a spontaneous emergence of semi-peripheral and peripheral neighbourhoods. This is the realistic result which reproduces the revitalisation of old parts of the city by "pioneers" in search of spaces for innovative activities.

Moreover, when the rent gap threshold is given in terms of "ratio" the average maintenance level is higher (case B - cycle 100, average CR = 2632; case A - cycle 100, average CR = 1005). Since the amount of available capitals is the same for both simulations, there is, in B, a higher number of investors operating on the market with a diffused improvement for the city as a whole.

Looking at the whole sequence of simulation B, is worth to note the cyclical dynamics of the maintenance levels and, in general terms of the spatial pattern of the system. At the beginning investors renovate some isolated semi-central areas; then the property owners strengthen and widen the investments northwards and southwards (cycle 30). Afterwards the gentrification process affects some peripheral neighbourhoods as well (cycle

50). Finally the city centre and north-western part of the city undergoes a new gentrification wave (cycle 60).

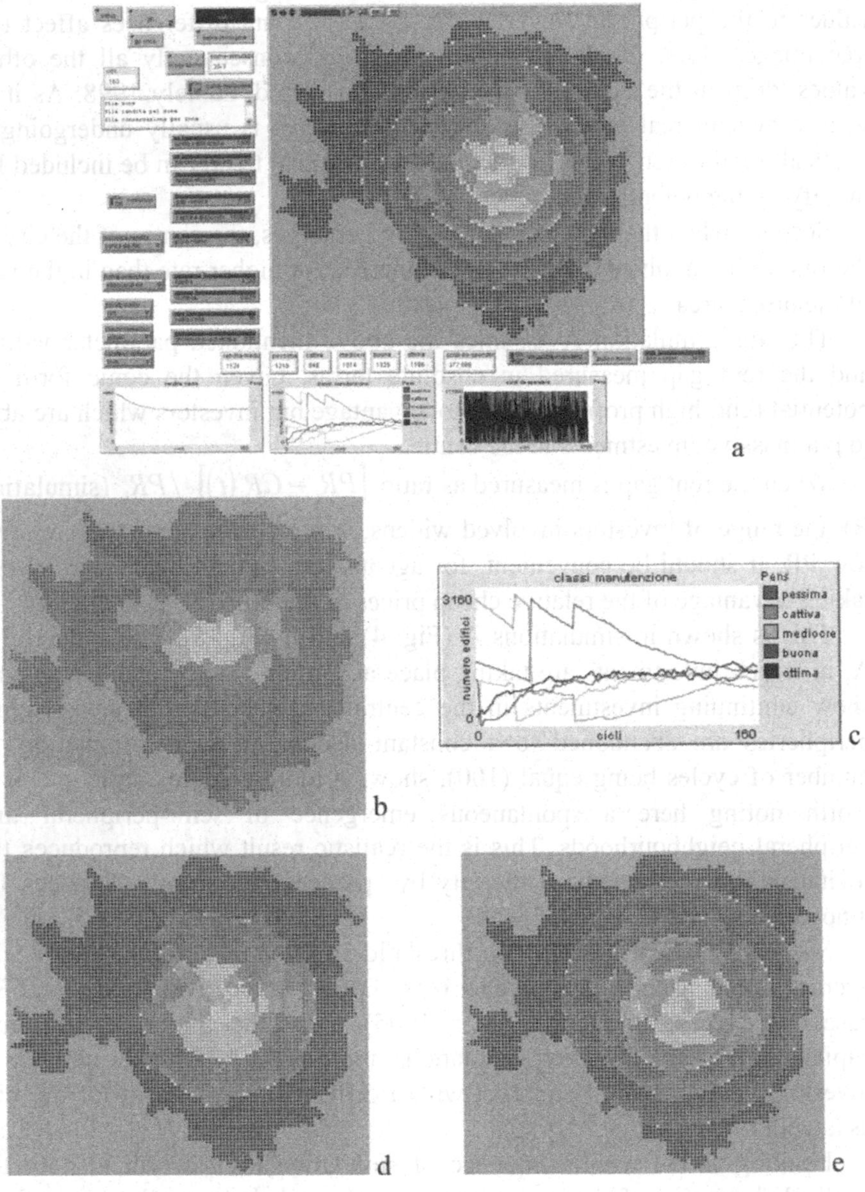

Figure 4. Simulation A – Investors rule: heighest rent gap (a)-The simulation environment of Netlogo; (b)-Initial state; (c)- the maintenance classes trends; (d)-cycle 40; (e)- cycle 100.

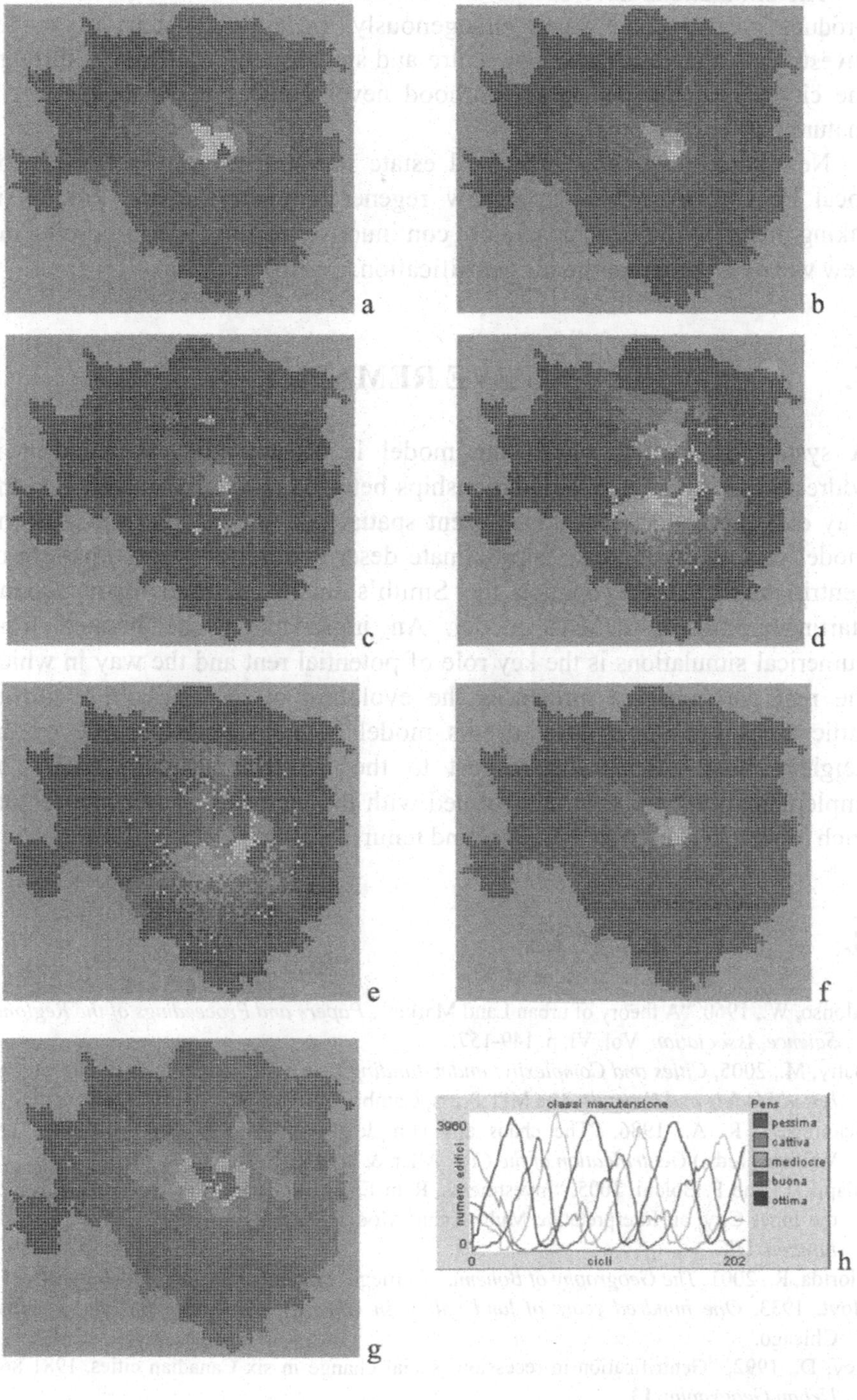

Figure 5. Simulation B – investors rule: highest ratio of rent gap over potential rent - cycles (a) 40, (b) 50, (c) 60, (d) 70, (e) 80, (f) 90 and (g) 100; (h) maintenance classes dynamics.

The simulations have shown that the real estate market logic is able to produce gentrification waves endogenously. In, approximately, 60 cycles investments restart in the urban centre and spread out progressively through the city. The gentrified neighbourhood never reaches a steady state or a mature equilibrium point.

New actors and subsequent real estate investments can destabilise the local housing market when a new regeneration/gentrification process is taking place. Social benchmarks are continuously moving upward producing new waves of neighbourhoods gentrification.

7. FEW CONCLUSIVE REMARKS

A systematic exploration of the model is already underway and it is addressed to investigate the relationships between model parameters and the way different values lead to different spatial pattern. In these first run the model was able to give an approximate description of the spatial pattern of gentrification. In our opinion the Smith's theory has provided a sound starting point for a MAS model. An important result obtained from numerical simulations is the key role of potential rent and the way in which the rent gap measure influences the evolution of the market. A further indication of the relevance of this model is the "ancillary" role of the neighbourhood effect with respect to the investors driving force. The implementation can be further refined with the inclusion of relevant factors such as age, urban fabric typology and tenure.

8. REFERENCES

Alonso, W., 1960, "A theory of urban Land Market", *Papers and Proceedings of the Regional Science Association,* Vol. VI, p. 149-157.

Batty, M., 2005, *Cities and Complexity: understanding Cities with Cellular Automata, Agent-based Models and Fractals*, The MIT Press, Cambridge, Massachusetts.

Beauregard, R. A., 1986, "The chaos and complexity of gentrification", in Smith and Williams (eds.) *Gentrification of the City*, Allen & Unwin, Boston, p. 35-55.

Diappi, L. and P. Bolchi, 2005, "Investments, Rent Fluctuations and Gentrification Waves in the Inner City: an Interpretative Multi-Agent Model", *paper presented at CUPUM 2005, London, UK.*

Florida, R., 2001, *The Geography of Bohemia*, Carnegie Mellon University, Pittsburgh.

Hoyt, 1933, *One hundred years of land values in Chicago*, University of Chicago Press, Chicago.

Ley, D., 1992, "Gentrification in recession: social change in six Canadian cities, 1981-86", *Urban Geography*, **13**.

Rose, D., 1984, "Rethinking gentrification: beyond the uneven development of Marxist urban theory, *Environment and Planning D*, **2**: 47-74.

Smith, N., 1996, *The new urban frontier: Gentrification and the Revanchist City*, Routledge, London and New York.

Smith, N., 1979, "Toward a theory of gentrification: a back to the City movement by Capital, not people", *APA Journal*, October, p. 538-548.

Urban Studies, 2003, Special Issue on Gentrification, **40**(12).

Wingo, L. Jr, 1961, *Transportation and urban Land,* Resources for the future, Baltimore, Maryland.

Wingo, L. Jr, 1961, "An economic model of the utilization of urban Land for residential purposes", *Papers and Proceedings of the Regional Science Association,* Vol. VII, p. 191-205.

Multi-Agent Model to Multi-Process Transformation

A housing market case study

Gerhard Zimmermann

University of Kaiserslautern

Keywords: Agent technology, User activity modeling, User activity simulation, Software engineering, Code generation, Software process

Abstract: Simulation is a means to help urban planners and investors to optimize inhabitant satisfaction and return on investment. An example is the optimal match between household preferences and property profiles. The problem is that not enough knowledge exists yet about dynamic user activity models to build reliable and realistic simulators. Therefore, we propose a modeling and software technique that produces simulator prototypes very efficiently for the development, test, and evaluation of many different user activity models, using executable models, code generation, and a domain specific software process. As a specific feature, the model is based on many agents acting independently from each other and that are mapped in several refinement steps into the same number of concurrent processes. The housing example is used as a case study to explain the process and show performance results.

1. INTRODUCTION

User activities in urban and building environments gain more and more interest in architectural design and evaluation tasks. Since user activities are non-deterministic and since model abstractions of these activities are not yet well understood or validated, tools are necessary to develop and test user activity models and to apply such models in design decision processes. Computer simulation is a means to implement such models and to execute, observe, and analyze user activities with different initial and boundary conditions.

Although a number of simulators exist for applications such as motorized and pedestrian traffic, models and simulators for other kinds of user behavior

Jos P. van Leeuwen and Harry J.P. Timmermans (eds.), Innovations in Design & Decision Support Systems in Architecture and Urban Planning, 203-219.

in urban and building environments still have to be developed and tested. One problem is that of finding the appropriate models for the desired applications. We specifically concentrate on dynamic models of many individual users instead of stochastic models of user groups because traffic simulations have shown that human behavior is modeled more realistically that way. Appropriate in this context means that because of the high complexity of such models the abstraction of the reality should be as simple as possible without missing essential details. Therefore, many experiments with different abstractions are necessary to arrive at models with the right amount of details.

This leads to the next problem, the implementation, execution, and variation of models. Advances in computing and software techniques in the last decade have provided us with a number of options to solve this problem as efficiently as possible. Efficiency is very important because during the development phase of models a large number of different experiments are necessary. Therefore, we propose and demonstrate a modeling and software prototyping process that is based on executable models and code generation tools. Executable models are formal models which include dynamic behavior and can be either interpreted as or compiled into executable computer programs. We will use a case study to demonstrate the process and show efficiency data.

2. STATE OF THE ART

Our work is based on three main areas: user activity modeling, simulation, and software engineering. We cannot cover all of them completely and will restrict ourselves to some representative examples.

Static user activity modeling in building environments was introduced by Eastman and Siabiris (1995) and by Eckholm (2001). In the IFC models user activities are not yet included. We also presented an effort to integrate user activities into a comprehensive building system model (Zimmermann, 2003) as a basis for extending it to user dynamics.

Models exist or are under development for pedestrian movements (Kukla et al. 2002) or in shopping centers (Borgers and Timmermans, 2005). The models are either based on cellular automata with a very restricted horizon and no differentiation of individuals or on agent systems with a much larger horizon and the possibility of giving different properties to each individual. Still, cellular automata exhibit much more realistic pedestrian behaviors than stochastic models and because of the regularity, synchrony, and simplicity can be executed with very large numbers of individuals. Agent models can implement much more complex behaviors and are in principle asynchronous. The possible size of such simulations has to be seen.

Models of other user activities are also under development. For our case study we adopted models from (Devisch et al. 2005) for the dynamic migration behavior of households in towns. Devisch is modeling individuals, households, realtors, and properties as agents and is implementing very sophisticated decision and negotiation models.

Simulation has become a mature tool in the area of building performance (e.g. ESP-r). Such simulators are based on modeling the physical environment by differential equations and numerical solutions at run-time. Also in the area of pedestrian simulation and visualization, for example for the purpose of emergency evacuation (Klüpfel, 2005), tools exist and are in use. These are based on cellular automata or agent technologies. Integrating the different simulation approaches is difficult. Therefore, we successfully experimented with performance simulators that are based on mapping physical entities such as walls and windows to autonomous objects and executing all objects concurrently (Zimmermann, 2002). This approach lends itself to integration with agent models.

The construction of simulators can be very time consuming tasks. This is true for many reactive systems and because of the typically low number of sold systems, development cost is an important factor. For this purpose intensive research has been conducted in the field of software technique for reactive systems. At the same time software engineering has made large progress, especially by better modeling techniques such as the *Specification and Description Language* SDL (Olsen et al. 1994) and the *Unified Modeling Language* UML. Both environments allow formal models of asynchronous dynamic behavior and tools are available for automatic code generation. This code may not be as computational efficient as manually programmed code, but the generation enhances process efficiency by magnitudes. We are using SDL for modeling and simulation purposes. Environments such as SIMULINK and Modelica are more orientated towards continuous simulation and not so suited for modeling user activities. Efficiency also depends on the applied software process. For the purpose of user activities special processes for reactive systems are of advantage, for example (Braek and Haugen, 1993). We have tailored such a process for building control systems and performance simulators (Zimmermann and Metzger, 2004) and will show in this paper that this process can also be applied efficiently to user activities in the case study.

3. MODELING ENVIRONMENT

The model used to describe the structure follows the principle idea of our object-based building system model (Zimmermann, 2003). The model has several levels of refinement. The top level in *Figure 1* represents the relation between the domains, in our case the *User Activity* and the *Functional Unit Domain*. The next level refines these domains to represent general types for urban models, for example *Group Role Type* in the *User Activity Domain* and *Organizational* and *Functional Unit Type* in the *Functional Unit Domain*. The *Application Domain Level* further specializes each *Domain Level Model* for a specific area, for example population migration.

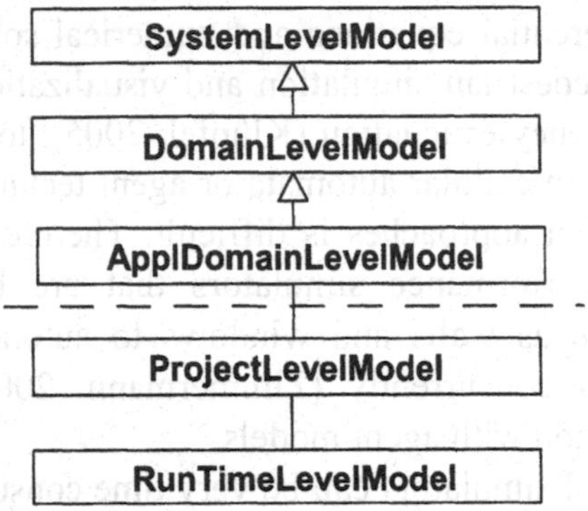

Figure 1. Modeling levels. Triangles depict inheritance, straight lines instantiation.

For a specific project like the one presented here, specific object types are derived from the types above the dashed line by instantiation. A *Household* type is an instantiation of the *Group Role Type* with the function of controlling and changing (by moving) the utility gained from the inhabited property. The *Property* type is an instantiation of *Functional Unit Type* and a *Market* type of the *Organizational Unit Type*. The function of a market is the management of properties and making offers on request. Finally, the simulator at run-time is a further instantiation of these types into individual objects. Objects can exhibit agent properties and we will also call them agents. *Figure 4* shows the object type diagram for our case study project as an example. For automation purposes we also use tabular representations of the object type structures.

Modeling the dynamic behavior is a more demanding task. We use several representation types for different refinement levels. With structured text we describe *Needs, Tasks*, and *Strategies* informally. Structured text means that identifiers are assigned to text blocks to be able to store, retrieve, and reuse the blocks in a structured way and to be able to define relations between blocks. This is important for consistency checks and for documentation and annotation purposes. In *Chapter 5* we will show some examples.

With *Needs* we describe what the system is supposed to do for its user. These *Needs* are then translated into *Tasks* and subtasks that are assigned to object types. From the *Tasks* we derive *Strategies*.

The first refinement of the *Strategies* is MTCs (Message Transition Chart). To understand these charts we have to introduce the model and software architecture of the target simulators in the language SDL (Olsen, 1994). All objects are mapped into threads that execute concurrently and communicate by messages. These threads act like agents and we will use this term from now on. Agents are internally modeled as extended finite state machines (EFSM). They are represented by state transition diagrams. Messages trigger state transitions. Messages are saved in an input queue for each agent. If a queue is empty, the agent becomes inactive. Time can be introduced by timers that also send messages at the set time.

Both the agent-message level and the EFSM level individually are not very complex when both are well structured. We do this by using aggregation hierarchies of agents in SDL to keep individual agent types reasonably simple. The difficulty to design and understand models arises when both levels are combined. This is necessary to understand the dynamic behavior of the whole system. MTCs are a means to abstract both levels as much as possible to understand the main dynamic actions and interactions of the agents. With MTCs we only model the messages that are exchanged between agents and the state transitions that are triggered by messages and create new messages. All other state changes, especially data transformations are neglected. MTCs do not have to be complete. Individual scenarios can be modeled as action-reaction chains. Further abstractions show agents as composed states. *Figure 5* and *Figure 6* show an example from the case study.

4. THE SOFTWARE GENERATION PROCESS

A software generation process PROBAnD (Metzger and Queins, 2002) was developed for reactive systems in general and especially for building control systems. The efficiency of the process is based on advanced modeling and software generation tools, but also on tailoring the process to the specific domain. Because of the latter, we can make use of predefined document and model types and architectures, specific process structures, and reuse partial descriptions and models.

We first adapted this process with small changes to building performance simulation (Zimmermann, 2002) and have now also used it for user activity modeling. Here we only describe the main steps of the process and explain it in more detail using examples from the case study in *Chapter 5*.

Figure 2 gives an overview. Starting point are the *Needs* that build the *problem description* and information about the *environment* such as layouts and other design information. Especially the environment structure is used to build an *object structure*, using the same terminology. This is important for the necessary discussions between modeler and customers, in this case urban planners. The mapping of *Needs* into *Tasks* now depends on the chosen *object structure* shown as a strong relation in *Figure 2.*

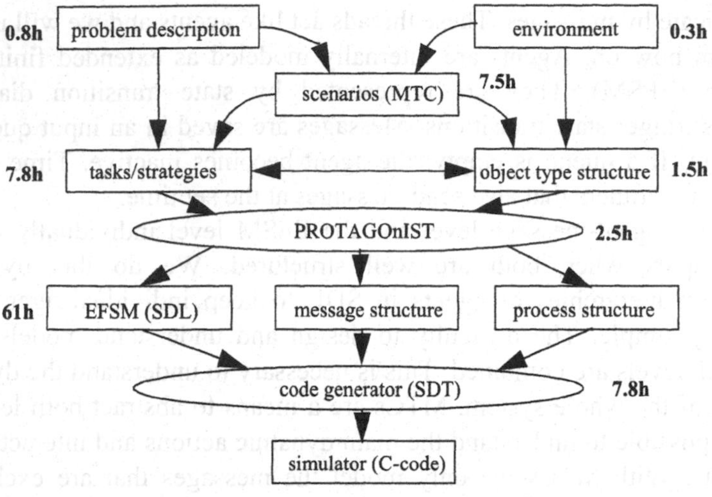

Figure 2. The software process. Numbers besides the boxes (documents) and ovals (tools) are recorded person hours for creating, modifying, testing, and debugging.

The modeling of the concurrent interaction between objects (agents) can be quite complex. The MTCs are a means to divide this task into *scenarios* which are orthogonal to the object structure. MTCs are optional and the direct path from the *problem description* to *strategies* is also possible.

Strategies are first described informally. A formal description is derived in several refinement steps. For this purpose we use two representations. The first is a textual formal language that describes all state transitions individually and is structured in tabular form. It is closely related to the task and MTC structure. The advantage of a textual over a graphical language is that reuse of text components is easier than of graphical components.

The step from the textual to the graphical representation of SDL can be done manually, but is error prone and time consuming. Therefore, we automated this by the tool PROTAGOnIST (Metzger and Queins, 2003). The graphical layout follows simple rules and is done automatically. SDL graphs can be further manipulated and analyzed by the commercial tool environment SDT (Telelogic).

The last step is the automatic generation of C-code using SDT. This also produces a run-time environment for the concurrent execution of agents and the message and timer administration. Finally the code is compiled into an executable program and can be run in simulation experiments. SDL provides different generators to create code for interactive debugging, real-time code, efficient code without debugging aids, and application code that can be run stand-alone without SDT licenses.

In *Chapter 5* we will demonstrate the process with examples. *Figure 2* shows efficiency data in the form of person hours spent on different document types for the case study. The data are automatically recorded by a simple document management and versioning environment, based on internet browsers. Adding all hours results in 89 person hours or 11 work days. The large number of 61 hours for SDL is mostly debugging and refinement time because of being to "hasty" in earlier phases. We can assume that one person can create such a simulator within three weeks. We achieved the same result with other applications as well. This is a very reasonable time frame. Creating initial and boundary data take at least this time and simulation experiments as well.

5. THE CASE STUDY

This case study was adopted from ongoing research at the Eindhoven University of Technology (Devisch, 2005) to be able to compare models, process, and results. It is not intended to copy all model details, nor to compete in any sense. All decision processes have been greatly simplified. Our goal is the demonstration of the software process and the efforts necessary to produce a simulator prototype and run some initial experiments.

5.1 The Problem

It is of interest for city planners to be able to predict the result of changes in the built environment for citizens and commercial enterprises regarding their satisfaction with the gained utility. It is also of interest to analyze the result on yield for owners and investors and on city tax income. Example changes are the construction of new properties, the change of the environment and quality of buildings, or of renting and buying prices. It is also of interest to predict changes in the population itself and the result on satisfaction and necessary changes in the built environment. This includes migration into and out of cities.

In this case study households are considered as basic entities. Households are created, change in size, and disappear, mostly because of biology and of decisions of individuals to join or separate. Other properties such as income, work place position, or preferences can also change over time. Households occupy properties in buildings. Properties can be owned or rented. The dynamics are mainly visible as moves of. Other changes are property improvements to increase satisfaction. Both events are determined by decisions of the households and by the property market. Therefore, we need models of the dynamic behavior of households and of property markets. Since we model all entities individually, they should not be uniform but display the whole spectrum of different behaviors. The behavior should also have some random components, modeling non-determinism.

One of the problems of such models is complexity. This is not so much a computational problem. If models have too many parameters, we have no way to determine a correct set of parameters by real life experiments. In our approach we therefore start with as simple models as possible and allow for extensions when necessary.

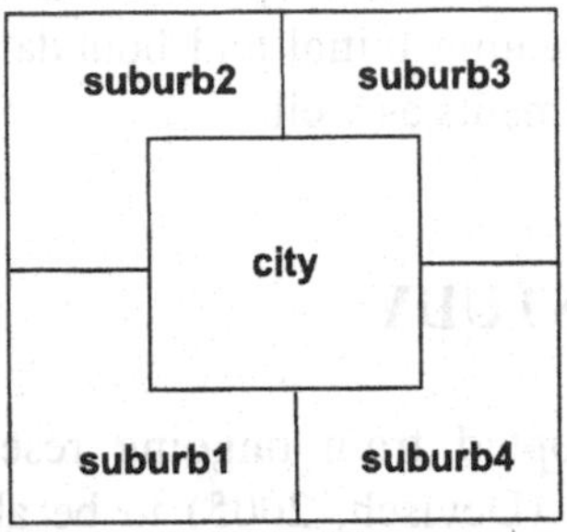

Figure 3. Sample town layout, outer dimensions: 4 x 4 km.

5.2 Basic Model

The basic model of a small town consists of five property markets, one in the center called city and four suburbs. *Figure 3* shows the layout. With this simple geometry tests for the assignment of positions to markets is simple and decision processes can still be tested. We also use a fixed assignment of one realtor agent per market as a further simplification. This agent manages all properties of its assigned market.

Households are not modeled as groups of individuals, but as basic agents. Households are grouped into 15 types according to the number of persons (1..5), employment status (student, employed, retired), and income level (low, high). Unlikely combinations are excluded.

Households are characterized by profiles and preferences. Profiles are data about the household, as for example income, current property, and position of workplace. Preferences describe how households value property profiles and how flexible households react to different property options. Household and property profiles are evaluated against the preferences to calculate the utility of a property for a household. Details of the data and the calculations are in *Section 5.3*.

Properties are characterized by property profiles. The profiles include typical data such as type, location, size, ownership, value and cost.

All households and properties can have different data. The data are stored in external files and are loaded into the simulator before the simulation starts (initial condition) and during simulation to communicate changes (boundary condition). The classification into household types is only a means to create sample data that reflect statistical data of real communities.

5.3 Utility Metrics

The goal of households in selecting properties and deciding on moves is the optimization of the utility it gains from using a property. Many factors influence the utility evaluation. Other goals such as minimizing the cost of providing the urban environment for the city administration and to optimize the return on investment for property owners can also be considered. In this case study we concentrate on utility metrics only, although the other goals can be observed as well using the resulting data.

In this case study we have selected eight criteria out of a much larger number of possible ones (see also Devisch, 2005) for the calculation of utilities. The following list contains a maximum of three choices for each criteria and short explanations:

1. urbanisation: city, suburb, rural. The first two correspond to the sections of our model town, rural is set equal to external.
2. tenure: renting and owning a property.
3. garden: no, small, large garden belonging to the property
4. type: apartment, row house, free standing house
5. size: <80%, 80-120%, >120% of the ideal floor space for a household which is a profile value.
6. distance: >10, 2-10, <2 km rectangular distance between property and centre of external household activities, for example the work place, the school, or a shopping centre.
7. remaining income: <50%, 50-70%, >70% of the total household income reduced by rent or sales price financing of the property.
8. resistance to move: single value that reduces the utility of a property in case a move to it is necessary. Can be extended to resistance to renovate.

Using a larger number of criteria and choices would in principal refine the utility calculation, but would also require a much larger number of realistic preference parameters to be determined by real life experiments and would not change the decision process in principal.

For each criteria i and choice j_i a preference value $b_i(j_i)$ typically for a household has been defined. The range of b_i is [0-100%] such that

$$b_i(1) + b_i(2) + b_i(3) = 100\% \tag{1}$$

The maximum values of a choice for each criteria show the relative main preference, the values of the other choices show the household's tolerance towards other choices. For example, if a household would prefer a free standing house, but would find a row house also ok, but not an apartment, it could express this as $b_4 = (0, 30\%, 70\%)$. In order to express the importance of each criteria in relation to the others, weight factors w_i are introduced.

We can now calculate the utility u for a household with known preference values and known j-vector as

$$u = \sum_{i=1}^{8} b_i(j_i) \cdot w_i \tag{2}$$

The resistance to move weight w_8 has to have a negative value in the case of a necessary move and zero otherwise.

We calculate three different utilities. u_{ideal} is the best value for a given preference matrix if for each criteria the best choice is selected. For the currently inhabited property its profile is compared against the household profile in order to determine the j-vector and to calculate u_{curr}. The same method is applied for offered properties, resulting in u_{offer}.

Next we need a value that shows the relative satisfaction s of a household with its current property or with offered properties to decide if a move is appropriate. We calculate the satisfaction by comparing the ideal utility with the current or offered one. A satisfaction factor sf describes the reduction of the ideal utility that is still seen as satisfactory. In our simulations, $sf = 75\%$ has been assumed to give a positive satisfaction. This leads to

$$s = (u_{curr} - sf \cdot u_{ideal}) / u_{ideal} \tag{3}$$

If this value is positive, the household is assumed to be content with its situation; if negative it will search for an alternative with a better satisfaction value. The possible range is of s is [0.25, -0.75].

Naturally, a household will first search for offers with a utility close to the ideal one. Therefore, three attempts are made with a minimum offer acceptance level of 90% of the difference between ideal and current utility and then going down in 8 steps of 10%. If still no offer is found, the search is aborted until a later date. This strategy may lead to a move into a property with negative satisfaction and also trigger a new search, but not immediately to allow the market to change.

The satisfaction of a household may change over time and trigger a search. Such changes can stem from changed preferences and profile values, as for example household size or new work place. The change may also stem from changed property profiles such as depreciating quality of it over time.

5.4 Model Structure

The structure of the model is static. All households are modeled as individual agents. We have limited the number to 500 to be able to generate and run simulation prototypes in the range of 10 minutes for a two year simulation time period. The five markets are also agents. A minimum of 500 properties are distributed to these markets but are not modeled individually. The aggregating agents are objects that can perform global bookkeeping, instrumentation, and data collection. Figure 4 shows the object type structure. The cardinalities define the number of objects that are aggregated by the object type. The object BigCity functions as simulation time management and user and file interface.

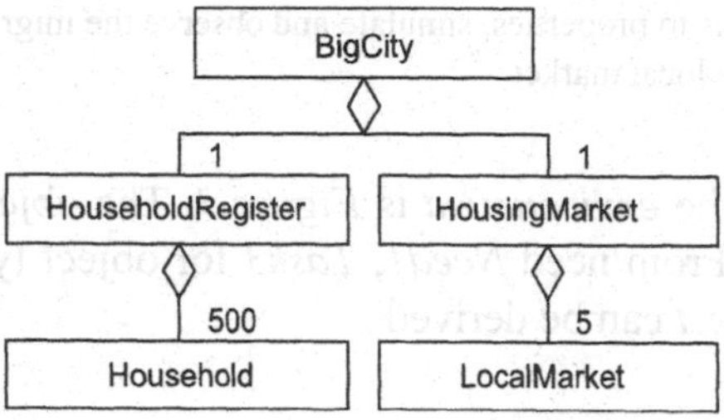

Figure 4. Object type structure, the diamond denotes aggregation, the numbers the cardinality.

5.5 Model Dynamics

The model functions as follows: Households with negative satisfaction send a request to the city and the suburb markets nearest to their center of activity. The request contains a floor area and a cost range and the tenure preference. Both markets select a maximum of three offers each from a list of free properties that match the request and send the property id to the household. The household the polls the property data, calculates offer utilities and sorts them accordingly. For offers that are acceptable an *accept* is sent to the market. Since the markets can send the same offer to many households simultaneously, acceptance messages are handled on a first-come-first-serve basis. If the property is still free, a confirmation is sent to one household, taking the property from the free list. When the household has moved, the old property is set on the free list. This is in short what happens.

All initial data and all dynamic changes are fed to the simulator from files that we derive from spreadsheets. This gives us the opportunity to run experiments with different sets of households and properties without changing the simulator. Result data are also stored in files and evaluated by spreadsheets. The amount of output data can be controlled by a simulation parameter. Real time simulation is possible and time compression is controlled by another factor.

5.6 Software Process

The process has been outlined in *Chapter 4* and we will follow it with examples of documents from the case study. Because of the limited space we have selected short document sections.

The *problem description* is composed of *needs* with identifiers, as in *Table 1*:

Table 1. Problem description.

Need1	Given an urban plan with all housing properties, and given an assignment of households to properties, simulate and observe the migration of households within the local market

An example for the *environment* is *Figure 3*. The object type structure is shown in *Figure 4*. From need *Need1*, *Task3* for object type Household and *Task5* for *LocalMarket* can be derived:

Table 2. Tasks.

Task3	Calculate satisfaction and request offers from near local markets	Household	Need1
Task5	Receive requests from households, find properties and send offer back	LocalMarket	Need1

As an option, MTCs can be created to design the interfaces between the objects (*Figure 5*) and the basic state transitions. The folded corner in the compound states mean that refinements exist. The diagram has been produced and edited with a tool environment DOME (dome).

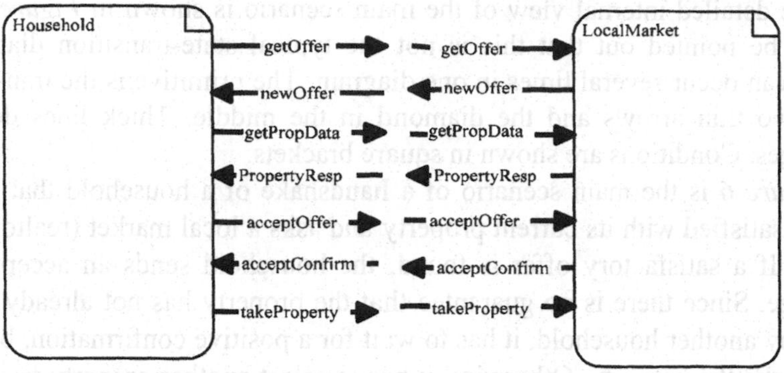

Figure 5. Household – LocalMarket MTC interaction diagram. Arrows are messages with message name. Message parameters not shown.

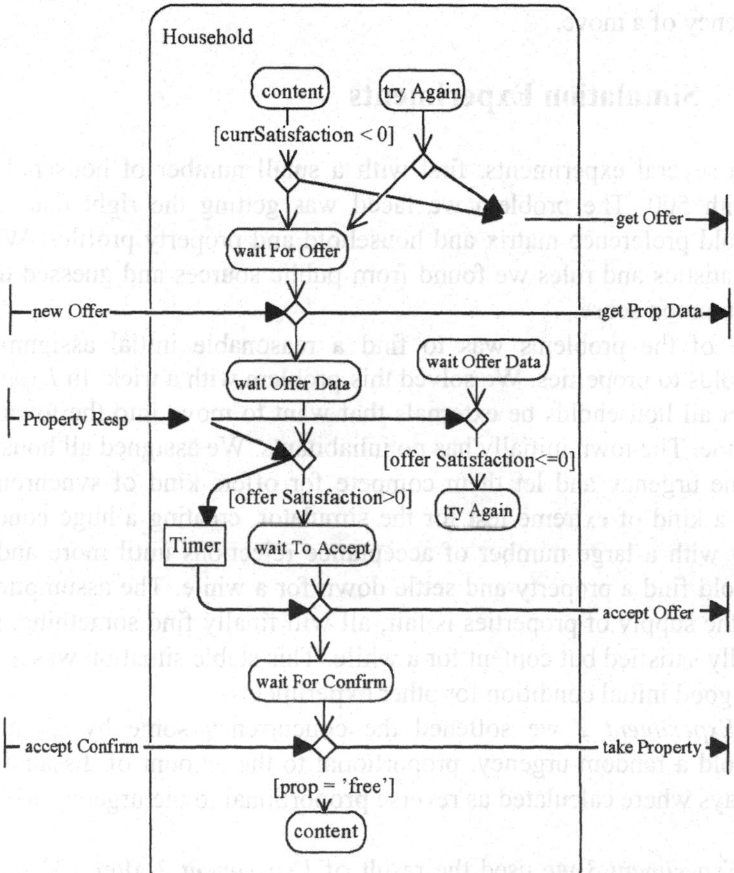

Figure 6. MTC of object type *Household*, scenario 'get offer'.

The detailed internal view of the main scenario is shown in *Figure 6*. It has to be pointed out that this is not the typical state-transition diagram. States can occur several times in one diagram. The primitive is the transition with two thin arrows and the diamond in the middle. Thick lines denote messages. Conditions are shown in square brackets.

Figure 6 is the main scenario of a handshake of a household that is no longer satisfied with its current property and asks a local market (realtor) for offers. If a satisfactory offer is found, the household sends an acceptance message. Since there is no guarantee that the property has not already been taken by another household, it has to wait for a positive confirmation, before it can take the property. Otherwise, it has to select another property from the offer or try again (not shown). Also not shown are additional built in delays, modeling the times necessary to evaluate and decide. The delay unit is one day, with a resolution of 1/1000. Delays can be fixed or made dependent on the urgency of a move.

5.7 Simulation Experiments

We ran several experiments, first with a small number of households and then with 500. The problem we faced was getting the right data for the household preference matrix and household and property profiles. We used what statistics and rules we found from public sources and guessed the rest using common sense.

One of the problems was to find a reasonable initial assignment of households to properties. We solved this problem with a trick: In *Experiment 1* we let all households be externals that want to move into the town at the same time. The town initially has no inhabitants. We assigned all households the same urgency and let them compete for offers kind of synchronously. This is a kind of extreme test for the simulator, creating a huge concurrent activity with a large number of acceptance rejections until more and more household find a property and settle down for a while. The assumption was that if the supply of properties is fair, all will finally find something, maybe not really satisfied but content for a while. This stable situation was assumed to be a good initial condition for other experiments.

In *Experiment 2* we softened the concurrency some by giving each household a random urgency, proportional to the amount of dissatisfaction. All delays where calculated as reverse proportional to the urgency with some offset.

In *Experiment 3* we used the result of *Experiment 2* after 100 days and kind of started a new phase. Now dynamic changes in household preferences and profiles and additional properties where introduced to observe a normal migration behavior. Still, also this experiment is not meant to give new insides into human behavior, but rather a validation of the models, decision processes and parameters.

The results from *Experiment 1* and *2* are nearly identical, although in *1* it takes 170 days before a stable condition is reached, compared to 65 days in *2*. We therefore only report on the first 70 days of *Experiment 2*. During this time 471 households found a property, all local markets together sent 3594 offers, each with a maximum of three properties. In addition, 1673 offer came back empty. 1823 accept messages were sent, but 1352 rejected as already occupied by a household that had been faster. The relation of 7 offers per household seems to be high, but it has to be taken into account, that the offer selection is only based on tenure, size and cost requests only and many are turned down as not satisfactory. The fact that 75% of the accepted offers are already occupied is not astonishing, given the initial condition. *Figure 7* shows how quickly the average satisfaction changes from negative to positive and in parallel the rent income grows. All properties together would reach 500 thousand Euro if occupied.

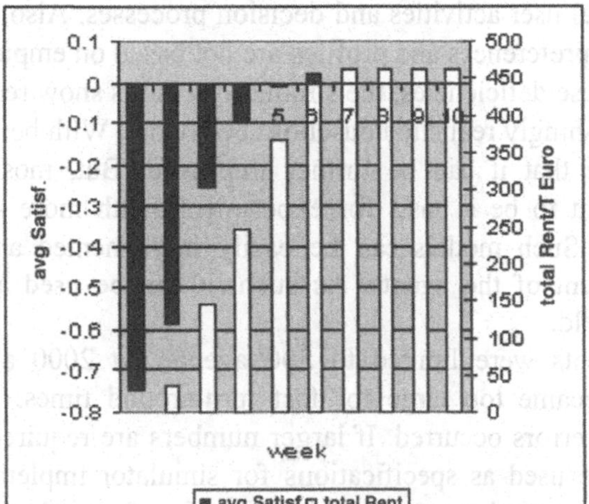

Figure 7. Change of average households and total rent paid during the first 10 weeks of Experiment 2.

In Experiment 3 the first 100 days are set up as in experiment 2 to assign properties to households. At 100 a new satisfaction evaluation starts with the goal of moves in the town to improve the utilities. In addition, over a period of 120 days 100 households get new random work locations which is also a trigger to reevaluate the satisfaction. First tests showed that under these boundary conditions no moves occur because the existing free properties are not attractive. Therefore, 80 additional free properties in the suburbs are added at day 130. The result is an average satisfaction of 0.04 at day 100, 42 additional moves with a final average satisfaction of 0.09 after 800 day. This is still far off the maximum of 0.25, but we did not try to adjust the housing market to the household preferences in the first place.

Common to all three experiments is the run-time of 10 minutes for a simulation period of 2 years. Memory requirements are about 5 MBytes. The simulation is executed with on a laptop with an 850MHz CPU.

6. CONCLUSION

It has been shown that a fully documented and tested prototype simulator for a housing case study can be developed in three person weeks, using the introduced software process, modeling techniques, and appropriate tools. It has also been shown that the agent based model is capable of simulating individual instance behavior with a large range of properties, using a small number of object types. The used models do not claim to be correct abstractions of real user activities and decision processes. Also, the parameters used for user preferences and profiles are not based on empirical study.

But despite these deficiencies, the simulation results show reasonable and sometimes astonishingly realistic household behaviors. With better statistical data we are sure that it can be further improved. But, most of all, the simulator is meant to be a basis for experiments with more sophisticated decision models. Such models can be easily implemented and tested by stepwise refinement of the agents. As such, it can be used as a tool for research in this field.

The experiments were limited to 500 agents. At 2000 agents model compile times became too large for fast turn-around times. Beyond that number compiler errors occurred. If larger numbers are required, the model documents can be used as specifications for simulator implementations in more efficient agent environments. We are looking forward to results from Devisch (Devisch 2005).

7. REFERENCES

Borgers, A. and H.J.P. Timmermans, 2005, "Modeling pedestrian behavior in downtown shopping areas", *CUPUM'05*, London.

Braek, R. and O. Haugen, 1993, "*Engineering Real-Time Systems. An Object-Oriented Methodology Using SDL*", Prentice Hall.

Devisch, O., T. Arentze, A. Borgers, and H. Timmermans, 2005, "An agent-based model of residential choice dynamics in non-stationary housing markets", *CUPUM'05*, London.

dome, http://www.htc.honeywell.com/dome/download.htm

Eastman, C.M. and A. Siabiris, 1995, "A generic building model incorporating building type information", *Automation in Construction*, **3**: 283-304.

Eckholm, A., 2001, "Activity objects in CAD programs for building design", *Computer Aided Design Futures*, Eindhoven.

ESP-r, http://www.esru.strath.ac.uk/Programs/ESP-r.htm

Klüpfel, H., 2005, "The simulation of crowd dynamics at very large events - Calibration, empirical data and validation", *PED2005*, Vienna.

Kukla, R., J. Kerridge, A. Willis, and J. Hine, 2002, "PEDFLOW: development of an autonomous agent model of pedestrian flow", *Transportation Research Record*, **1774**: 11-17

Metzger, A. and S. Queins, 2002, "Specifying building automation systems with PROBAnD, a method based on prototyping, reuse, and object-orientation", OMER - Object-Oriented Modeling of Embedded Real-Time Systems, *Lecture Notes in Informatics*, P-5, p. 135-140.

Metzger, A. and S. Queins, 2003, "Model-based generation of SDL specifications for the early prototyping of reactive systems", *Telecommunications and Beyond: The Broader Applicability of SDL and MSC, Springer Lecture Notes LNCS 2599*, p. 158-169.

Olsen, A. et al. 1994, "*Systems Engineering Using SDL-92*", North Holland.

Telelogic Tau SDL Suite, http://www.telelogic.com

Zimmermann, G., 2002, "Efficient creation of building performance simulators using automatic code generation", *Energy and Buildings*, **34**: 973-983.

Zimmermann, G., 2003, "Modeling the building as a system" *8th International IBPSA Conf. Proc.*, Rio de Janeiro, p. 707-714.

Zimmermann, G. and A. Metzger, 2004, "A software generation process for user-centered dynamic building system models", *Proc. ECPPM2004*, Istanbul.

Research on New Residential Areas Using GIS

A case study

Selma Celikyay

Zonguldak Karaelmas University

Keywords: Decision support systems, Ecological analysis, Geographical information systems, Residential areas, Spatial analysis

Abstract: Planning is a decision-making process which is about “the future”. In each scale of planning process, spatial rules of the social life are formed. In this process, firstly series of spatial analyses should be practised. Throughout the world, spatial planning strategies which focus on the sustainable development adapt an ecological approach and both the regional and urban planning processes are based upon ecological bases. Under the guidance of this notion, also in Turkey, spatial planning strategies should be urgently reviewed and any level of planning process should be directed to ecological bases. Furthermore, in all these steps, natural resources and ecological characteristics should be taken into consideration. In the city of Bartin, where Bartin River flows through, a case study has been carried out regarding the above mentioned planning strategies. The case study has three stages. These stages also frame the data, analysis and evaluation stages. In the case study, a combination of McHarg’s ecological evaluation method and Kiemstedt’s usage value analysis in planning has been employed. With the help of ecological analyses, in the rural areas that have not been settled yet, the potential of the natural resources has been examined for the new residential areas. As a result, in the city of Bartin, the potential residential areas have been defined on the unsettled regions. What is more, concerning the subject, a map has been formed on the scale of 1/25 000. As a result of the case study, it has been concluded that in Bartin city because of the physical planning which ignores the potential of the natural resources, some of the existing residential areas have been chosen improperly.

Jos P. van Leeuwen and Harry J.P. Timmermans (eds.), Innovations in Design & Decision Support Systems in Architecture and Urban Planning, 221-233.

1. INTRODUCTION

In a broader term, planning is the formation of spatial arrangements so as to improve the social welfare and to supply the needs. Land use decisions which have been made during the process of physical planning determine the relations of human/life/nature and their interrelations with each other. With regard to this relation and interaction, while man is provided with the new and better life conditions, the formation and sustainability of a healthy environment should be a "must" in order not to ruin the life conditions of the future generations (Celikyay, 2005a).

Nowadays, in the world, the strategies of residential planning gain a new perspective which utilizes natural resources and potentials, and focuses on the sustainable residential development. Furthermore, it is observed that both regional and residential planning are focused on the ecological bases. All over the world the principle that provides the harmony of the socio-economical development with the geographical aspects of the residential area and preserves the ecological balance has been accepted (Atabay, 1998). Since the use of natural resources without total consumption is one of the major concerns of the sustainable development, the major goal of planning should be preserving the ecological balance. The most damaging activities that are being done for the urban development, the pollution in the cities, all these brought the necessity of taking the primary precautions related to the preservation of nature in urban planning strategies. In Turkey a planning process which takes environment into consideration should be developed since the environmental problems are getting worse (Celikyay, 2005a).

Since land use decisions related to residential areas influence economic activities, these activities also influence the physical environment, the topography on which these activities take place, soil, and natural biotopes. These activities constantly affect ecological structure and thus environmental problems occur (Atabay, 1991). In this case, when the subject is discussed under the principle of benefit-cost theory, it is seen that these decisions provide economical benefits in the short run; but in the long run they eventually cause the destruction of the natural environment and ecological collapses that lead to ecological costs attributed to the whole society. Hence, in both settled and unsettled areas, ecological planning must be taken into consideration in order to preserve and develop the urban and rural environment (Celikyay, 2004). Built environment improving together with the economical growth has created a great deal of damage on the natural resources. Thus, the land uses should be determined by rational and environment-sensitive planning process. Potentials of the natural resources have to be investigated by ecological threshold analysis (Celikyay, 2005b).

2. MATERIAL AND METHOD

2.1 Material

In the process of saving and analysing the natural resources data the following computer software and programmes have been applied.

2.1.1 Geographical Information System

GIS is an effective computer based tool that has been employed in the study of area-based analysis. The analyses which have been carried out by GIS provide the occasion of evaluating the area in an appropriate manner.

GIS give the opportunity of the input, storage, preservation, analysis, output and presentation of the data. In the study, the multi layer processes, surface analyses, and grid analyses have been carried out with GIS.

2.1.2 ArcView 3.2

The GIS software that has been utilized in the study is ArcView 3.2. The processes such as the digitalization of all data related to the research, integration of the data with the given grid system, determination of the dominant surface type with grid analysis by calculating the area in each grid are all done by ArcView 3.2 software.

2.1.3 AutoCAD

In the stage of ecological analysis of the study, AutoCAD programme was used to question the factors within the grid size on the 1x1 km^2.

2.1.4 Potential Value Analysis Programme

A computer programme has been prepared for the study. The numerical weights of the ecological factors and the numerical values of the ecological sub-factors in the table of the potential evaluation formed according to the method used in the study are entered as data in a computer programme called 'Potential Value Analysis Programme'. In the case area, the storage of the data, which has been questioned by grid analysis, the calculation of the total potential values, the selection of the new potential areas for new settlements are all conducted with the help of this programme.

2.2 Method

Ecological planning is based upon the analysis of the areas, which have natural resources, in terms of ecology. The method of McHarg (1992) employed in this study is also based on the determination of the utilization potentials of an area in terms of natural resources and the use of this analysis in physical planning. The study consists of data, analysis and evaluation stages.

2.2.1 Data

The input data of natural resources related to case area have been digitized. Through the application of GIS, related to natural resources the following maps have been obtained.

- Topographic structure
- Geological Structure
- Hydrology
- Soil Structure
- Soil Type
- Flora
- Existing Land Uses

2.2.2 Analysis

The stage of analysis consists of the research on whether the natural potential of the unsettled areas in the case area is suitable for the residential land use or not, and the analysis of the ecological factors according to this aim. Land resources are the ultimate natural sources that provide man, animal and plants with the suitable life conditions and produce various things needed for sustainable and easier life. When land resources are under the danger of deterioration as a result of natural events and inappropriate land uses, they cannot sustain their functions properly. Hence, the land as a limited source has significant economical, social, and ecological roles in the realization of the sustainable development; in order to achieve all these roles in a balanced way, its components and their responses to different impacts should be considered. This can be achieved through a comprehensive study on soil, climate, water, and geological structure and their characteristics that create the land resources.

An appropriate mathematical research and evaluation method has been developed by making use of McHarg's (1992) and Carl Steinitz's (1996) landscape evaluation methods, Golany's (1976) method on the analysis of urban settlement, Kiemstedt's method of "Usage Value Analysis" –

translated by Köseoğlu (1982) in order to define natural resources data and to determine potential residential areas. The mathematical method is preferred because it makes the grouping and analysis of the data related to the case area easier and it also makes the determination of the appropriate and inappropriate ones simple. In the process of analysis and evaluation, if mathematical method was not used, it would be necessary to frame a map for each criterion, as a result during the process of analysis there would be so many maps. The integration of these maps would be another problem. However, with the help of the mathematical method employed in the study, only a map has been framed as the result of the analysis.

2.2.2.1 Mathematical Analysis

Within this method, the natural and the ecological factors' suitability thresholds, related to residential areas, have been studied and the evaluation criterions have been formed. The numerical weights of the ecological factors determined as the evaluation criterions have been decided according to their suitability for residential land uses. The sub-factors of each ecological factor have been determined, and these sub-factors are given numerical values according to their importance in the determination of the residential land use.

With the help of Kiemstedt's usage value analysis method, the formula of "the suitability value analysis of the natural potential" has been formed:

$$RLU_{PV} = Factor1_{PV} + Factor2_{PV} + Factor3_{PV} + Factor\ (n)_{PV} \quad (1)$$

RLU= Residential Land Use

PV= Potential Value

The potential of a land for residential use is the sum of the potential values of the ecological factors related to the land. This expression has been mathematically formulated:

$$RLU_{PV} = \Sigma Factor_{PV} = g_1.e_1 + g_2.e_2 + g_n..e_n. \quad (2)$$

$Factor_{PV} = g.e$

RLU_{PV} = Total potential value related to residential land use

PV = Potential value of the ecological factor related to residential land use

g = The ecological factor's level of importance in determining the residential land use

e = The numerical value of the ecological factor (functional value)

2.2.2.2 Grid Analysis

Research on the suitability of natural resources potential in the case area for residential land uses has been carried out through the grid analysis. The case area has been divided into the grids of 1x1 km^2 according to geographical coordinates on maps drawn to a scale of 1/25 000 and thereby geographical grid system has been developed. Through the grid analysis, the presence of ecological factors in grid units has been researched according to the mathematical method with the help of geographical information system (GIS). Firstly the data of the natural resources has been digitised and then, whole data has been questioned separately for the residential land use according to the formula of "the suitability value analysis of the natural potential" through the programme of the potential value analysis.

In accordance with the method, ecological factors have been determined in order to research on the suitability of natural resources for residential land uses. Numerical values of the ecological factors and ecological sub-factors have been determined. Considering not only one sub-factor but also several sub-factors existing in these grids, the dominant type method has been used in the assessment of ecological sub-factors researched in the grid system. Through the dominant type method, numerical value of grid units has been assigned according to the type of the largest surface value and sub-factors covering the largest area, in grids have been taken into consideration as the dominant characteristic related to the grid unit in question while the other sub-factors have been eliminated and have not been taken into consideration.

A computer programme has been developed and employed in order to calculate the numerical values of potential values related to grid units with the help of grid analysis. According to the presence of these sub-ecological factors in grid units, total potential value of each grid unit has been calculated by this programme called 'Potential Value Analysis Programme'.

2.2.3 Evaluation Process

According to the interpretations of numerical values of grid units obtained through the grid analysis, the grids which received a value less than the minimum potential value indicated in the natural factor evaluation table related to residential areas through the computer programme developed for this purpose have been determined as the areas lack of potential and extracted from the evaluation.

The grid units having numerical value between minimum potential value and mid-potential value are the second degree potential areas, the grid units that have numerical value above mid-value degree are determined as the first degree potential areas. In this way, a map illustrating two optional residential areas has been acquired.

3. CASE AREA: BARTIN SETTLEMENT

The city of Bartin is situated in the northern part of Turkey and in the West Black Sea Region. The city takes its name from Bartin River, which was called as "Pharthenius" -meaning the God of Water and The Young Virgin- in the ancient times.

The city once was a trade centre since Bartin River provides 12 km waterway, related to that, in time the city became a residential area with the emergence of the need for accommodation. As a trade route, Bartin River has a great influence on the establishment and development of the residential areas. The city of Bartin was an important commercial and cultural centre in the nineteenth century, and today the settlement still keeps its historical characteristic and its importance as a commercial centre. The case area includes three sites: urban site in which characteristic of its being a historical trade centre are dominant, archaeological and natural site. By the Ministry of Culture and Tourism, the areas alongside the river are declared as the first degree of natural site area.

3.1 Existing Land Uses

The settled areas in the centre of the city of Bartin have historical characteristic, and within this historical structure there are some wooden constructions as the types for the traditional architecture, a fountain, a Turkish bath, and two inns. The heights and the scales of these wooden houses are the distinguishing aspect of the urban settlement. So in the areas, houses have three floors, but outside of these areas, in the developing areas, it can be observed that houses can have 3 floors to 5 floors.

In application plan of 1970, it can be seen that there are some residences on some of the first class agricultural lands in spite of the inappropriate conditions resulting from the erosion risk, high subsoil water and earthquake movements. Because of these negative conditions, old residences are placed on slightly sloppy hillsides which consist of eosen layers over the level of 20m.

In the developing residential area- outside of the settled areas- the housing areas started to be built as they were suggested in 1980's plan on the scale of 1/5000. These housing groups having 4 or 5 floors were built by cooperatives. Some of these residential areas have fourth class land use ability and they are on 12-20% gradient hillsides. Some of the residential areas are on valuable agricultural lands that have first class land use ability. In the areas outside of the district's borders, residences having 2 or 3 floors can be observed. These residential areas also have first class land use ability and they are set on the most fertile agricultural lands (Figure 1).

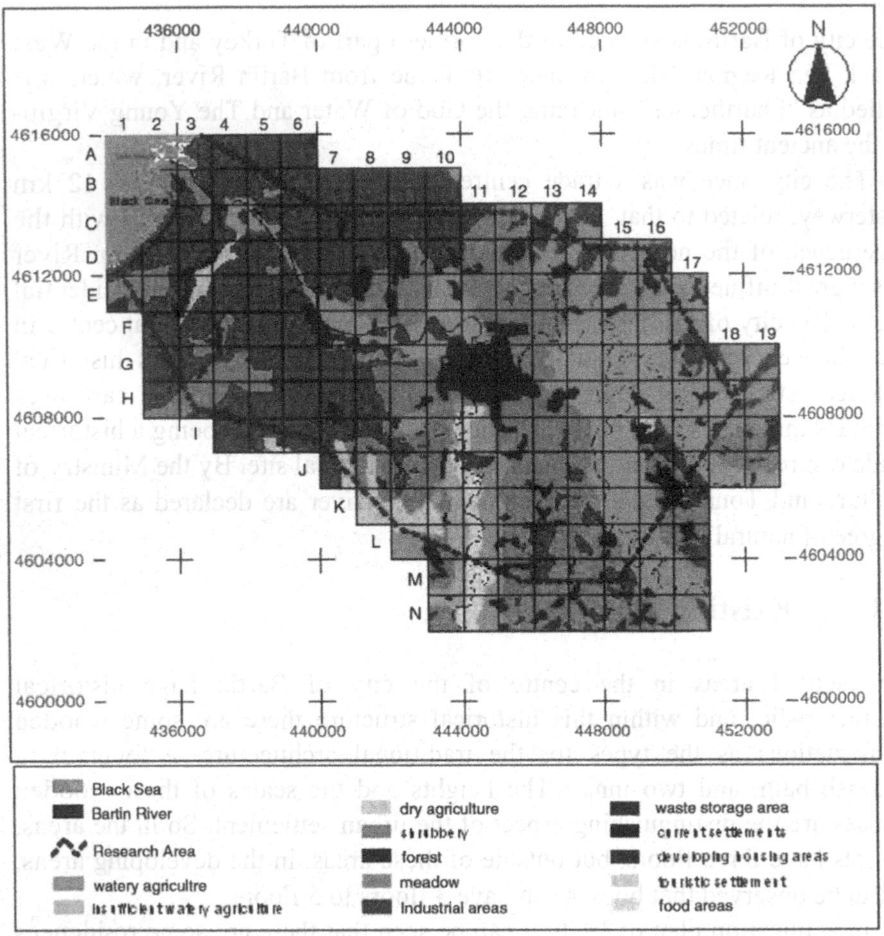

Figure 1. Existing land uses.

3.2 The Research on the New Residential Areas Suitable to the Potential of the Natural Resource

The ecological factors employed as the criterions for the determination of the residential areas are ability classes of the lands, slope, erosion and aspect.

In order to evaluate the land use related to residential areas, the following formula has been developed by adding the following elements up: potential value related to ability classes of the lands, slope potential value, erosion potential value and aspect potential value (Table 1).

$$PRA_{pv} = \text{Ability Class}_{pv} + \text{Slope}_{pv} + \text{Erosion}_{pv} + \text{Aspect}_{pv} \quad (3)$$

PRA = Potential residential areas

Table 1. Assessment criterions related to the potential residential areas.

Grid unit no.				
Ecological factors	Ability Class	Slope	Erosion	Aspect
Factors' values (g)	**3**	**3**	**2**	**2**
Sub-factors' values in grid area (e)	Max: 4 Min: 1	Max: 4 Min: 1	Max: 4 Min: 1	Max: 4 Min: 1
Total values (g.e)	g.e	g.e	g.e	g.e
Total potential value of grid unit (∑g.e)	Max. value: 40, Min. value: 10, Mid-value: 25 $\sum g.e = PRA_{PV}$			

The suitable residential areas should be those outside of the fertile agricultural lands. With regard to this principle, it is concluded that the most important factor in the determination of the residential areas is the ability classes of the lands. Its sub-factors are the fourth class, fifth class, sixth class and seventh class lands.

Table 2. Assessment criterions related to the ability class factor of the lands.

Grid unit no.				
Ecological factor	Ability Class of Land			
Factor's value (g)	**3**			
	Sub-factors			
Sub-factors	IV.Class	V.Class	VI.Class	VII.Class
Sub-factor's value (e)	4	3	2	1
The value of grid (e)				
Potential value of grid unit (g.e)	AC_{pv}=			

The classification of the suitability of the land ability classes according to the residential areas has been carried out by giving the highest value (4) to the fourth class lands, giving high value (3) to the fifth class lands, giving mid-value (2) to the sixth class lands, and finally by giving low value (1) to the seventh class lands (Table 2).

Table 3. Assessment criterions related to the slope factor.

Grid unit no.				
Ecological factor	Slope			
Factor's value (g)	**2**			
	Sub-factors			
Sub-factors	0-6%	6-12%	12-20%	20 +%
Sub-factor's value (e)	4	3	2	1
The value of grid (e)				
Potential value of grid unit (g.e)	$Slope_{pv}$=			

The classification of the suitability of residential uses according to slope of the land is done by giving the highest value (4) to the slope until 6%, giving high value (3) to the slope between 6% and 12%, giving mid-value (2) to the slope degree between 12% and 20%, and finally by giving low value (1) to the 20% and higher slope (Table 3).

In the classification of the sub-factors- related to erosion factor- so as to determine the suitable lands for residential purposes, the following evaluation has been applied. The lands where no or little erosion takes place are given the highest function value (4), the lands with mid-degree erosion risk are given high value (3), the lands that have high risk of erosion are given mid-function degree (2), and the lands that carry the highest erosion risk are given low value (1) (Table 4).

Table 4. Assessment criterions related to the erosion factor.

Grid unit no.				
Ecological factor	Erosion			
Factor's value (g)	2			
	Sub-factors			
Sub-factors	The highest	High erosion	Mid-erosion	Little or no
Sub-factor's value (e)	1	2	3	4
The value of grid (e)				
Potential value of grid unit (g.e)	$Erosion_{pv}$=			

The aspect as a significant factor makes hillsides take little or much sun light during a day time, so this factor should be taken into consideration in the choice of the residential areas. Since mostly in Turkey south (S), south-east (SE), south-west (SW) and west (W) aspects receive much sunlight, these exposures are hotter than the other ones. North (N), north-east (NE), north-west (NW) and east (E) aspects are shadowy and cooler since they receive little sunlight. The appropriate choice for the settlements in order to reduce the dependency of people on exhaustible energy, and to make them benefit from solar energy and natural air conditioning, shortly to benefit from the natural sources in a maximum level. To achieve these, the choice of the settlement area should be parallel to the climate of that area. Hence, north hill sides are not preferred since they receive low sun light degree. The top of the south-east and east hill sides are suitable for the settlement when the relationship between climate appropriateness and settlement is considered. In the evaluation of the appropriateness of aspects to settlements, aspects of south-west, south and south-east take the highest function value (4), west and east aspects are given high function value (3), north-west and north-east aspects are given mid-function value (2), and finally north aspect takes low function value (1). The ecological factors for determining the evaluation criterions of the residential areas, the ecological sub-factors used as the thresholds for settlement areas and their weights, finally their functional values are illustrated in Table 5.

Table 5. Assessment criterions related to the aspect factor.

Grid unit no.				
Ecological factor	Aspect			
Factor's value (g)	3			
	Sub-factors			
Sub-factors	SW, S, SE	W, E	NW, NE	N
Sub-factor's value (e)	4	3	2	1
The value of grid (e)				
Potential value of grid unit (g.e)	$Aspect_{pv}$=			

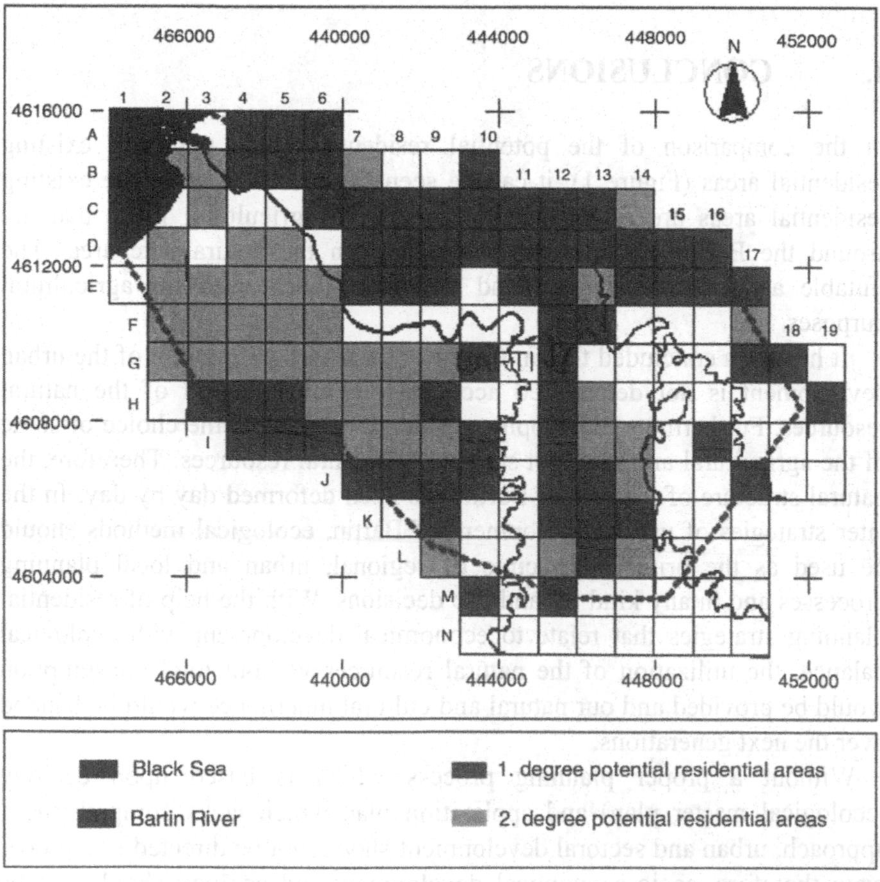

Figure 2. Potential areas related to the new residential uses.

3.3 Results of the Analysis and New Residential Areas

In order to determine the potential residential areas, after the questioning of the factors-given above- with grid analysis, new potential residential areas are determined for residential purposes. As a result of the grid analyses, the fourth class lands having slope between 12-20%, the lands having mid-erosion degree, and aspects of south, south-east and east were determined as first class potential areas that are suitable for residential uses (Figure 2).

The hill sides around the settled areas are the places that have high potential in terms of residential land uses. Contrary to the settlements on the first class agricultural lands, and the places having risks in terms of level of the subsoil water and earthquake, these areas have characteristic of strong ground in terms of geological structure.

4. CONCLUSIONS

In the comparison of the potential residential areas with the existing residential areas (Figure 1), it can be seen that while some of the existing residential areas are established on the fertile agricultural lands that are around the Bartin River at the same time on the natural site area. The suitable areas for residential land uses have been used for agricultural purposes.

It has been concluded that in the city of Bartin the direction of the urban development is not determined according to the potential of the natural resources. Furthermore, developing residential areas and the choice of some of the agricultural areas are not suitable for natural resources. Therefore, the natural structure of the city of Bartin has been deformed day by day. In the later strategies of urban development of Bartin, ecological methods should be used as the primary principle of regional, urban and local planning processes and in any kind of land use decisions. With the help of residential planning strategies that relate to economical development with ecological balance, the utilization of the natural resources without total consumption would be provided and our natural and cultural inheritance would be handed over the next generations.

Without a proper planning process which is based upon ecology (ecological master plan) and application plan which lacks comprehensive approach, urban and sectoral development should not be directed to a proper way; therefore, socio-economical development and ecology should not be reconciled. Since ecological development is not interconnected with ecology, its existing and future negative impacts on the natural resources would not make the sustainable development possible and it would be impossible to establish a protection-use balance. Hence, with the land use decisions made during the physical planning process, a healthy environment and its sustainability should be provided without destroying the life-conditions of the future generations. Ecological planning should be the basic principle in the settled and unsettled areas of Bartin, and in the process of the protection and development of the urban and rural lands.

In the process of urban development in which natural resources are being destroyed, it should be considered that natural resources are limited and they can be used up; so, socio-economical development strategies should be focused on ecological bases. It should be considered to provide the balance between conservation and usage of the natural resources.

GIS is the primary tool in forming a wide range of data bases in the planning processes, in the analyses of both natural resources and ecological threshold analyses, in both determination and sustainability of the appropriate ways of land use with regard to the potential of the natural resources.

In order to provide sustainability, both ecological approach should be the major principle and geographical information systems should be used as the major tool in planning process.

5. REFERENCES

Atabay, S., 1991, "Dogal cevreye uyumlu planlama", *Cumhuriyet gazetesi, Kasım 5.*

Atabay, S., 1998, *Ekolojik temele dayalı bölge planlama,* Yıldız Teknik Üniversitesi, 98.083, Istanbul.

Atabay, S., 2000, *Cevre duyarlı planlama,* YTU.SBP.MF., Istanbul.

Celikyay, S., 2004, "Ekolojik planlamanın gerekliligi", in *Proceedings of V.Ulusal Ekoloji ve Cevre Kongresi,5-8 Ekim, Bolu-Abant,* p. 561-567.

Celikyay, S., 2005a, "Çevre düzeni planlarında stratejik çevresel etki değerlendirmesi", I.Çevre ve Ormancılık Şurası, 21-24 Mart, Antalya, http://www.sura.cevreorman.gov.tr

Celikyay, S., 2005b, "Determination of the land uses by the ecological analysis", *Proceedings of ITAFE'05, International Conference on Information Technology in Agriculture, Food and Environment, Adana,* p. 419-423.

Geissler, E., 1991, "Ekolojik acıdan peyzaj planlama", *Cevreye Uyumlu Planlama Aracları ve Politikaları Sempozyumu, April 29-30,*Yıldız Üniversitesi Mimarlık Fakültesi Yayın No: MF-SBP 92.040, Istanbul, p. 25-48.

Golany, G., 1976, *New-Town planning, principles and practise,* John Wiley and Sons, New York.

Köseoglu, M., 1982, *Planlamada kullanım değeri analizi,* Ege Üniversitesi Yayınları, İzmir.

Lozano, E, E., 1990, *Community design and the culture of cities,* Cambridge University Press, New York.

McHarg, I, L., 1992, *Design with nature,* The American Museum of Natural History, Doubleday/Natural History Press, Doubleday & Company, Inc., Garden City, New York.

A Comparison Study of the Allocation Problem of Undesirable Facilities Based on Residential Awareness

A case study on waste disposal facility in ChengDu City, Sichuan China

K. Zhou, A. Kondo, A. Cartagena Gordillo[1], and K. Watanabe
The University of Tokushima, Tokushima, Japan
[1]*Yokohama National University, Yokohama, Japan*

Keywords: Facility location, Undesirable facilities, Stochastic method, ChengDu City

Abstract: The purpose of this paper is to propose a model, which can be used for allocation planning of undesirable facilities by analysing citizen's awareness. As the endurance distance is regarded as a variable, the relation between the distance from the residential location to a waste facility and residential endurance rate is a problem of probability and statistics. Three kinds of stochastic methods are compared in this study.

1. INTRODUCTION

Many studies of facility location theory are concerned with location of facilities such that some distance objective, expressing nearness to residents is optimized. Distance minimization is however not a good performance indicator for all facilities, because some facilities such as waste disposal give residents feelings of dissatisfaction. Location problems for these kinds of facilities require new methodologies with corresponding solutions.

The aim of this research project therefore is to suggest such methodology for addressing the location problem of undesirable facilities. It is based on the concept of residential awareness. Disposal facilities serve as units of analysis. We introduce the concepts "endurance distance" and "endurance rate" as criteria for the allocation of such facilities, and propose methods to measure these new concepts. First, the Weibull distribution is employed as a modelling approach. With the aim of improving the accuracy of the model,

Jos P. van Leeuwen and Harry J.P. Timmermans (eds.), Innovations in Design & Decision Support Systems in Architecture and Urban Planning, 235-250.

two other methods are introduced in sections 6 and 7. The chapter is completed with a conclusion.

2. EXISTING STUDIES

In the context of the waste disposal facility location problem, Ejima, et al. (2000), using a linear planning method, searched for the optimal group setting when the consumed energy for garbage transfer reached a minimum. Nishimura et al. (2002) paid attention to one kind of trade-off relationship between dioxin generated by incinerators and NOx emitted from garbage collection trucks. From the viewpoint of locating an incineration facility at the place where total transportation distance and NOx emissions reach their minimum, he gave a solution based on a mathematical planning formulation. Plastria (1996) gave a critical overview of research on the location of undesirable facilities in continuous space. The most popular way of handling undesirability for a single facility is to minimise the highest effect on a series of fixed points with the principle of locating the undesirable facilities as far as possible from all sensitive places. Thus, distance or time, were mainly used as important variables in the study of facility location. However, the psychology element of facility users was not given enough attention.

Our objective in this research project therefore is to analyse location problems of undesirable facilities by using a model based on probability theory, which considers residential awareness. The reason why we put importance on residential awareness for studying location problems is because people need a reasonable location for an undesirable facility location. That is also why public involvement has become very necessary nowadays in the field of urban planning. From the viewpoint of emphasizing public involvement, Baba (2002, 2003) described a public involvement assessment framework for the location problem of undesirable facilities, and discussed the effect and possibility of public involvement in locating an undesirable facility. But in those two papers, he did not use any facility location methodologies.

So in this research, considering the above, we choose garbage transfer stations and final disposal facilities as research objects due to their high level of variety. Until now, it has been a task to determine how to reflect residential awareness into undesirable facilities' location planning. From such a perspective, we carry out a case study in Chengdu city of Sichuan province in China, to achieve this objective.

3. ENDURANCE RATE MODELLING FOR THE UNDESIRABLE FACILITIES LOCATION

3.1 Definition of Endurance Distance

For purely desirable facilities, residents hope the facility will be located shorter than a certain distance, thus the maximum of this hoped for distance can be defined as a satisfied distance for desirable facilities (Yoshitaka, et al., 1986). Contrary to the purely desirable facilities, for purely undesirable facilities, residents hope the undesirable facility can be located farther than a certain distance, which means the residents can endure the location of the undesirable facility if the facility is located farther than that distance. Then the minimum of this desired distance can be defined as endurance distance for undesirable facilities. And, when an undesirable facility is located at a certain distance, the rate of residents who could endure the facility location is defined as endurance rate.

3.2 Distribution of Endurance Distance

First, we consider that for undesirable facilities, all people have an endurance distance in their mind. Here, we define the endurance distance as w, because w is a variable that depends on each person's different choice, w can be considered as a stochastic variable. Then we can assume that w follows a probability density function $f(w)$. The relationship between endurance distance w and the probability density function $f(w)$ can be shown in the graph of $f(w)$, plotted at upper part in *Figure 1*. The probability density function $f(w)$ has the property

$$\int_0^\infty f(w)dw = 1.0 \tag{1}$$

So, when a facility is located at a certain distance x from a residential location, if the residential endurance distance $w > x$, then the residents can accept that facility location. If we define the endurance rate as $P(x)$, $P(x)$ is the grey part in the upper graph in *Figure 1*, and can be found by equation (2). The relationship between $P(x)$ and $f(w)$ can be described using *Figure 1*. The relationship between x (the distance from residential location to a facility) and $P(x)$ (the rate that residents could endure the facility location) is also shown in the lower graph in *Figure 1*. From the relationship between x and $P(x)$ shown in *Figure 1*, we can see that the farther the distance x from a residential location to a facility, the higher the endurance distance becomes. When the distance to the facility is 0m, the endurance rate is 0%, and conversely, when the distance to the facility becomes infinite, the endurance rate becomes close to 100%.

$$P(x) = \int_0^x f(w)\, dw \tag{2}$$

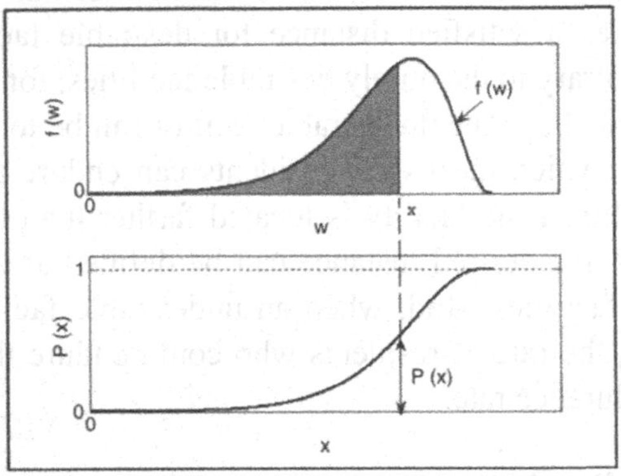

Figure 1. Relationship between the endurance rate and distance to a facility.

3.3 Assumption for Distribution of Endurance Distance

In this part, we estimate the distribution function f(w) which is used for expressing the residential endurance rate in equation (2). Referring to the existing literature (Yoshitaka, 1972), the Weibull distribution defined in equation (3) was chosen as the distribution testing result, where α and m are scale and shape parameters. During earlier work, we found that the Weibull distribution had some interesting characteristics: 1. It is a distribution with good elasticity. The shape changes following changing the parameters. 2. The distribution function is completely integrative, which allows easy parameter estimation. Putting equation (3) into equation (2), equation (4) is obtained. Equation (5) was obtained by taking the logarithm at both sides of equation (4). Furthermore, we take the logarithm of both sides of equation (5) and obtain equation (6).

$$f(w) = \frac{m}{\alpha} w^{m-1} e^{-\frac{w^m}{\alpha}} \tag{3}$$

$$P(x) = \int_0^x \frac{m}{\alpha} w^{m-1} e^{-\frac{w^m}{\alpha}}\, dw$$

$$= 1 - e^{-\frac{x^m}{\alpha}} \tag{4}$$

$$\log_e(1-P(x)) = -\frac{x^m}{\alpha} \tag{5}$$

$$\log_e[-\log_e(1-P(x))] = m\log_e x + \log_e \frac{1}{\alpha} \tag{6}$$

Because equation (6) is a linear function, it is possible to estimate parameters α and m employing the least-squares method. If we have real data for endurance distance w and frequency distribution $f(w)$ of w, which is the threshold distance where the residents hope a waste facility can be located, and given a distance x_i (i=1, 2, 3...), we can find the proportion of people whose endurance distance w is shorter than x_i. Based on that data, parameters α and m can be estimated by using regression analysis.

4. QUESTIONNAIRE SURVEY IN CHENGDU CITY

For estimating parameters in equation (4), data concerning the endurance rate $P(x)$ is collected for Chengdu City. It is the capital city of Sichuan province, China (*Figure 2*). Recently, construction of a highway and city-loop road is in progress as part of the national program of "West Great Development Strategy" (*Figure 3*). Future development is expected.

In recent years, following increases in quality of life and thus waste production, waste disposal problems has become more serious. Inadequate numbers of waste disposal facilities has been given more attention by a larger number of citizens. It is thought that reasonable facility location would influence urban planning in Chengdu.

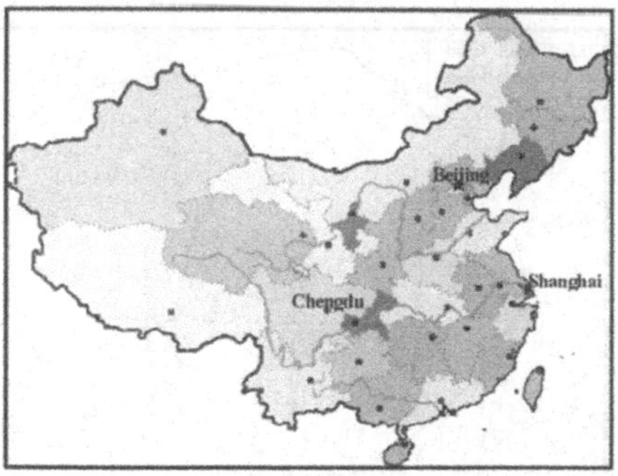

Figure 2. The location of Chengdu City.

According to the Chengdu city environmental and health office (1999), for garbage collection and disposal, each family puts their garbage at a garbage container nearby. Then the garbage is taken to a garbage transfer station. Most of the garbage transfer stations are located along back lanes in residential districts, avoiding main roads. Garbage collected from each garbage transfer station is transported to final disposal facilities which are 34 km away from the inner city of Chengdu. After recyclable waste is taken out, the rest is finally disposed in a landfill.

For getting the data, a survey on residential awareness about undesirable facilities was carried out in February 2004. The main question was: At least how far should a waste facility be located to your home? According to the endurance distance, a few alternatives were given in advance. Then respondents choose their desired endurance distance from the alternatives or a certain number they considered adequate. The choices, for garbage transfer stations and for final waste disposal facilities, were from 1km to 10km and from 5km to 30km respectively. For both facilities there was the option: "If there's no endurance distance you considered, please write down a distance you can endure". Data analysis was based on the endurance distance which residents chose or wrote. A simple explanation concerning the present conditions of waste disposal in Chengdu was given before the questions.

With the cooperation of the Chengdu Jinjiang Shahe Community Working Committee, the questionnaire was distributed to 350 households which were randomly sampled based on identification cards of 3,085 local families. We got 323 valid answers collected directly from the corresponding households. The valid response rate was 92.3%. The respondents can be classified by gender as 52% male and 48% female. According to age, 4% were in their teens, 27% were 20-29, 24% were 30-39, 31% were 40-49, 11% were 50-59, and 3% were 60 and over.

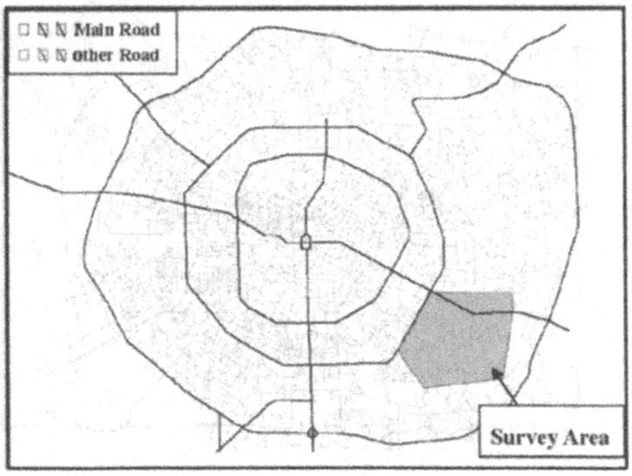

Figure 3. Object area of the research (Himeno et al., 2004).

5. ENDURANCE RATE FUNCTION

5.1 Endurance Distance and Endurance Rate

First, from the survey results, we calculated the distance to garbage transfer stations and final waste disposal facilities and corresponding residential endurance rates. Distance to facilities and the proportion of people who considered it an expectable endurance distance obtained from the survey are shown in *Figures 4 Figure 5*.

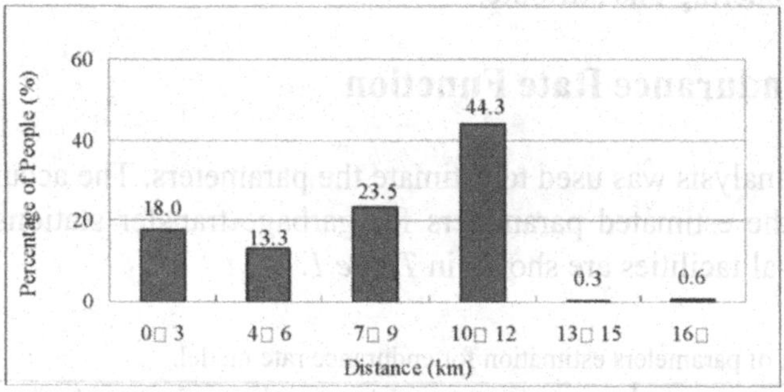

Figure 4. Distance to garbage transfer stations and corresponding residential endurance rate.

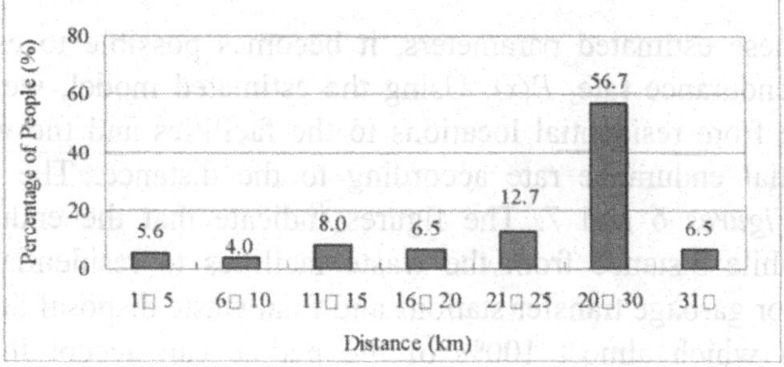

Figure 5. Distance to final waste disposal facilities and corresponding residential endurance rate.

5.2 Average Endurance Distance Classified by Attribute

Residents average endurance distance, classified by attribute, was calculated based on the survey data. For garbage transfer stations, the average endurance distance was classified by age, and most of the residents whose age ranged from 20 to 59 gave an answer of around 7 km. Teen's average endurance distance is 5.86 km and for people in their sixties' it is 9.93 km. Although there is a difference of 4.07 km, we cannot say there is a significant difference between these two values with a 5% confidence level. The final disposal facility gives quite similar results. For average endurance distance classified by sex, there is also no big difference between males and females according to both garbage transfer stations and final waste disposal facilities. From the above analysis of average endurance distance classified by attribute, we assume the same model to analyse all residents' awareness without considering this attribute.

5.3 Endurance Rate Function

Regression analysis was used to estimate the parameters. The accuracy of the model and the estimated parameters for garbage transfer stations and final waste disposal facilities are shown in *Table 1*.

Table 1. Result of parameters estimation for endurance rate model.

facilities	garbage transfer stations	final waste disposal facilities
parameter α	38.7(9.971)	522.5(19.312)
parameter m	1.809(9.383)	1.781(16.968)
R^2	0.880	0.947
numbers of sample	14	18

where the numbers between parentheses represent the value of *t*.

Given these estimated parameters, it becomes possible to estimate the residential endurance rate, *P(x)*. Using the estimated model, we calculated the distance from residential locations to the facilities and the variation of the residential endurance rate according to the distance. The results are shown in *Figures 6* and *7*. The figures indicate that the endurance rate increases while distance from the waste facilities to residential locations increases. For garbage transfer stations and final waste disposal facilities, the distances at which almost 100% of the people can accept the facilities location are more than 20 km and 80 km respectively.

As *Figures 6* and *7* show, for the garbage transfer stations, at a distance of 10 km (for final waste disposal facilities, it is 30 km), there is an error between the calculated value from the model and the survey data. This is probably an effect of people's tendency to round off numbers. Another possibility is simply influenced by people's psychology, namely, people chose the farthest distance from the provided alternatives in the questionnaire survey because they have undesirable feeling regarding those facilities. If the above considerations are right or can be proven, for e.g., the survey data may change when the alternatives are changed to be smaller scale or added longer distance, the ratio of people who chose 10 km / 30 km would be lower. In such cases, the Weibull distribution method would indicate a better fit. It is also possible that the survey result would have been different in the case that there was no advance explanation about undesirable facilities or in the case that the explanation was more "positive". As these considerations cannot be proven in this paper, we will provide other solutions in later sections to improve accuracy of the model. Nonetheless, from the viewpoint of a government that constructs waste disposal facilities, the cost for collection and transport has a big influence on location decisions.

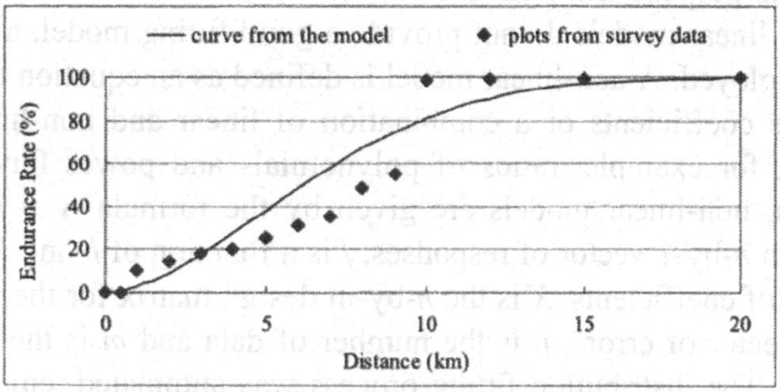

Figure 6. The residential endurance rate for garbage transfer stations.

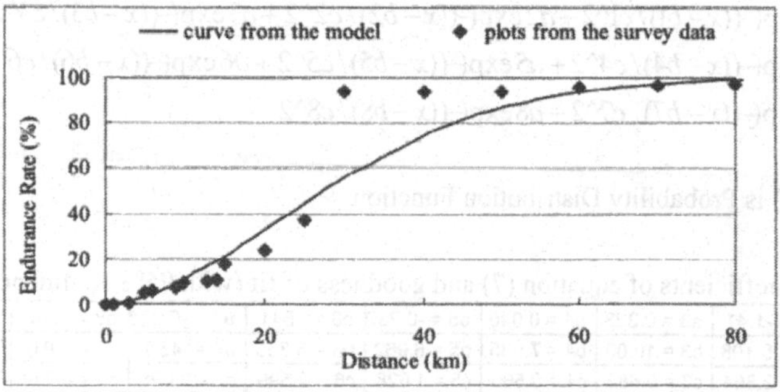

Figure 7. The residential endurance rate for final waste disposal facilities.

As a next step, we consider a value which is helpful for facility location planning from the viewpoint of residents. For example, when the residential endurance rate is 80%, suitable distances for garbage transfer stations and final waste disposal facilities are 9.8 km and 43 km. This result is useful for facility location planning.

6. NON-PARAMETRIC DISTRIBUTION METHOD

In the following 2 sections, we discuss two other stochastic methods. First, a Non-Parametric distribution method will be introduced; then a 3-Parameter Loglogistic distribution will be discussed. We only discuss the two methods applied to the garbage transfer facility due to page limitations.

6.1 Non-Parametric Distribution

The objective of this section is to find a close-form expression for the survey data. For this purpose the well-known Least Squares method is employed. Because the linear models do not provide a good fitting model, a non-linear model is employed. A non-linear model is defined as an equation that is non-linear in the coefficients or a combination of linear and non-linear in the coefficients, for example, ratios of polynomials and power functions. In matrix form, non-linear models are given by the formula: $y = f(X, \beta) + \varepsilon$, where y is an n-by-1 vector of responses, f is a function of β and X, β is a m-by-1 vector of coefficients, X is the n-by-m design matrix for the model, ε is an n-by-1 vector of errors, n is the number of data and m is the number of coefficients. The distribution fitting process was automated, employing the commercial software Statistics Fitting Toolbox from Matlab. Based on the survey data, we estimated the model shown in equation (7). Its coefficients are shown in *Table 2*.

$$\begin{aligned} y(x) = {} & a1\exp(-((x-b1)/c1\text{^}2 + a2\exp(-((x-b2)/c2\text{^}2 + a3\exp(-((x-b3)/c3\text{^}2 + \\ & a4\exp(-((x-b4)/c4\text{^}2 + a5\exp(-((x-b5)/c5\text{^}2 + a6\exp(-((x-b6)/c6\text{^}2 + \\ & a7\exp(-((x-b7)/c7\text{^}2 + a8\exp(-((x-b8)/c8\text{^}2 \end{aligned} \tag{7}$$

where, $y(x)$ is Probability Distribution Function

Table 2. The coefficients of equation (7) and goodness of fit (with 95% confidence bounds).

a1 = 0.001	a2 = -1.41	a3 = 0.323	a4 = 0.040	a5 = -0.237	a6 =1.541	a7 = -0.222	a8 = 0.067	SSE: 0.0002
b1 = -1.62	b2 = 5.108	b3 = 10.00	b4 = 7.935	b5 = 6.962	b6 = 5.359	b7 = -45.73	b8 = 1.015	R^2: 0.9997
c1 = 56.90	c2 = 2.364	c3 = 0.682	c4 = 0.565	c5 = 1.628	c6 = 2.599	c7 = 24.950	c8 = 0.744	RMSE: 0.0012

6.2 Cumulative Distribution Function

The cumulative distribution function $h(\lambda)$ of the endurance rate is estimated as equation (8). Its coefficients are shown in *Table 3*. The function of equation (8) is shown in *Figure 8*. Using this model, the endurance rate expressed in percentage can be found for any distance. When residential endurance is 80%, a suitable distance for garbage transfer stations is 10.1 km.

$$h(\lambda) = \int_0^\lambda y(x)\,dx$$
$$= d1erf(e1\lambda + g1) + d2erf(e2\lambda + g2) + d3erf(e3\lambda + g3) + d4erf(e4\lambda + g4) +$$
$$d5erf(e5\lambda + g5) + d6erf(e6\lambda + g6) + d7erf(e7\lambda + g7) + d8erf(e8\lambda + g8) + h1 \quad (8)$$

where *erf(.)* is the error function

Table 3. The coefficients of equation (8) (with 95% confidence bounds).

d1 = 0.64e-1	d2 = 0.44e-1	d3 = -2.96	d4 = 0.195	d5 = 0.2e-1	d6 =-0.342	d7 = 3.549	d8 = -4.91
e1 = 0.18e-1	e2 = 1.344	e3 = 0.423	e4 = 1.466	e5 = 1.770	e6 = 0.614	e7 = 0.385	e8 = 0.401
g1 = 0.28e-1	g2 = -1.364	g3 = -2.161	g4 = -14.66	g5 = -14.04	g6 = -4.28	g7 = -2.06	g8 = 1.833
h1= 5.363							

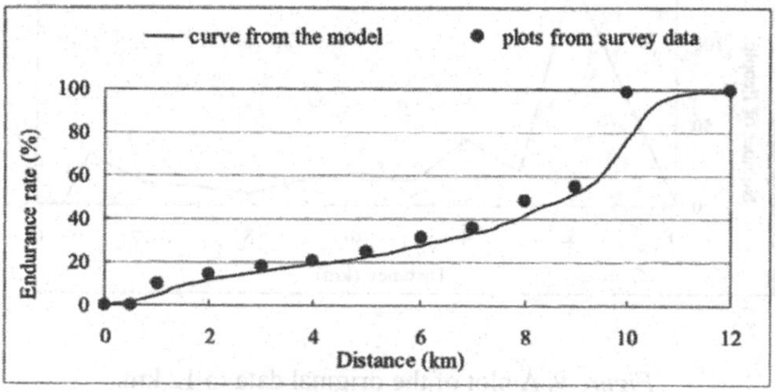

Figure 8. Distribution of the endurance rate *h(λ).*

7. 3-PARAMETER LOGLOGISTIC DISTRIBUTION

7.1 Analysis of Data

In this section, we propose a procedure based on data analysis. *Figure 9* shows a plot of the original data. Values greater than 12 km are excluded from the original data for the following reasons: 1. Only 0.9% of the original data have values greater than 12 km. 2. If we divide the original data into two halves, from 0 to 12.5 km, and from 12.6 km to 25 km, and define a density coefficient

$$Density\_Coefficient = \frac{\textbf{Number of People}}{\textbf{Distance\_Interval}}$$

we can see that the Density Coefficient for the first half is 25.6 persons/km. For the second half it is 0.07persons/km. it is most important to emphasize the negative effects they would have on mathematical modelling. For these reasons, data greater than 12 km can be excluded for modelling.

Then we need a distribution function that can provide an approximate to the data with the restrictions applied. There exist two options for distribution functions: 1. Parametric distributions, they have equations for the probability distribution functions. 2. Non-Parametric distributions, they have no fixed equations that describe the probability distribution functions, and can have many peaks. They are more precise.

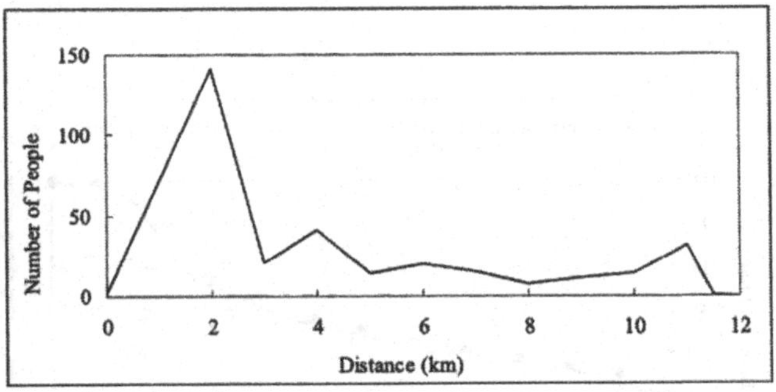

Figure 9. A plot of the original data to 12 km.

7.2 Flipping the Data

Because most of the parametric distribution functions have their peak value in the lower values rather than the higher ones, we can flip the peak of the data from the higher to the lower values. Then, the resulting values NewData are plotted in *Figure 10*.

$$g(x) = f(12 - x) \tag{9}$$

The idea of this procedure is to find the probability distribution for the NewData in *Figure 10*, which will be much more useful because the peak is on the left side of the abscissa, and then flipped back into the resulting function. Let us assume that the distribution function found for the New Data is *h(x)*. Then, we flip this function and get the final result *j(x)*, where

$$j(x) = h(12 - x) \tag{10}$$

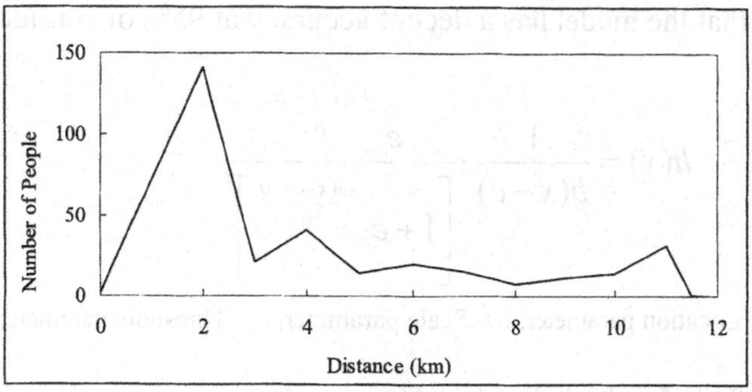

Figure 10. NewData.

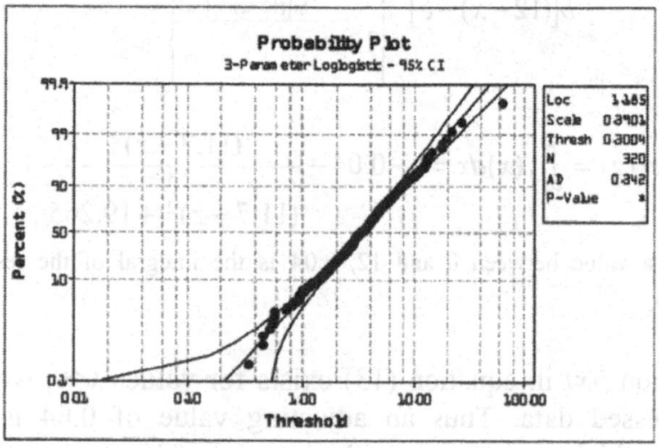

Figure 11. Result of parameter estimation and test of goodness of distribution (Where, Loc, Scale & Thres mean parameters of *a*, *b* & *c*; *C1* means NewData shown in *Figure 10*).

7.3 Distribution Analysis of NewData

Employing Least Squares, a 3-Parameter Loglogistic distribution function as in equation (11) was found. It was selected based on the result of a goodness of fit test for fourteen distributions by employing MINITAB. The resulting values for the parameters and the goodness of fit are shown in *Figure 11.* This Probability Plot helps us to determine whether a particular distribution fits the data. There are 3 lines, the middle line is the model and the sidelines are the margin of 95% confidence. If the distribution fits the data, the plotted points will roughly form a straight line; the plotted points will fall close to the fitted line; the Anderson-Darling statistic will be small. It can be appreciated that the model has a decent accuracy at 95% of confidence.

$$h(x) = \frac{1}{b(x-c)} \cdot \frac{e^{-\frac{\ln(x-c)-a}{b}}}{\left[1+e^{-\frac{\ln(x-c)-a}{b}}\right]^2} \tag{11}$$

where, a = Location parameter, b = Scale parameter, c = Threshold parameter

A modified Loglogistic function is shown in equation (12). The endurance rate can be obtained in equation (13).

$$j(x) = \frac{1}{b[(12-x)-c]} \cdot \frac{e^{-\frac{\ln[(12-x)-c]-a}{b}}}{\left\{1+e^{-\frac{\ln[(12-x)-c]-a}{b}}\right\}^2} \tag{12}$$

$$sr(z) = \int_0^z j(x)dx = 1 - 0.04 - \frac{(11.7-z)^{\frac{5}{2}}}{(11.7-z)^{\frac{5}{2}} + 19.265} \tag{13}$$

where z is a value between 0 and 12, 0.04 is the integral of the resulting function from -∞ to 0.

The function *j(x)* in equation (13) exists for values *x*<0, which is unreal for the processed data. Thus an adjusting value of 0.04 is included in equation (13) which corresponds to the integral of *j(x)* from -∞ to 0. For this reason, the endurance rate never reaches 100%. A corrective coefficient can be applied to the equation of the endurance rate *sr(z)*. The result is defined by equation (14). The corrected function *newsr(z)* is illustrated in *Figure 12*.

Based on this model, when the residential endurance rate is 80%, a suitable distance for garbage transfer stations is 9.9 km.

$$newsr(z) = CC \cdot sr(z)$$
$$CC = \frac{1}{sr_{\max}(z)} = \frac{1}{0.96} = 1.04 \tag{14}$$

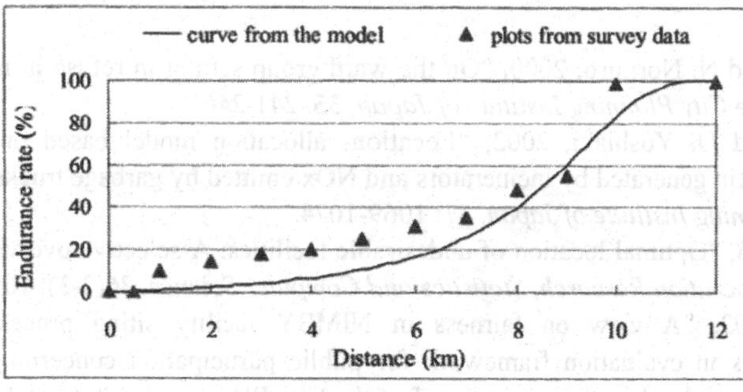

Figure 12. Distribution of the endurance rate *sr(z)*.

8. CONCLUSIONS

Contrary to traditional situations where nearness to a facility is valued, in this research, we address undesirable facilities for which a farther location is more expected. Regarding those kinds of facility locations, we established an endurance rate model. This allows us to consider the relationship between distance to facilities and residential endurance rates. We also made a comparative study for three different models derived from three probability distributions.

The results of a comparative study show that the differences between the models are small. The advantage of the model employing the Weibull distribution is its simplicity; it has only 2 parameters and can be used for the both kinds of facilities though the accuracy was not good enough. The Non-Parametric model is a better model even though many parameters were needed for describing the data in more detail.

Where an undesirable facility is located, it is possible that some people live in an effected area (distance to the facility is shorter than their endurance distance). If we locate undesirable facilities far away from urban areas to avoid negative feelings that the facilities would bring, it is also possible to cause other environmental problems (such as gas emission increase due to transport, or influence to ecosystems due to natural development). To balance these kinds of opposing forces with new corresponding solutions would be objective of our future research.

9. REFERENCES

Masayasu, E. and N. Norihiro, 2000, "On the ward group setting in refuse in inner Tokyo", *Journal of the City Planning Institute of Japan*, **35**: 241-246.

Masashi, N. and O. Yoshiaki, 2002, "Location- allocation model based on the tradeoff between Dioxin generated by incinerators and NOx emitted by garbage trucks", *Journal of the City Planning Institute of Japan*, **37**: 1069-1074.

Plastria, F., 1996, "Optimal location of undesirable facilities: A selective overview", *Belgian Journal of Operation Research, Statistics and Computer Science*, **36**(2-3): 109-127.

Kenshi, B., 2002, "A view on fairness in NIMBY facility siting process – primary considerations in evaluation framework for public participation concerning distributive justice and procedural justice –", *Journal of the City Planning Institute of Japan*, **37**: 295-300.

Kenshi, B., 2003, "Each actor's role on decision making process – examining a possibility of hybrid public participation on NIMBY facility siting issue –", *Journal of the City Planning Institute of Japan*, **38**: 217-222.

Yoshitaka, A. and A. Kondo. 1986, "The optimal distance between residential location and urban facility", *Journal of the City Planning Institute of Japan*, **21**: 295-300.

Yoshitaka, A., 1972, *Research of methodologies regarding systems approach of urban facility location planning*, PhD dissertation, the university of Kyoto.

Satoshi, H., A. Kondo, K. Zhou and T. Wada, 2004, "Evaluation of Public Service Facilities based on Satisfactory Distance in Chengdu City, Sichuan, China", *JSCE*, **21**(1): 239-246.

Chengdu City Environmental and Health Office, 1999, "Document of Chengdu City Environmental and Health Office", 88, in Chinese.

Chengdu Jinjiang Shahe Community Working Committee, 2004, "Document of Chengdu Jinjiang Shahe Community Working Committee", *Family Register 2004*, in Chinese.

Decision-Making on Olympic Urban Development

A multi-actor decision support tool

E.W.T.M. Heurkens

Faculty of Architecture, Delft University of Technology, The Netherlands

Keywords: Combination of sub-solutions, Group decision-making, Olympic urban development, Optimum interorganisational design, Preference measurement modelling

Abstract: Subject of study is the possible organisation of the Olympic Games of 2028 in the Netherlands, as seen from an urban development viewpoint. The project focuses on the decision-making process in the initiative phase. Aim of the project is the development of a decision support tool for the complex, interdisciplinary decision-making process which should result in an optimum interorganisational design. The methodology used to find the optimum choice is the combination of sub-solutions. Preference Measurement modelling based on a multi-criteria decision analysis is the technique employed. The group decision is a choice out of a number of Olympic urban development combinations, which is made in such a way that the preferred combination is the 'best' among the possible candidates for all relevant stakeholders.

1. INTRODUCTION

1.1 Problem Setting

The ongoing political discussions about the possibility of the future organisation of the Olympic Summer Games in the Netherlands, given the 'success' of the 2004 Athens Olympics, eventually resulted in an investigation proposal by the Dutch Olympic Committee (NOC*NSF) in March 2005. The most important conclusion was the announcement of 2028 as the most likely earliest possible year for organising the Dutch Olympics. This was underlined by several arguments.

Jos P. van Leeuwen and Harry J.P. Timmermans (eds.), Innovations in Design & Decision Support Systems in Architecture and Urban Planning, 251-262.

This paper describes the development of a decision support model for the design preparation of the Olympic urban facilities. The project's main question was as follows: "In what way can the complex decision-making process of the 2028 Dutch Olympic urban development be designed, such that it supports the stakeholders of the Olympic urban development project."

We will first clarify the problem with the use of three fields of expertise in which the project can be placed.

1.2 Integrated Urban Development

Urban development processes are no longer characterised by an unambiguous course. Changing social demands and needs necessitate adjustments to the built environment and spatial composition. The adjustments of urban areas refer to restructuring, renewal, transformation, and replacement of functions (Bruil, 2004). Consequently, Integrated Urban Development has become one of the focus points of our department.

Through urban development, essential shifts emerge in future economic, spatial, and social-cultural structures. More and more, urban managers are confronted with tuning and directing on different levels, development phases, policy sectors, and fields of expertise. Furthermore, tuning and directing often take place in complex decision-making processes in interorganisational networks. The current and future urban redevelopment tasks require a steering condition which unites different sorts of knowledge, insights, and skills (Bruil, 2004). It is this managerial context in which we can place the project's problem.

1.3 Olympic Games

2028. This seems lacking in ambition, but for the author this target only states the efforts necessary for a successful Dutch Olympic Games.

First of all there is not one universally suitable 'Olympic urban development product'. Studies made at different former and future Olympic urban developments show that the quantitative and qualitative aspects of Olympic facilities and locations have a wide variety. The product or development task is therefore complex. Sport stadiums for example can be newly built or renovated for the Olympic occasion alone. Then there is the post-Olympic usage of these buildings, the infrastructural and social-economic impact, and the best possible allocation of the Olympic facilities.

The most important requirement of the International Olympic Committee (IOC, 2003) is the accessibility of Olympic sites. To determine suitable Olympic cities, we made an accessibility study in which different means of transport and different levels of approach (national, conglomerate, regional) resulted in the Dutch possible solution of Amsterdam and/or Rotterdam, or the Randstad as metropolitan area (Van Susteren, 2005) as host cities.

Secondly, the decision-making process towards, during and after the urban development of the Olympic facilities is complex, not in the last place due to the size and the investment costs of the development. Numerous organisations and fields of expertise are involved, all with their own means, interests, and goals for the project.

1.4 Decision-Making Environment

Before entering the decision-making process on the Olympic urban development, one has to consider the context we are in now. The NOC*NSF is currently in the study phase of the project. A go or no-go decision has not yet been made. This project is aimed to be implemented in the initiative phase, which would take place around the year 2010.

The project's aim is to decide where, and in what spatial cohesion, the Olympic facilities could be allocated to three locations in one of the two cities. The actors involved are the provincial government, the local government, the municipal development agency, and the sport federation NOC*NSF. These actors are policy-makers within the fields of urban planning and sports infrastructure. These experts work together to achieve the optimum group design for the Dutch Olympic urban development. The decision made is the determination of the preferable situation for the Olympic urban development.

2. BASIC PREMISE: INTERORGANISATIONAL DESIGN

2.1 Individual versus Collective Optimum

Integrated Urban Development asks for tuning and steering in complex decision-making processes in interorganisational networks. Activated parts of networks are called policy arenas (Teisman, 1998). Within these arenas we find different actors, or individuals, or representatives of different types of organisations. I shall assume that, in the decision-making process on the Olympic urban development, each actor individually and constantly strives to improve his part of the design, and thus to achieve his individual optimum. The project team as a whole will also continually strive to achieve the best group result possible. This is referred to as the optimum interorganisational design (Van Loon, 1998), the final product of the decision-making process.

2.2 Optimum Interorganisational Design

The final product of the decision-making should be an optimum design. There are widely varying interpretations and definitions to be found in the literature. The optimum interorganisational design solution can be found within the planning conception category, concerning the optimum choice. This conception is an elaboration of one aspect of the design conception: the optimum combination of sub-solutions. Planners refer to the optimum choice from alternative possibilities. The optimum interorganisational design then can be defined as:

> … the design which has been selected by an explicitly defined procedure from alternatives which fall within mathematically defined constraints accepted by those involved. (Van Loon, 1998).

The methodology used to find the optimum choice, or most preferable combination of sub-solutions, is founded in Open Design theory (Van Gunsteren & Van Loon, 2000). This methodology tackles the combinatory explosion, which emerges in complex and interdisciplinary decision-making processes, such as the Olympic urban development.

In Open Design three groups of methods are developed:

1. the means of combination of sub-solutions;
2. the means of the production of design information;
3. the means of the quantification of design decisions.

In this project the combination of sub-solutions is applied.

3. METHOD: COMBINATION OF SUB-SOLUTIONS

3.1 Structuring the Combination Process

In design methodology, the structuring of the combination process and the limiting of the number of sub-solutions are generally done on a hierarchic basis: one of the parties involved determines the number of sub-solutions which may be designed and the sequence of the combinations. Architects have introduced a process sequence into design based on the combination of sub-solutions (Hamel, 1990). The analysis-synthesis work structure of the architectural engineer is shown in *Figure 1*.

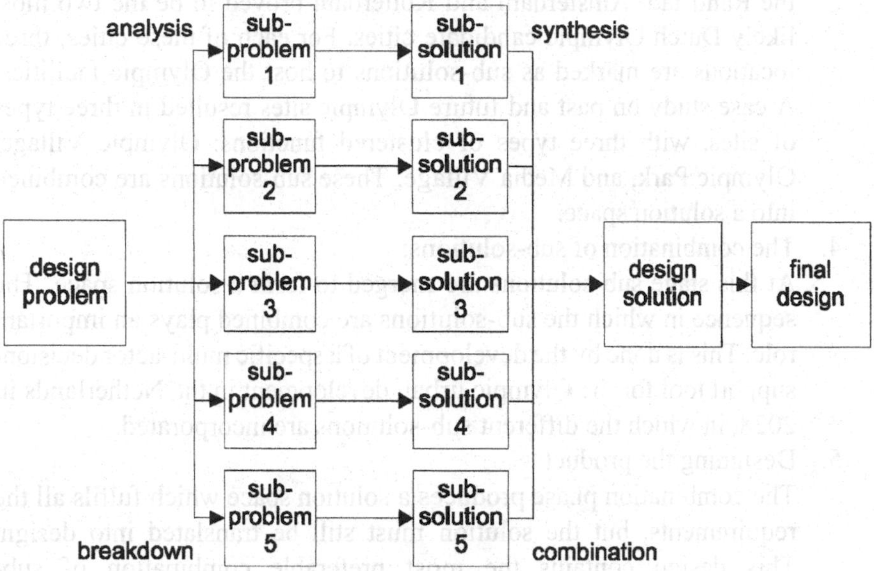

Figure 1. The architectural engineer's analysis-synthesis work structure.

Applied to this project, the process is as follows.

1. The gathering of information:
   The first step in the design process is to study the commission. This provides information on what is desired, in the form of design specifications. Literature references such as the IOC Candidature Acceptance Procedure, The Metropolitan Debate and the Rise of the Network Society are used to determine the design problem. The final problem is the design of the decision-making process on the preferable Olympic urban development situation in all its components.
2. Breaking the commission into its constituent parts:
   As there is never one solution to a design problem, the commission is split up into a series of smaller problems which can be solved individually. The first sub-problem is the determination of the possible Dutch Olympic candidate cities. The second sub-problem is the determination of the possible Olympic locations within these cities. The third problem is the determination of the possible Olympic urban developments.
3. Designing different solutions for these smaller problems:
   Decisions are taken on how these smaller problems are to be solved. The first problem is solved by an attainability study on four cities in

the Randstad. Amsterdam and Rotterdam proved to be the two most likely Dutch Olympic candidate cities. For each of these cities, three locations are marked as sub-solutions to host the Olympic facilities. A case study on past and future Olympic sites resulted in three types of sites, with three types of clustered functions: Olympic Village, Olympic Park, and Media Village. These sub-solutions are combined into a solution space.

4. The combination of sub-solutions:
   At this stage sub-solutions are merged to form a solution space. The sequence in which the sub-solutions are combined plays an important role. This is done by the development of a specific multi-actor decision-support tool for the Olympic urban development in the Netherlands in 2028, in which the different sub-solutions are incorporated.
5. Designing the product:
   The combination phase produces a solution space which fulfils all the requirements, but the solution must still be translated into design. This design contains the most preferable combination of sub-solutions on the most preferable locations in an Olympic candidate city. This is done with the use of the preference measurement technique, explained in further detail in the next chapter.

3.2 Interorganisational Optimisation of the Combinations of Sub-solutions

To apply the method we made the following assumptions. In the first place we have assumed that all the Olympic urban functions have to be newly developed, despite existing infrastructure, stadiums, housing, and commercial buildings. We also have assumed that there are no limitations regarding financial means, the involved actors should have enough opportunities to invest in the Olympic urban development. Furthermore, we have assumed that the selected sites have enough capacity for the Olympic urban development. Based on these assumptions the total permitted solution space is defined. A formal representation of the total solution space, from which the optimal interorganisational design can be made, is shown in *Figure 2.* Since the method is a combination of sub-solutions, and in this study all the combinations are feasible, the solution space contains all possibilities.

In general, the boundaries of the solution space feature implicitly in the combination method. Its limits and form are not defined. It is only possible to derive the boundaries 'retroactively', after each step in the combination process, from the 'position' of and the 'link' between the sub-solutions. In technical terms, the sub-solutions are 'points' in a solution space that has not been delineated beforehand, and is therefore always vague in outline.

As we have assumed that all the combinations are feasible, this retroactive aspect is not applicable in this study, as the boundaries are defined quite explicitly. This is expressed in *Figure 2*.

The study tries to find the preferable solutions within the feasible solution space. To do so we apply the Preference Measurement method.

4. PREFERENCE MEASUREMENT

4.1 Method

Scepticism about the usefulness of computer modelling in architecture and urban planning is often based on the argument that soft variables like beauty cannot be measured. This is actually a misconception. The beauty itself can indeed not be measured, but the preference of stakeholders for one design in comparison with other designs can be established without much difficulty. Preference Measurement can be seen as the key to incorporating soft variables (Binnekamp, Van Gunsteren, Van Loon, 2005).

To measure preference correctly, measurements have to be taken relative to two arbitrarily chosen reference points. What is measured is the ratio of difference. This operation is independent of the chosen origin (zero-point) and selected unit of measurement.

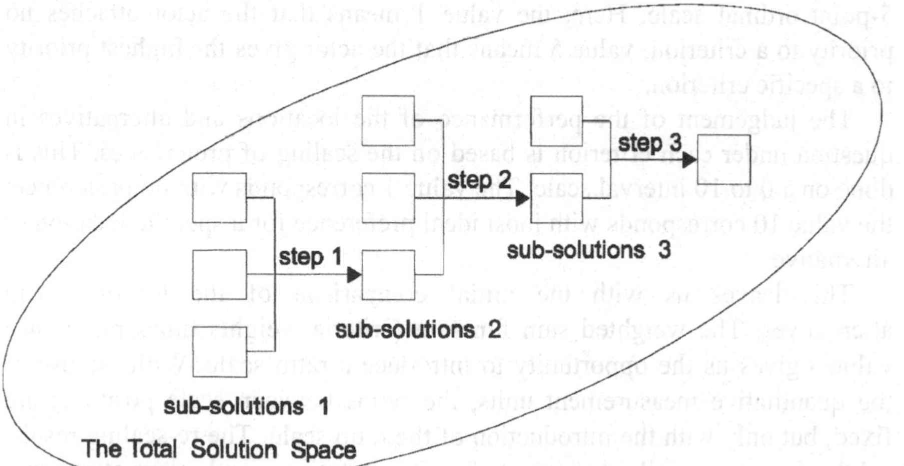

Figure 2. Sub-solutions inside the permitted solution space.

The method can be used for a great variety of choices or decisions. It is particularly useful in extremely complex cases involving many alternatives and many criteria. An example of such a case is the decision-making of the possible future Olympic urban development in the Netherlands.

For the Olympic urban development, a multi-criteria model is used to determine the preferences of the relevant actors within the Olympic policy arena. The preferences for the types of locations and way of developing these locations (with Olympic functions) are measured. This type of decision-making, finally resulting in an optimum interorganisational design, lends itself to an approach whereby different alternative solutions of Olympic urban development per candidate city are generated. Among them is the optimum interorganisational design or choice.

4.2 Modelling

Preference is modelled with the use of a multi-criteria decision analysis model. The decision is a choice out of a number of alternatives, and the choice is made in such a way that the preferred alternative is the 'best' among the possible candidates. The decision maker does not only have the task to judge the performance of the alternatives in question under each criterion, he/she also has to weigh the relative importance of the criteria in order to arrive at a global judgement (Lootsma, 1999).

The criteria are subdivided into location specific criteria and alternative specific criteria. The weighing of the relative importance of the criteria is done by giving priorities to criteria. This is done by entering values on a 5-point ordinal scale. Here, the value 1 means that the actor attaches no priority to a criterion, value 5 means that the actor gives the highest priority to a specific criterion.

The judgement of the performance of the locations and alternatives in question under each criterion is based on the scaling of preferences. This is done on a 0 to 10 interval scale. The value 1 corresponds with no preference; the value 10 corresponds with most ideal preference for a specific location or alternative.

This leaves us with the initial comparison of the locations and alternatives. The weighted sum function (criteria weights times preference values) gives us the opportunity to introduce a ratio scale. With the use of the quantitative measurement units, the ratios between scale positions are fixed, but only with the introduction of the ratio scale. The re-scaling results solely in an overall preference for the locations and alternatives per individual actor.

The next crucial step is the determination of the optimum individual design, in order to come up with an optimum collective design. Therefore we have to combine the preferences on locations and alternatives to end up with different solutions for Olympic urban developments per candidate city. Here the weighted sum of the scaled preferences of locations and alternatives results in all possible sub-solutions, all points within the total solution space. We refer to these as combinations.

Finally, this individual decision-making process results in a group decision-making process, with the use of the technical capabilities of the Preference model. The individual scaled preferences of the combinations are compared. We can again use the weighted sum for this. The power of individuals within the group (their weight) is set to be equal.

The decision-making process is shown in *Figure 3*.

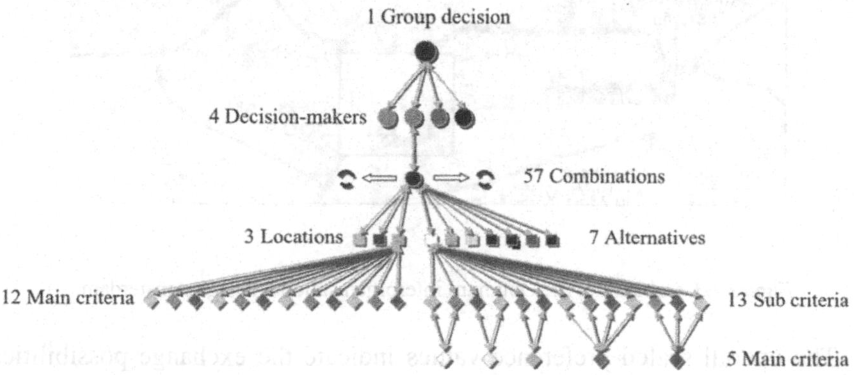

Figure 3. Structure of the decision-making process for a single city.

5. RESULTS

Several workshops with students were organised to come up with an optimum interorganisational design. The three combinations (out of 57 for Amsterdam) with the highest group preference are presented in Table 1.

Table 1. Preferred optimum interorganisational designs for Amsterdam.

Alternative	Combination	G. Preference	Location N	Location SW	Location SE
A04 & A08	M16	P = 0.99	OP & MV	OV	
A04 & A08	M19	P = 0.97		OP & MV	OV
A04 & A08	M21	P = 1,00		OV	OP & MV

The combinations M16, M19, and M21 are most ideal group preferences for the Amsterdam situations. They all belong to combinations which are converted from alternative 4 and 8, which means that the group is quite unanimously satisfied about the combined development of the Olympic Park and Media Village and the separate development of the Olympic Village on specific locations in Amsterdam.

The table can be translated into an optimum interorganisational design, the preferred Olympic urban development plan for Amsterdam is shown in *Figure 4*.

Figure 4. Combination 21: Optimum interorganisational design Amsterdam.

The overall scaled preference values indicate the exchange possibilities of the group's preferences. What if an individual actor is not really satisfied with the end result? Therefore, we can take a closer look at output scaled preferences of the individual actors for each of the three solutions. It might appear that three of the four actors show a high satisfaction on one combination but, one actor is lagging behind. In that case we are comparing the individual optima. With this analysis we can identify the coalitions and oppositions within the group of students. After confronting the different actors with this analysis, the negotiation phase takes place. The (dis) agreements on one specific solution can be tracked within the model's variables. The argumentation behind given preference values forms the base of the negotiation process. This could lead to an adjustment of the actor's goals, which could led to reviewing one's individual preference values, resulting in a repetition of the group decision-making process.

Once all individual stakeholders played by students agree on the crucial individual variables, the group of students identifies one solution to be the optimum interorganisational design. The choice on the most preferable

Olympic urban development is made. The first step in the decision-making process has been taken; the base for the next step is set.

6. CONCLUSIONS

The project's main question was as follows: "In what way can the complex decision-making process of the 2028 Dutch Olympic urban development be designed, such that it supports the stakeholders of the Olympic urban development project." To address this question we have designed a Preference Measurement model for the Olympic urban development process. In the initiative phase, this tool provides a quick scan of the agreements and disagreements of the different stakeholders on the complex product.

The most important characteristic of the decision-support tool is the possibility to incorporate soft variables, a characteristic that answers the common need for structure in the orientating phase. An architectural engineer, or process manager, could use this instrument to bring together the most relevant policy makers in order to determine the optimum interorganisational design. This design creates more effectiveness and transparency in the process. Furthermore, the tool monitors and compares different alternative solutions generated throughout the decision-making process. An earlier experiment in practice with this method was conducted by Feuth (2003). These studies suggest that this method can have practical value for decision making in complex planning projects.

7. REFERENCES

Binnekamp, R., L.A. van Gunstern and P.P. van Loon, 2005, *Open Design, cases and exercises preliminary edition*, Delft University Press, Delft.

Bruil, A.W., F. Hobma, G.J. Peek and G. Wigmans, 2004, *Integrale gebiedsontwikkeling, Het stationsgebied's-Hertogenbosch*, SUN, Amsterdam.

Boersma, S.K.T., 1989, *Beslissingsondersteunende Systemen; een Praktijkgerichte Ontwikkelingsmethode*, Academic Service, Schoonhoven.

Castells, M., 2000, *The Rise of the Network Society*, Blackwell, Oxford.

Frieling, D.H., 1998, *Het Metropolitane Debat*, THOTH, Bussum.

Feuth, J., R.P. de Graaf, and L.A. van Gunsteren, 2003, "OR in public debate, the case of military airport 'Valkenburg'," *45th Annual Conference of the Canadian Operational Research Society*, June 2003.

Gunsteren, L.A. van, and P.P. van Loon, 2000, *Open Design*, Eburon, Delft.

Hamel, R., 1990, *Over het Denken van de Architect, een Cognitief Psychologische Beschrijving van het Ontwerpproces bij Architecten*, AHA Books, Amsterdam.

International Olympic Committee, 2003, *Candidature Acceptance Procedure, Games of the XXX Olympiad in 2012*.

Loon, P.P. van, 1998, *Interorganisational design, a new approach to team design in architecture and urban planning*, Publicatiebureau Bouwkunde, Delft

Lootsma, F.A., 1999, *Multi-Criteria Decision Analysis via Ratio and Difference Judgement*, Applied Optimization, Vol. 29, Kluwer Academic Publishers.

Susteren, A. van, 2005, *Metropolitan World Atlas*, 010 publishers, Rotterdam.

Teisman, G.R., 1998, *Complexe besluitvorming, een pluricentrisch perspectief op besluitvorming over ruimtelijke investeringen*, Elsevier, Den Haag.

Usage of Planning Support Systems

Combining three approaches

Guido Vonk[1,2], Stan Geertman[1], and Paul Schot[2]
[1]*Urban and regional Research center Utrecht (URU)*
[2]*Copernicus Institute*
Utrecht University
g.vonk@geo.uu.nl; s.geertman@geo.uu.nl; p.schot@geo.uu.nl

Keywords: Planning support systems, Usage, Instrumental quality, User acceptance, Diffusion

Abstract: Although a wide range of Planning Support Systems (PSS) exists, their actual utilization in planning practice, to support planners in doing their planning tasks, stays behind. This is problematic since many see PSS capable of aiding planners to handle the complexity of their planning tasks. Our current study explains under usage of PSS from three different angles: the instrument, the user and the transfer of the instrument towards the user. The main conclusion is that usage of PSS is hampered by lack of awareness of and experience with PSS in planning practice as well as by instrumental quality problems and hampered user acceptance and diffusion. The main recommendation to enhance usage of PSS is that it should be made transparent which PSS types should be used for what planning tasks, by which kinds of users, in which kinds of organizations and under which external conditions.

1. INTRODUCTION

Regional planning of land use concerns the development of long-range strategic-level structure plans that organize land use functions. It is one of the most complex tasks of public organizations. Regional land use planners firstly need to deal with a wide range of objectives that occur at different spatial and temporal scales. Secondly, they need to handle large amounts of information of various qualities to serve as a basis of solutions for planning problems. Thirdly, they need to weigh and add up all this information to

Jos P. van Leeuwen and Harry J.P. Timmermans (eds.), Innovations in Design & Decision Support Systems in Architecture and Urban Planning, 263-274.

synthesize planning solutions with a high degree of synergy and that leave little conflict. Fourthly, they need to communicate this all in such a way that their plans receive the support required to make them feasible (Faludi 1973; Friedman 1987; Forester 1989; Hall 2002; Archibugi 2004). Planners demand support to handle these aspects of complexity of planning, but have so far only utilized few of the opportunities for computer support (Stillwell, Geertman et al. 1999). The majority of planners use little more than generic office software to support them in their daily work that tends to be extended with simple spatial capabilities.

Since long, geo-information technology developers have focused on supporting planners dealing with various aspects of complexity. In particular support of information management and scenario analysis have received a great deal of attention. Nonetheless, the large-scale urban models from the 1960s and 1970s have failed to meet expectations and have failed to become widely accepted as planning support instruments (Lee 1973; Batty 1979; Openshaw 1979; Lee 1994). The Geographical information systems (GIS) from the 1980s and 1990s have also not become a great success in supporting planners (Croswell 1991; Innes and Simpson 1993; Stillwell, Geertman et al. 1999). Although many planners use them for basic information functions, most GIS are general-purpose tools that make a poor match to the demands and capabilities of planners in the planning process. A new generation of geo-information technologies known as Planning Support Systems (PSS) is much more dedicated to the demands and capabilities of planners in planning processes (Geertman and Stillwell 2004). These PSS may be better suitable to assist planners in handing the ever-increasing complexity of planning. They have been defined as a subset of geo-information technologies, dedicated to support those involved in planning to explore, represent, analyze, visualize, predict, prescribe, design, implement, monitor and discuss issues associated with the need to plan (Batty, 1995). PSS bring together the functionalities of GIS, models and visualization and take the form of "information frameworks" that integrate the full range of information technologies useful for supporting the specific planning context for which they are designed (Geertman and Stillwell, 2003b; Klosterman, 1997). Inventories show that PSS cover a wide diversity of tools that are readily available for planning support purposes that have not yet become widely applied in planning practice (Brail and Klosterman 2001; Geertman and Stillwell 2003).

In our studies we have taken three different approaches to explain the problem of many PSS not reaching planning practice. The approaches overlap in the sense that they all look at the same problem but each approach emphasizes slightly different aspects of the problem. The 'instrument' approach, explains the problem mainly from the instrumental quality of the PSS, thereby focusing particularly on fitness for use and user friendliness of the PSS (Vonk, Geertman et al. 2006). The 'user approach' explains the problem from the extent of user acceptance of PSS, thereby focusing on a broader set of factors related to the accepting environment (Vonk, Geertman et al. 2005). The 'transfer' approach explains the problem from the extent of diffusion, thereby focusing particularly on the flow of information on and experiences with PSS from sender to receiver (Vonk, Geertman et al. 2006).

In the following we will first describe the three approaches that have been followed to study usage of PSS in general and subsequently describe the explanations found in following these approaches.

2. THEORETICAL FRAMEWORK

2.1 Instrument Approach

The first of three approaches to our problem explains usage of PSS in planning practice from characteristics of the PSS instruments themselves. It does so by focusing on those characteristics of the instrument that determine their instrumental quality. The underlying assumption is that poor instrumental quality of PSS hampers users from using PSS. This approach takes PSS themselves more or less as a dependent variable. It emphasizes in what sense they should change in order to enhance usage. We define instrumental quality as consisting of a judgment of: a) how well the instruments are capable of carrying out the tasks that they were made for; and b) how well they fit to the capabilities and demands of intended users. Googhue and Thompson (1995) showed the importance of these characteristics as determinants of usage of information technologies in their model of task-technology fit (Goodhue 1995; Goodhue and Thompson 1995; Dishaw and Strong 1999; Dishaw, Strong et al. 2002). In terms of this model, under usage of PSS is explained by insufficient fit of PSS to user characteristics and planning task characteristics in comparison with other options that have a better fit. The latter options then may have a relative advantage over using PSS in terms of instrument quality, depending on costs of the options.

2.2 User Approach

The second approach to our problem explains usage of PSS in planning practice from characteristics of the user, focusing on characteristics that determine acceptance of PSS. This user approach is related to the instrument approach as it incorporates user perceptions of instrument characteristics as determinants of acceptance. Furthermore the approach incorporates a much broader set of acceptance factors to explain usage than the instrument approach. The underlying assumption of the user approach is that non-acceptance hampers potential users from using PSS. This approach takes the user as a dependent variable. It emphasizes in what sense users should change in order to enhance usage of PSS. We define the acceptance process as "the process through which an individual or other decision-making unit passes from first knowledge of an innovation, to forming an attitude toward the innovation, to a decision to adopt or reject, to implementation of the new idea, and to confirmation of his decision" (Rogers 1995). We see this acceptance process influenced by user characteristics, instrument characteristics, organizational characteristics, characteristics of the social environment, characteristics of the external environment and facilitating conditions. These factors that influence acceptance have been framed in the 'Technology Acceptance Model'. Since its first publication in 1986, the model has been refined numerous times and applied for a broad range of information technologies (Davis 1986; Compeau and Higgins 1991; Mathieson 1991; Keil, Beranek et al. 1995; Dishaw and Strong 1999; Karahanna and Straub 1999; Venkatesh and Davis 2000; Frambach and Schillewaert 2002; Venkatesh, Morris et al. 2004; Beaudry and Pinsonneault 2005). In terms of this model, under usage of PSS is explained by a hindered acceptance process due to problems with acceptance influencing factors.

2.3 Transfer Approach

The third approach to our problem explains usage of PSS in planning practice from characteristics of the transfer of PSS towards planning practice. It does so by focusing on those characteristics of the transfer that determine PSS diffusion. Innovation diffusion has been defined as "the process by which an innovation is communicated through certain channels over time among members of a social system" (Rogers 1995). It is concerned with the transfer of an innovation into a practice context, through the acceptance by individuals, groups and organizations. The approach is different from the user approach as it emphasizes the course of the innovation in its diffusion among users instead of a single users acceptance process. The assumption that underlies this approach is that diffusion problems hamper users from using PSS. This approach takes the transfer processes as a dependent variable. It emphasizes in what sense they should change in order to enhance usage of PSS. Diffusion is regarded as a process that takes the innovation from the system developers towards widespread usage in practice over the various levels of aggregation: individual, group, organization, and branch of organizations. In diffusion, the aggregation of individuals within groups, groups within planning organizations, and planning organizations that have adopted the innovation follows a path such as described by the innovation adoption curve (Rogers 1995). The curve describes that a group of so-called 'innovators' are the first individuals, groups or organizations to see opportunities and are most likely to perceive the complexity of adoption as a challenge or perceive to be capable of handling the complexity, followed by 'early adopters', 'early majority', 'late majority' and finally the 'laggards' who cannot but accept the innovation after having been confronted with it all over by individuals, groups and organizations who adopted the innovation before they did. The fact that PSS are not used widespread in planning practice indicates that their diffusion has not evolved beyond the early stages.

2.4 Conceptual Framework

Figure 1 shows the three approaches to explain the under usage of PSS in planning practice.

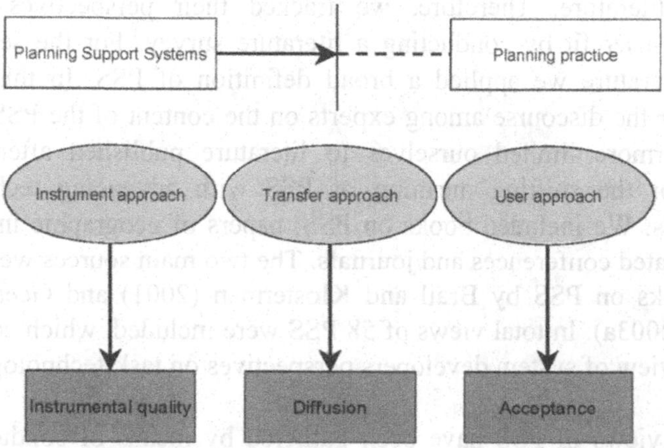

Figure 1. Three approaches to explain usage of PSS.

3. METHOD

To achieve our aim, we combine user perspectives, system developer perspectives and expert perspectives. We realize that these three groups are not fully distinct in their knowledge and experience and account for this in our analysis. The knowledge and expertise of these three groups has been gathered between June and December 2003. User views of PSS have been gathered by holding a series of interviews among 43 employees of 12 highly comparable Dutch regional planning organizations commanded with the task of regional strategic land use planning, which they do by developing plans for water, traffic, environment, economic etc. and integrating these into a comprehensive structure plan for the area governed. In particular we interviewed three archetypes of users that currently have an important role in using and evaluating PSS: the geo-information specialist, the planner, and the manager. We expect these to be capable of providing us with a good and representative overview of user perspectives on PSS-technology in most of the developed world, particularly since evidence suggests the existence of more or less similar planning organizational environments in these societies and since we do not focus on specific aspects of the planning style. In the end, most of the participants were geo-information specialists (15), planners (12) and managers (3), but also people with strongly related specializations such as environmental planning, economic planning, social planning and general IT joined in (13). The interviews were carried out in groups, during 12 sessions of several hours each.

System developer views of PSS have been quite well recorded in scientific literature. Therefore, we tracked their perspectives on task-technology-user fit by conducting a literature survey. For the selection of suitable literature we applied a broad definition of PSS. In this way we account for the discourse among experts on the content of the PSS concept. We furthermore limited ourselves to literature published after 1998 to account for the shifting meaning of PSS with advancing technological possibilities. We included books on PSS, papers of geographic information science related conferences and journals. The two main sources were the two edited books on PSS by Brail and Klosterman (2001) and Geertman and Stillwell (2003a). In total views of 58 PSS were included, which constitute a good overview of system developers perspectives on task-technology-user fit of PSS.

Expert views of PSS have been gathered by means of conducting two worldwide web-surveys. Via several PSS-related listserv e-mail networks 800 PSS-interested persons were asked to participate. The first survey had 96 respondents; the second had 40 respondents. 86 respectively 30 of these respondents were considered experts, since they indicated to know at least 2 PSS from a list. The majority of the expert respondents were university researchers and employees of public planning bureaus dealing with planning support in their work. Although many users from planning practice were asked to participate, among the respondents they were a minority. The first survey consisted of a series of bottlenecks that potentially block widespread usage of PSS, to be judged on importance by the experts. The second survey consisted of open questions on strengths, weaknesses, opportunities and threats of PSS, as well as closed questions to express the perceived fit of a range of combinations of planning task, PSS and user and to state experience with these combinations. Respondents could judge the importance of potential bottlenecks as well as the fit by selecting from 'not useful', 'neutral', '(very) useful' and 'don't know'.

We combine the findings of the literature survey, the interviews and the web-survey to find what underlies the under usage of PSS in planning practice. The underlying explanations are found from the results by analyzing the bottlenecks that block widespread usage of PSS in general. We identify bottlenecks in instrumental quality by evaluating PSS on the fit to planning tasks, fit to user competences and relative advantage. We identify bottlenecks for acceptance by evaluating PSS on perceived user friendliness and usefulness, their users' awareness of, experience with and intention to use these instruments, and the presence of social influences and organizational facilitators that affect acceptance. We identify bottlenecks in diffusion by evaluating PSS on their take up from the organizational environment, and feed forward and back through the organization by means of initiation, decision and implementation. These bottlenecks were interpreted in terms of our conceptual framework.

4. RESULTS

4.1 Instrument Approach

Results show that one of the primary reasons for the under usage of PSS in planning practice is that PSS technology is in an early stage of development. The large diversity and little standards associated with this development stage cause large differences in instrumental quality between instruments.

Results furthermore show that there exists a large dichotomy between PSS demanded in practice and supplied by system developers: while practice demands rather simple PSS for exploratory tasks such as making an inventory of conditions, the majority of PSS focus on more analytical tasks, especially modeling. These on their turn are seen as making a poor match with the demands of planning practice. The instrumental quality of simple instruments is considered acceptable while that of advanced instruments is generally considered to be poor. Results suggest that simple instruments have a relative advantage over doing it all by hand, while for many currently existing advanced instruments the advantage is doubtful at the least.

Results furthermore show that planning practice regards PSS as hardly suitable for direct usage by decision makers themselves. Decision making itself is seen as a game of politics and power, in which PSS hardly have a place. PSS are seen as more promising to support the other planning stages that require information management, communication and analysis rather than politics.

4.2 User Approach

Results of the user acceptance approach show that there exists a large diversity of bottlenecks blocking widespread acceptance of PSS in planning practice, the main of which are lack of awareness of the existence of PSS and for which purposes they can be used, lack of experience with PSS, which makes users unaware of the benefits of PSS and the conditions under which they can be used and low intention to start using PSS among possible users. Also high scoring bottlenecks are insufficient user friendliness and usefulness, the absence of the required organizational facilitators and social influences and data quality and accessibility problems (Vonk, Geertman et al. 2005).

Results clearly show that while system development is continuing at a rapid pace, development has remained largely unnoticed by the intended users. If the planning community remains unaware, it will not acquire experience and develop a demand side, which causes insufficient pay-off in the investments in PSS development. Furthermore a process of improvement of existing tools by learning from practice will remain at a low pace. In terms of a product lifecycle, the product will not get a chance to mature and to reach the point after which its development and proliferation becomes self-enforcing. If no marketing action is taken, it is therefore likely that PSS will not get a good chance to prove its worth as a means for improving spatial planning practice.

4.3 Transfer Approach

Results show that many managers and planners are hardly aware of the existence and potential of PSS and have so little affinity with these technologies that they cannot develop a good strategy. Geo-information specialists are often the only ones in the organization capable of initiating adoption and implementation from the bottom up. Their diffusion-oriented actions are often motivated by an experienced opportunity.

Results furthermore show that lack of opportunity for innovation allowed by the management and lack of the required personal characteristics often causes geo-information specialists to be unable to initiate adoption and implementation of PSS. Furthermore, exploratory activities of those geo-information specialists that do possess the required personal characteristics to be initiators are often repressed instead of nurtured.

Results also show that regional planning organizations often exploit management-supported strategies on geo-information technology diffusion. These strategies often hold back significant steps in diffusion, since they are based upon a persistent negative image of geo-information technology that exists among many managers. Geo-information specialists that do take up their role as initiators of diffusion, often face a wall when trying to convince managers of the worth of new developments in geo-information technology such as PSS. Showing examples in real projects has proven to be a good means of convincing managers, but their preparation requires innovation time, which geo-information specialists often do not have. This traps diffusion in a stalemate.

Results show that geo-information specialists themselves are hardly ever able to reach spatial planners and cooperate with them. If they do, they often encounter a discrepancy between planners' questions and geo-information specialists' offers that obstructs successful cooperation. This hampers development of useful innovations that are applied in planning practice, since these are likely to evolve from cooperation of geo-information specialists and spatial planners (Vonk, Geertman et al. 2006).

5. INTERPRETATION

Figure 2 shows the results of our current study in terms of the theoretical framework of our study. It shows: 1) the problem of many Planning Support Systems being blocked from reaching planning practice and being widely used; 2) the perspectives to look at this problem that correspond with an instrument approach, a user approach and a transfer approach; 3) the main focus of these approaches: instrumental quality, user acceptance and diffusion; and 4) the main reasons explaining under usage of PSS in general that were found in our studies using these approaches.

6. CONCLUSIONS AND RECOMMENDATIONS

We conclude that usage of PSS in general is hampered by a broad range of bottlenecks. These bottlenecks are related to instrumental quality as well as user acceptance and diffusion issues. Results show that lack of *experience* with PSS, lack of *awareness* and lack of *instrument quality* of PSS are the main bottlenecks blocking user acceptance. Results show that the effects of these main bottlenecks on usage are enhanced by hampered *user acceptance* and *diffusion*.

The main recommendation to enhance usage of PSS is that it should be made transparent which PSS types should be used for what planning tasks, by which kinds of users, in which kinds of organizations and under which external conditions. More in particular, to enhance primarily *awareness* of PSS we recommend spreading the news of the existence and potential of PSS in planning practice. Awareness generation should not to stop after a single rejection since innovation in PSS may be a timely process of human and organizational adaptation. To enhance primarily *experience* with PSS we recommend applying best practices of PSS application that will maximize chances that the gained experiences will be positive. To enhance primarily *instrumental quality* of PSS we recommend system developers and geo-information specialists to improve communication with practice in order to be capable of actively analysing the tasks that may be supported by PSS and the application environments. To enhance primarily *user acceptance* and *diffusion* of PSS we recommend managers to adopt the management paradigm of the learning organization.

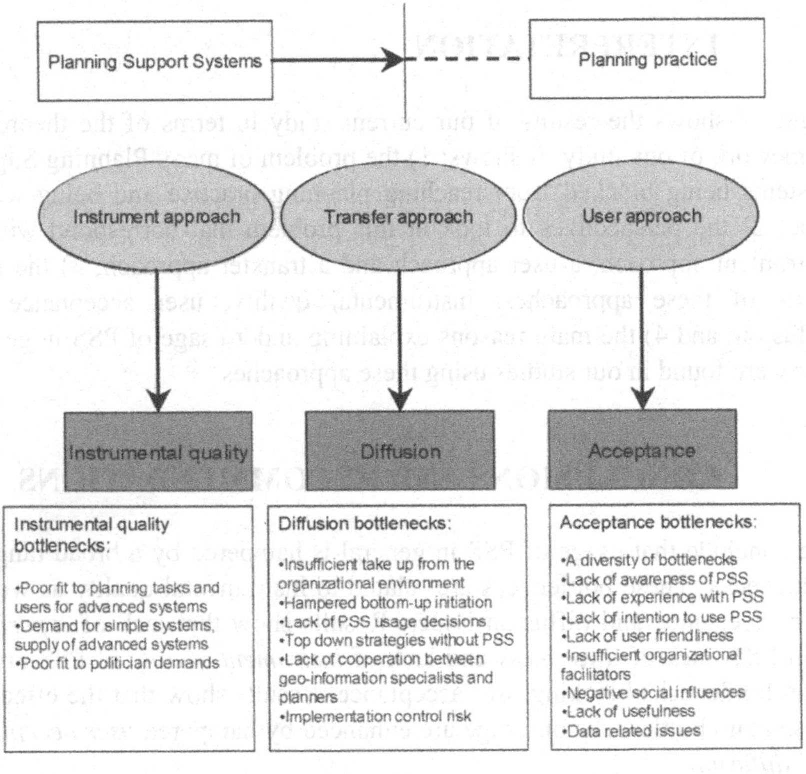

Figure 2. Framework explaining under usage of PSS from three different approaches.

7. REFERENCES

Archibugi, F., 2004, "Planning Theory: Reconstruction or Requiem for Planning," *European Planning Studies* **12**(3): 425-445.

Batty, M. (1979). "Progress, success and failure in urban modelling," *Environment and Planning A* **11**: 863-878.

Beaudry, A. and A. Pinsonneault, 2005, "Understanding User Responses to Information Technology: A Coping Model of User Adaptation," *MIS Quarterly* **29**(3): 493-524.

Brail, R. and R. Klosterman (eds.), 2001, *Planning Support Systems, integrating geographic information systems, models and visualization tools*, Redlands California, Esri Press.

Compeau, D. and C. Higgins, 1991, "A social cognitive theory perspective on individual reactions to computing technology," *12th Annual conference on information systems (ICIS)*, New York, USA, Association for Information Systems.

Croswell, P., 1991, "Obstacles to GIS Implementation and Guidelines to Increase the Oppurtunities for Success." *URISA Journal* **3**(1): 43-56.

Davis, F., 1986, *A technology acceptance model for empirically testing new end-user information systems: theory and results*, Cambridge Massachussets, MIT.

Dishaw, M. and D. Strong, 1999, "Extending the technology acceptance model with task-technology fit constructs," *Information & Management* **36**(1): 9-21.

Dishaw, M., D. Strong, et al., 2002, "Extending the Task-Technology Fit Model with Self-Efficacy Constructs," *Eighth Americas Conference on Information Systems, AMCIS 2002*, Dallas, USA, Association for Information Systems.

Faludi, A., 1973, *Planning Theory*, New York, Pergamon Press.

Forester, J., 1989, *Planning in the face of Power*, Berkeley, University of California Press.

Frambach, R. and N. Schillewaert, 2002, "Organizational innovation adoption, A multilevel framework of determinants and opportunities for future research," *Journal of Business Research* **55**(2): 163-176.

Friedman, J., 1987, *Planning in the public domain; From knowledge to action*, Princeton, Princeton University Press.

Geertman, S. and J. Stillwell (eds.), 2003, *Planning Support Systems in Practice, Advances in Spatial Science*. Berlin, Springer Verlag.

Geertman, S. and J. Stillwell, 2004, "Planning Support Systems: an inventory of current practice," *Computers Environment and Urban Systems* **28**(4): 291-310.

Goodhue, D. L., 1995, "Understanding user evaluations of information systems," *Management Science* **41**(12): 1827-1844.

Goodhue, D. L. and R. L. Thompson, 1995, "Task-Technology fit and Individual Performance," *MIS Quarterly* **19**(2): 213-235.

Hall, P., 2002, "Planning: millennial retrospect and prospect," *Progress in Planning* **57**: 263-284.

Innes, J. and D. Simpson, 1993, "Implementing GIS for planning," *Journal of the American Planning Association* **59**(2): 230-236.

Karahanna, E. and D. Straub, 1999, "The psychological origins of perceived usefulness and ease-of-use," *Information & Management* **35**(4): 237-250.

Keil, M., P. M. Beranek, et al., 1995, "Usefulness and ease of use: field study evidence regarding task considerations," *Decision Support Systems* **13**(1): 75-91.

Lee, D., 1973, "Requiem for large-scale models," *Journal of the American Planning Association* **39**(3): 163-178.

Lee, D., 1994, "Retrospective on large-scale urban models," *Journal of the American Planning Association* **60**(1): 35-40.

Mathieson, K., 1991, "Predicting user intentions: comparing the technology acceptance model with the theory of planned behaviour," *Information Systems Research* **2**(3): 173-191.

Openshaw, S., 1979, "A methodology for using models for planning purposes," *Environment and Planning A* **11**: 879-896.

Rogers, E., 1995, *Diffusion of innovations*, New York, Free Press.

Stillwell, J., S. Geertman, et al., 1999, "Developments in Geographical Information and Planning," in J. Stillwell, Geertman, S. and S. Openshaw (eds.) *Geographical Information and Planning*, Berlin, Springer-Verlag: 3-22.

Venkatesh, V. and F. Davis (2000). "A Theoretical Extension of the Technology Acceptance Model: Four Longitudinal Field Studies," *Management Science* **46**(2): 186-204.

Venkatesh, V., M. Morris, et al., 2004, "User Acceptance of Information Technology: Toward a Unified View."

Vonk, G., S. Geertman, et al., 2005, "Bottlenecks blocking widespread usage of Planning Support Systems," *Environment and Planning A* **17**: 909-924.

Vonk, G., S. Geertman, et al., 2006, "New technologies stuck in old hierarchies: an analysis of the diffusion of geo-information technologies in Dutch planning organizations," *Public Administration Review* **66**(Forthcoming).

Vonk, G., S. Geertman, et al., 2006, "A SWOT-analysis of Planning Support Systems," *Environment and Planning A* **38**(Forthcoming).

DEVELOPMENT OF DESIGN SUPPORT SYSTEMS

Managing and Deploying Design Knowledge

Sieving Pebbles and Growing Profiles

Capitalising on knowledge embodied in design practice

Cristina Cerulli
School of Architecture, University of Sheffield, UK

Keywords: Knowledge management, Design knowledge, Recommender systems

Abstract: This paper discusses tools and strategies to support the capturing and use of knowledge embodied in design practice. The tools described are part of a system conceived as a suite of modular tools, KMan, to be developed and adopted incrementally and independently, in the framework of a wider organizational knowledge management strategy. The concept of knowledge pebbles is introduced to describe collections of heterogeneous data formats to form a unit of potential knowledge In the final section of the paper an overview of all the tools part of the system is presented and some of the tools described and illustrated in more detail. In particular the Brief Management Tool, the Generic Search Tool and the Pebble Creation Tool are discussed.

1. RESEARCH FRAMEWORK AND APPROACH

This paper reports on some of the outcomes of ongoing research started with an EPSRC funded project on the issues of knowledge creation and management within construction industry projects (Cooper, Lawson et al. 1998; Cerulli, Peng et al. 2001; Cerulli, Peng et al. 2002; Cooper, Cerulli et al. 2005) and continued through a doctoral research, combined with architectural practice, looking at Knowledge Management tools for construction design practices (Cerulli 2004).

The research framework within which tools and strategies here discussed were designed and identified was fundamentally practice focused, although initially technology driven. Dealing with real construction project processes and complexities was at the core of the research approach and a significant number of leading architectural practices and construction organizations

Jos P. van Leeuwen and Harry J.P. Timmermans (eds.), Innovations in Design & Decision Support Systems in Architecture and Urban Planning, 277-291.

contributed to the study through hosting ethnographic field studies, and participating in interviews, workshops and advisory panels.

In terms of system design the approach adopted within this study has evolved from an effectively top-down approach, centred around the ADS prototype, to a bottom-up approach that culminated with the design of the KMan tool, discussed below. The ADS Project (Cooper, Lawson et al. 1998), was essentially technology driven: a set of technologies had become available, namely java and engineering components, and a tool was developed to explore the potential of those technologies within the construction industry, after identifying the problems that needed addressing within that domain.

Although every attention was paid to real practice through careful gathering of user requirements, this approach created an intrinsically top-down system. With a system like ADS a hierarchically structured organization would customise the system to reflect its practices and needs and the information and knowledge captured by the system would be structured according to the company's ethos and directives expressed by the company's executives. For this system to successfully work in practice there needs to be an alignment between the aspirations and intentions of executives, who have a larger input in the briefing, commissioning and funding of the system, and those of the non executive members of the organization, who will, potentially, populate and use the system. This alignment, although achievable, does not necessarily constitute the norm within large design organizations and therefore alternative approaches are required to facilitate the undertaking of knowledge management initiatives by architectural and engineering practices.

As the research progressed it became gradually apparent that Information Technology *per se* is not sufficient to address those problems of the construction industry that this research set out to provide solutions for and that an alternative, practice driven, approach was required. Rather than looking for suitable applications of the latest technology within the chosen domain, the practice driven approach, adopted for the KMan tool, aims to let the tools emerge from practice by providing a system structure that supports emergence. In this model, using collaborative filtering techniques, the organization's hierarchical structure loosens up and clusters of users emerge, based on user's affinities and interests and not on company roles. Similarly the information and knowledge stored into the system is gradually structured according to users' feedback and it becomes increasingly easier for users to find good quality material that is relevant to them.

The difference in approach between the ADS System and the KMan tool is epitomised by the techniques they use: while ADS makes use of taxonomies, structures that presuppose an organization from the top down, KMan uses recommender systems, typically bottom up structures.

2. WHAT KNOWLEDGE

The boundaries and definitions of information, knowledge and data can be subjective and change over time with circumstances. A collection of information does not necessarily constitute knowledge but has the potential to become such when the contextual information retrieved is used to infer and construct knowledge (Cerulli 2004).

In the KMan tool, knowledge is mainly inferred from contextual information made easily accessible and filterable; the tool aims to manage Data, Information and Knowledge, without trying to fit particular items into any of these categories. An item might be input into the system as Information at a point in time A and be retrieved later with other information to form knowledge at a point in time B. Or, alternatively, something that is pure data for an experienced designer X, for instance, the dimensions of standard steel members, can later become the basis for knowledge inferral for a novice user Y that might read this as a summary of steel construction solutions.

The KMan system aims to:

- Capture, manage and index for easy retrieval information already available. This functionality of the tool falls in what is widely accepted as information management but the author argues that effective information management and, hence, retrieval can be the basis for knowledge creation and design activities support.
- Provide a framework within which to gather knowledge by attaching metadata to existing documents to incorporate feedback on their quality or usefulness.
- Provide a place to record and share opinions, ideas and experiences by submitting them to the system.
- Support the emergence of communities of practice as defined by Lave and Wenger in their influential work "Situated Learning: Legitimate Peripheral Participation" (Lave and Wegner 1991).
- Support the creation of a shared repertoire of knowledge amongst users. Creating this type of shared knowledge warehouse for a design organization has the potential of:
    - o Reducing time spent solving routine design problems. Part of the proposed shared repertoire could be a collection of successful details to be re-used in future projects. The re-use through adaptation of previous design solution does not necessarily hinder innovation as it is often argued (Grint,

Case et al. 1995; Willmott and Wray-Bliss 1995); on the contrary, it could even facilitate it by speeding up routine design tasks and, thus, freeing up time for ad-hoc inventive and pioneering design (Henderson and Clark 1990; Veshosky 1998).

- Allowing novice members of the organization to benefit from experts' knowledge. One way of doing this is, for instance, making available experts' personal lists of *favourites*, regarding exclusively their area of expertise. These *favourites* can be, for instance, internet bookmarks or a collection of documents that are considered very topical and of good quality.
- Fostering intra-project knowledge transfer and learning. Design problems and technical solutions developed for one project, as well as the performance feedback about those solutions, can be easily made available to the whole organization to potentially inform the design of other projects, contemporary or future.

Stenmark (Stenmark 1999; Stenmark 2001; Stenmark 2002), building on Polanyi's theories (Polanyi 1966), maintains that although we might not always be able to articulate explicitly what we are interested in, we intuitively know it when we see it and suggests that there is scope for KM systems that use recommender systems to make use of tacit knowledge without needing to make it explicit.

> "By identifying certain documents as interesting the user could tell an agent based retrieval system to maintain a dynamic profile that represents a certain limited perspective on the user's tacit knowledge without requiring explicitly defined keywords or manually updated records". (Stenmark 2001)

In the light of the results obtained by Stenmark's research at Volvo and given the quantity and diversity of information to be managed, captured and retrieved within a construction design process the KMan tool uses recommender system techniques to help users to:

- Overcome low signal to noise ratio and to sieve relevant information. As mentioned above in responding to the need of filtering relevant information the approach chosen for the KMan system is to avoid unnecessary formalization, and to opt for a strategy that is less demanding on the user in terms of cognitive overload such as the use of recommender systems.
- Capture and make use of some of the tacit knowledge embedded in people, without having to make it explicit.

3. PEBBLES OF KNOWLEDGE

Within the ADS system heterogeneous data formats were pulled together into the complex dataset *ADS Design Decision Record*, described elsewhere (Cerulli, Peng et al. 2001; Cerulli 2002; Cerulli, Peng et al. 2002; Cerulli 2005). The different types of data constituting the ADS Decision Record were: CIMM management attributes, ADS taxonomical attributes, CAD transaction data, natural language description of design intents/rationale, affected and pending design decisions and hyperlinks to other related documents.

The KMan set of tools keeps this notion of the collection of heterogeneous data formats to form a unit of *potential* knowledge. The term *potential* is used here because, as mentioned above the author endorses that a collection of information does not necessarily constitute knowledge but it has the potential to become such once it is retrieved by a user who will infer and construct knowledge from the provided contextual information (Gruber and Russell 1996).

These *units* of potential knowledge are likened to pebbles on a beach: each pebble is an individual unit and has its individual colour and properties, nevertheless the collection of a large number of pebbles can be read as a portion of beach, just as a collection of information and data can, when interpreted, be read as knowledge or used to construct it.

Moreover if equipped with the right sieve, or filtering tool, one could filter out unwanted pebbles or keep only the desired ones. One, for instance, could decide to filter the whole beach and get rid of all blue pebbles or to keep just the yellow ones that are bigger than one millimetre; similarly one could filter the whole information stored in the system to eliminate decisions prior to a certain milestone or to highlight only the scheduled actions that have not been attended to.

What is a *pebble*? It can be any piece of information to which someone attaches some sort of interpretation, semantic, comment or link. Examples of a pebble could be a folder containing site pictures, a research report on shading techniques, an informative e-mail or a CAD intelligent engineering component.

Once *pebbles* are created their rating will start evolving on the basis of user's feedback. Users will be encouraged to flag out particularly good pieces of information, to provide feedback on the individual *pebbles* and to make links between various *pebbles*. At any time users can flag out pebbles that they consider particularly good and useful. The introduction of a ranking system will also allow accommodating negative feedback, so that poor quality material can be filtered out. The overall pebble rating displayed for each pebble, when retrieved, should be the average of all the rating provided by the different users, but the individual comments should also be made available for reading, if desired.

As knowledge is dynamic and contextual each pebble should be able to evolve over time by being modified and edited. This requires a versioning system to keep a record of the evolution of the pebble over time.

Users can also link several pebbles by defining dependencies and relationships. Regarding the way these links are defined, one possibility is to use structured link-types that could be, for instance, similar to IBIS categories. This strategy would call for an undesired action of formalization (Shipman and Marshall 1999) which would essentially bring on board the limitations of the argumentation approach to design rationale capturing (Shipman 1993; Shipman and McCall 1997). The other possibility is to refrain from defining the nature of these links and simply indicate the presence of a generic link between items. These generic links would be unstructured but they could be indexed by date or author to make them searchable. For KMan tool the latter possibility was considered more adequate because, while still providing a pointer whose nature can be inferred when required, it is less demanding on users, thus avoiding a *cognitive overload* (Shipman and Marshall 1999).

The system also enables the links between pebbles as well as their rating/feedback to evolve over time.

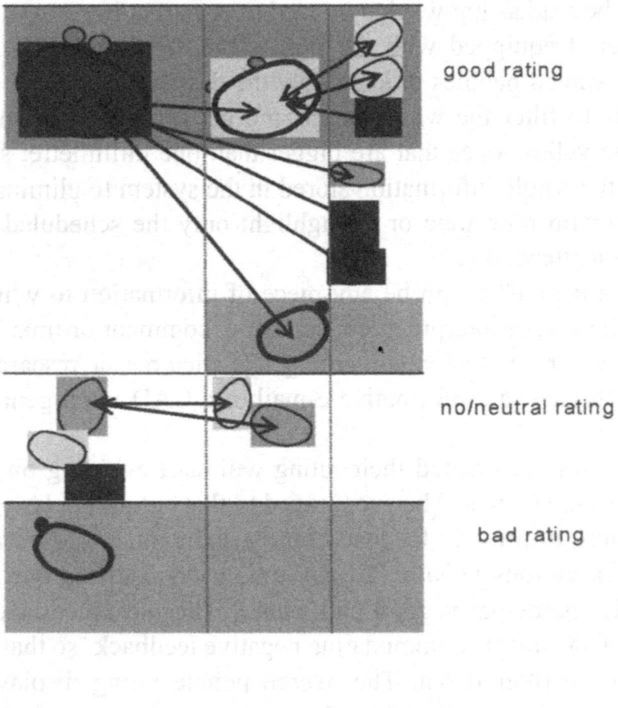

Figure 1. Retrieved pebbles are displayed according to their rating.

Users will be able to define within the search tool what type of pebbles they would like to filter: they could decide, for instance to search for all items related to *photovoltaic* with a positive rating of three and above. In the case of un-filtered search the system will first show the pebbles with good rating, in descending order, together with all their related pebbles, including those with negative rating; secondly pebbles with no or neutral rating will be displayed and, finally, those with negative rating.

The company Intranet was chosen as software environment to host the Information and Knowledge Management functionalities because of its widely spread use in UK construction design practices (Carrillo, Robinson et al. 2002) and for the flexibility that Intranets offer by allowing to easily integrate information from various sources and in heterogeneous formats.

4. THE KMAN SUITE OF TOOLS

In line with the practice based nature of this research the proposed tool was devised using a real architectural practice as target user while, at the same time, every attempt was made to keep the system generic enough to be valid for other organizations within the construction industry domain. The practice chosen this exercise is Reid Architecture (RA hereafter). RA was chosen because of its high interest in knowledge related issues applied to architectural design practice and because the author had the opportunity to work for that organization both as an architect and as adviser for the company's KM development programme.

RA was at the time working to implement a system that, building on and integrating other systems in use within the practice, would assist and facilitate IT support to knowledge creation and management. This study contributed to the specification of that system.

The KMan suite comprises several tools, summarised in *Table 2* below. In terms of System Architecture each tool could be an independant application and seamless integration could be achieved through the Intranet.

Some of the applications could be developed by partially reusing software created for other research projects or customising software available off the shelf.

In the sections below the Briefing Management, the Pebble Creation and Generic Search Tools are described in more detail, with a focus on the Graphical User Interface (GUI).

Table 1. Synoptic Table of knowledge capturing tools integrated in KMan.

KNOWLEDGE CAPTURING TOOLS	
Tool	**Function**
Profile Keeping Tool	To capture user preferences as a by-product of their tacit knowledge and, via a recommender system, make that knowledge available to others
Pebble Creation Tool	To flag out and attach DR information to any type of information. Heterogeneous
Brief Management Tool	To record and track the requirements' development and their corresponding elements of the solution
Document Archival Tool (Qdoc+)	Improved Qdoc Interface to provide opportunity to add comments, links or pictures when saving documents
Design Review Recording Tool	To capture the information and knowledge made explicit in a semi-structured way during design reviews
Meeting Minutes Generator Tool	To facilitate production of meeting minutes and gathering contextual information
Project DB	To contain standard information about projects
Thematic Homepages	To allow quick compiling of mini-sites featuring specific topics
Feedback Bar	Provides user with opportunities to express their feedback on the system recommendations and on the quality of documents

Table 2. Synoptic Table of knowledge retrieving tools integrated in KMan.

KNOWLEDGE RETRIEVING TOOLS	
Tool	**Function**
Generic Information Search Tool	The Generic Search Tool is conceived to be the starting point of any search. Quick and Advanced search.
Easy Qdoc Retrieval	To improve the retrieval of documents archived using the Qdoc software.
Project Homepages	An intranet site where pointers to all the information concerning one specific project are provided and tools to navigate and view that information can be launched.
Project Directory	a subset that a Contacts Database containing the contacts for a specific project team.
Drawing Management Tool	drawing browsing and searching facility with preview and drawing history management.
Project History tool	To build and navigate historical project information from the desired perspectives.

4.1 Pebble Creation Tool

In section 3 the notion of pebbles as units of potential knowledge was introduced. This notion is very generic and can assume several formats; in other words a pebble can be any piece of information to which some Design Rational like information is attached. A pebble can therefore be, for instance, an idea for a marketing event, a proposal for an exhibition in the office gallery space, a folder containing good examples of factory buildings or a record of a design decision. Meeting minutes and design review records are pebbles too, but, given their recurrence and importance, an ad hoc tool has been created to generate them.

The interface of the Pebble Creation Tool as proposed contains the following components:

- author and date, transparently captured by the system;
- List of Participants. Participants' names can be selected by ticking the project directory list.
- Circulation. As for the Participants, names can be easily selected from the project directory. By ticking them from a comprehensive list.
- Category list. Users can choose from a drop down menu a category applicable to the pebble that is being created. Examples of these categories are: product information, interesting building, event, suggestions. Should users feel that none of the categories provided are suitable for the pebble they are creating, by selecting OTHER they will be given the option to define a new category.
- Brief Description [Title]. Text box with limited number of characters. This is where the user will summarise the content of the pebble by giving it a sort of heading. This field is compulsory and it is the minimum requirement of textual input needed to create a pebble.
- Explanation [Design Rationale]. Free text box for text of unlimited length. Here is where users that are willing to do so can provide detailed unstructured explanations on the object of the pebble.
- Linked files. User can provide links to contextual information related to a pebble. As mentioned in Chapter 5 Gruber and Russell found that rationale explanations are more often constructed by interpreting stored information available than by retrieving exhaustive explanations (Gruber and Russell 1996, 330). Access to hyperlinks to relevant items will help inferring design rationale and knowledge, thus reducing the amount of contextual information that need to be input at the time of the pebble creation.
- Related images. This functionality allows including illustrative images to the heterogeneous data set that constitutes a pebble.

- Qdoc Save. This button allows saving the Pebble. By clicking this button the user is presented with two options: one is to save and commit to the system and the other is to save in a draft folder for future editing.
- Create Report. This button creates a printer friendly report of the Pebble created.

It is possible to envisage that after a test period during which users confront themselves with the structure of the generic pebble creation tool, user feedback would lead to an evolution of the interface for this tool in which data input forms become category specific. For instance by selecting the category "interesting building", some fields specific to that type will appear in the pebble creation interface such as "architect", "date", "location", "area", "cost" etcetera.

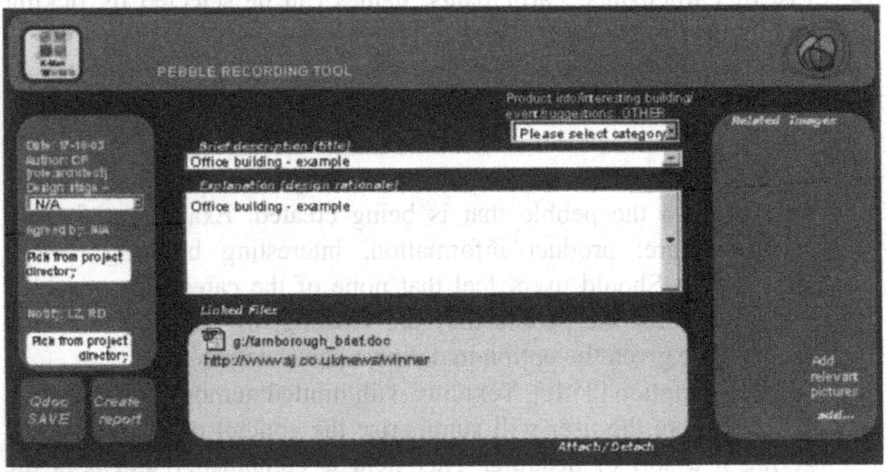

Figure 2. Pebble Recording Tool Interface.

4.2 Brief Management Tool

The Brief Management Tool (BMT) proposed in the KMan suite is a device to record and track the development of requirements, elements of the problem space, and their corresponding components of the solution.

BMT has both knowledge capturing and retrieving capabilities. This tool, by allowing all actors involved in a project to view how requirements map onto solutions, at least partially, and vice versa, has a substantial potential to improve current practice (Figure 3).

Clients could, for instance, understand that their request for a *flexible building* was addressed by the design team by choosing a *steel structure with easily replaceable composite panels* and by choosing a *building footprint* that, although sub-optimal for the first phase of the project, will offer optimum land use for the future phases of project development.

Similarly the BMT tool could allow a new member of the design team that was not part of the initial design process, to find out that the reasons for choosing composite panels were the client's request for flexibility, the fire ratings requirements and the design team desire for an *industrial slick aesthetic*.

The interface of the BMT Tool as pictured in Figure 3 contains the following components:

- author and date, transparently captured by the system;
- Requested by list, names can be picked by selecting them from the project directory list.
- Notify. This function will send an automated notification to all the actors that the author names as interested or affected by that item. As for the Requested by names can be easily selected from project directory.
- Notify author of changed status. This function sends a notification to the author of an item each time the item is modified.
- New Requirement text input box. Here is where the user can input a new requirement into the briefing database using free text to describe it.
- Add new requirement. Once the item description is complete by clicking Add the item will be added to the requirements list displayed in the central part of the screen.
- Browse Requirements List. In the central part of the BMT interface the requirements submitted to the system can be viewed and browsed. Users can Edit/move up/move down and add new versions for each item and view the whole list using a scroll bar.
- Requirements Summary. In this section on the right hand side of the BMT is displayed all the information regarding the requirement item selected in the requirement browser section, on the left hand side. This summary is has four components:
    1. *generic information* about the requirement such as when it was last modified, by whom, who was notified of the changes and what is the current status of the requirement (e.g. ongoing, addressed, to be addressed etc.).
    2. *full textual description* of the selected requirement. Users can edit this description and view previous versions of the same item.

3. *linked files*. Where files linked to that item can be define and viewed to provide contextual information for the selected requirement.
4. *list of elements of the solution*. In this section all the elements of the solution that satisfy the selected requirement are listed. By selecting one of these elements, a list of all the requirements that that specific component will appear.

- Create Report. This button creates a printer friendly report of the BMT.
- Qdoc Save. This button allows saving the content of the briefing database. As every Qdoc save operation it is possible to save and archive or save in a draft folder.
- Toggle REQUIREMENT/SOLUTIONS view. This button allows switching from a requirement centred interface to a solution-focused interface. The BMT interface to navigate the components of the solution space has the same structure and features of the requirements interface.
- Items from meeting minutes that are labelled as Design Brief will by automatically added by the system to the requirements/solutions database.

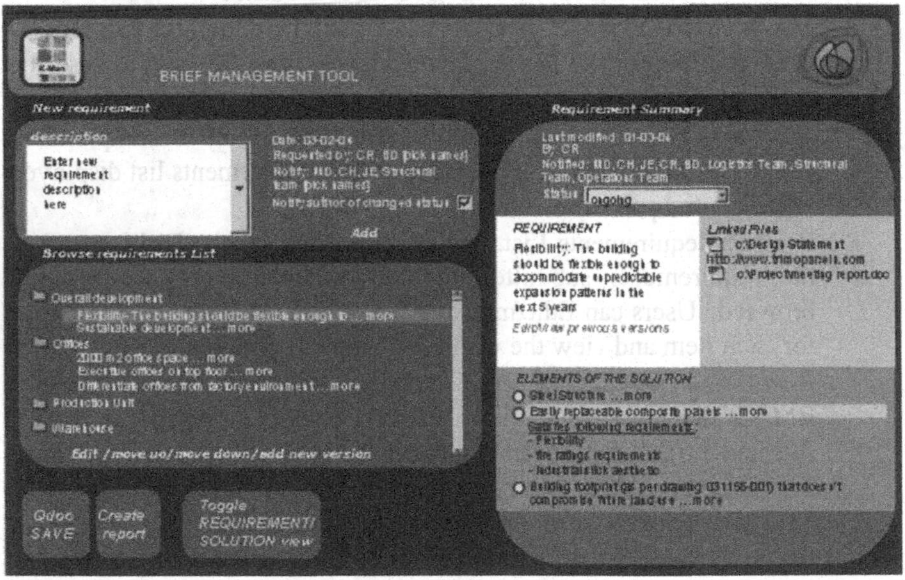

Figure 3. Brief Management Tool Interface.

4.3 Generic Search Tool

The *Generic Search Tool* is conceived to be the starting point of any search and it consists of a *quick search* functionality and an *advanced search* one. The search results are displayed taking into account the quality score of the various items and system recommendations for the specific user based on his profile.

After selecting the appropriate search strategy and parameters the user can click on the "search" button and the results of the search are displayed in a new window (Figure 4). The search results are split into three tiers:

- Projects. A list of relevant projects carried out within the practice is displayed in the first tier. In the scenario illustrated in Figure 4, displaying the results of a search for the word "Airport" the projects listed are Heathrow Terminal 1, Gatwick South Terminal, Edinburgh Air Traffic Control Tower and Farnborough Airport Competition. By selecting one item from this list, say Farnborough Airport, on the right hand side a preview summary of the project is displayed containing an image and information like project name, date, client, cost and construction materials. By clicking on the "open" button another window will be open the project homepage with a bottom frame for feedback on accuracy of the recommendations.
- Related Pebbles. Items are displayed in descending order of ranking. Again by selecting one item a preview/summary is displayed on the right hand side. Even without opening any of the items users can express interest in specific items and by doing so will cultivate their profile. If the item is opened, like for the projects, the item is displayed in a separate window with a bottom bar to provide feed-back on the quality of the recommendations.
- Contacts, with a list of the people in the practice involved in the area object of the search. Beside each name a list of the major projects they took part in for the area in question.

5. CONCLUSIONS

This paper gave a partial account of the research that led to the design and specification of the KMan suite of tools. In particular the concept of pebbles of knowledge, implicit in the KMan system, was introduced and some of the individual tools that compose KMan were described in more detail.

KMan was designed to address the problems related to the knowledge management and transfer within construction projects. Although designed for flexibility and incremental implementation, it is relatively development-intensive and therefore expensive and not affordable for small to medium practices. Further research is now being carried out into creating similar tools and functionalities to those of KMan, using freely or cheaply available software, like, for instance wikis.

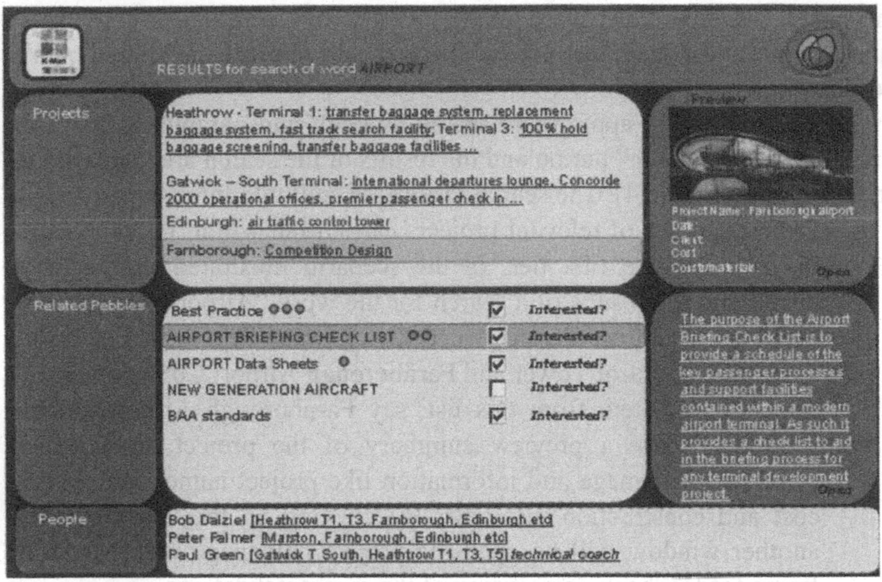

Figure 4. Search Results Interface.

6. REFERENCES

Carrillo, P. M., H. S. Robinson, et al., 2002, *Survey of Knowledge Management in Construction,* Department of Civil and Building Engineering, Loughbourough Univerisity.

Cerulli, C., 2002,"Tools for Design Decision Support: the ADS Prototype and its Successors." *DSIage2002 Decision Making & Decision Support in the Internet Age*. Doctoral Research Consortium, Cork, Ireland.

Cerulli, C., 2004, "Supporting the Knowledge Management of Design Decision Making In Architecture." *School of Architecture*. Sheffield, University of Sheffield: 257.

Cerulli, C., 2005, "Towards a Framework for Knowledge Management and Learning across Projects in Construction Design Processes: The ADS and KMan Prototypes." *The Fifth International Conference on Knowledge, Culture and Change in Organisations*, Rhodes, GR, Common Ground.

Cerulli, C., C. Peng, B. Lawson, 2001, "Capturing Histories of Design Processes for Collaborative Building Design Development: Field Trial of the ADS Prototype." *CAAD Futures*, Eindhoven, The Netherlands, Eindhoven University of Technology, pp. 427-437.

Cerulli, C., C. Peng, et al., 2002, "Design Rationale and Information Management in the Construction Domain: the outcome of the ADS Project and suggestions for future research." *Design Research Society – Common Ground Conference*, London.

Cooper, G., B. Lawson, et al., 1998, *Advanced Decision Support for the Construction Domain* (ADS).

Cooper, G. S., C. Cerulli, et al., 2005, "Tracking Decision-Making During Architectural Design." *Electronic Journal of Information Technology in Construction* (ITCON) **10**: 125-139.

Grint, K., P. Case, et al., 1995, "Business process reingeneering reappraised; the politics and technology of forgetting." In W. Orlikowski, G. Walsham, M. R. Jones and J. I. DeGross (eds) *Information technology and changes in organizational work.* London, Chapman and Hall: 62-88.

Gruber, T. R. and D. M. Russell, 1996, "Generative Design Rationale: Beyond the Record and Replay Paradigm." In T. P. Moran and J. M. Carroll (eds) *Design Rationale: Concepts, techniques and use.* Hillsdale, NJ, Lawrence Erlbaum Associates: 323-349.

Henderson, R. M. and K. B. Clark, 1990, "Architectural Innovation: The reconfiguration of Existing Product Technologies and the Failure of Established Firms." *Administrative Science Quarterly* **35**: 9-30.

Lave, J. and E. Wegner, 1991, *Situated Learning: Legitimate Peripheral Participation.* Cambridge, UK, Cambridge University Press.

Polanyi, M., 1966, *The Tacit Dimension.* Garden City, New York, Doubleday & Company, Inc.

Shipman, F., 1993, *Supporting Knowledge Based Evolution with Incremental Formalization.* Deparment of Computer Science. Boulder, CO, University of Colorado.

Shipman, F. M. I. and C. Marshall, 1999, "Formality Considered Harmful: Experiences, Emerging Themes, and Directions on the Use of Formal Representations in Interactive Systems." *Computer-Supported Cooperative Work* **8**(4): 333-352.

Shipman, F. M. I. and R. McCall, 1997, "Integrating Different Perspectives on Design Rationale: Supporting the Emergence of Design Rationale from Design Communication." *Artificial Intelligence in Engineering Design, Analysis, and Manufacturing* (AIEDAM) **2**(April 1997): 141-154.

Stenmark, D., 1999, "Asynchronous Brainstorm: An Intranet Application for Creativity." *WebNet 99*, Honolulu, Hawaii, AACE Press.

Stenmark, D., 2001, "Leveraging Tacit Organisational Knowledge." *Journal of Management Information Systems* **17**(3): 9-24.

Stenmark, D., 2002, "Information vs. Knowledge: The Role of Intranets in Knowledge Management." *35th Hawaii International Conference on System Sciences*, Hilton Waikoloa Village, Island of Hawaii.

Veshosky, D., 1998, "Managing Innovation Information in Engineering and Construction Firms." *Journal of Management in Engineering* **14**(1): 58-66.

Willmott, H. and E. Wray-Bliss, 1995, "Process reingeneering, information technology and the transformation of accountability; the remaindering of the human resource?" In W. Orlikowski, G. Walsham, M. R. Jones and J. I. DeGross (eds) *Information technology and changes in organizational work.* London, Chapman and Hall: 62-88.

Concept Formation in a Design Optimization Tool

Wei Peng and John S. Gero
Key Centre of Design Computing and Cognition
University of Sydney, Australia

Keywords: Situated agent, Concept formation, Knowledge, Design optimization tool, Design & decision support systems

Abstract: This paper presents how a situated agent model can wrap around a design optimization tool and construct concepts from interaction between the agent, the design problem and the use of the tool. The agent develops its structure and behaviour specific to what it is confronted with – its experience. As a consequence, designers can integrate their expertise with the learning results from the agent to develop design solutions. We present preliminary results.

1. INTRODUCTION

Design optimization is concerned with identifying optimal design solutions which meet design objectives while conforming to design constraints. The design optimization process involves some tasks that are both knowledge-intensive and error-prone. Such tasks include problem formulation, algorithm selection and the use of heuristics to improve efficiency of the optimization process. Many design optimization tools focus on gathering a variety of mathematical programming algorithms and providing the means for the user to access them to solve design problems. For example, Matlab Optimization Toolbox 3.0[1] includes a variety of functions for linear programming, quadratic programming, nonlinear optimization and nonlinear least squares, etc. Choosing a suitable optimizer becomes a bottleneck in a design optimization process, in which efficiency is related to how many

[1] http://www.mathworks.com/products/optimization/

Jos P. van Leeuwen and Harry J.P. Timmermans (eds.), Innovations in Design & Decision Support Systems in Architecture and Urban Planning, 293-308.

optimization cycles are involved in solving a design optimization problem. Designers rely on their experience to carry out this task. The limitation of such a manual process is that a sub-optimal design solution may result and hence an inefficient design produced. To increase efficiency, knowledge-based design optimization systems (Balachandran, 1988) have been used with the aim of leveraging design tasks that require human expertise. These knowledge-intensive programs respond reflexively to design variables based solely on their predefined knowledge, and are not able to adequately cope with the dynamics of the design process. These systems keep repeating themselves irrespective of their interactions with the design environment. The knowledge and functions are encoded in a "hard-wired" manner during the development stage.

Our objective here is to construct a computational model that is able to capture the knowledge of using the design tool, and hence to aid the tool's future use. This paper introduces an approach that uses a situated agent to wrap around a design optimization tool and to construct concepts from interactions between the agent, the design problem and the use of the tool. We demonstrate how concepts can be learned in a situated agent-based design optimization tool.

2. CONCEPT FORMATION

Concept formation is essential for an agent to develop its experience in a dynamic environment. Many researchers consider categorization the essence of a concept and its formation. Concept formation has been regarded as a process of incremental unsupervised acquisition of categories and their intentional descriptions (Fisher and Pizzani, 1991). Based on this theory, a broad spectrum of computational models has been developed, including inductive learning methods, explanation-based learning approaches and connectionist algorithms. However, theories of concept formation that merely focus on categorization are not able to address the complexity of the world (Bisbey and Trajkovski, 2005). A concept lacking an understanding of why the object, entity or event has its particular properties is called a protoconcept (Vygotskii, 1986; Bisbey and Trajkovski, 2005). We use the idea that learning a concept inherently involves understanding its influence on its environment. We believe that in the dynamic activity of designing, concepts that incrementally capture the knowledge of a design process are formed as a consequence of "situatedness" (Gero and Fujii, 2000). Interactions in designing play critical roles in shaping the design practice in which new design experiences can be learned.

2.1 Situated Concept Formation

Situatedness involves both the context and the observer's experiences and the interactions between them. Situatedness (Clancey, 1997) holds that "where you are when you do what you do matters" (Gero, 1998). A concept formation process can be regarded as the way an agent orders its experience in time, which is referred as conceptual coordination (Clancey, 1999).

A concept is a function of previously organized perceptual categories and what subsequently occurs, *Figure 1(a)*. A concept is generally formed by holding active a categorization that previously occurred (C1) and relating it now to an active categorization C2 (Clancey, 1999). *Figure 1(b)* illustrates a scenario of such a situated concept learning process in which sensory data is augmented into a Gestalt whole. Perceptual category C1 groups sensory sequence "S1 → S2" and activates the agent's experience to obtain similar organizations. The agent's experiential response (E1) represents the agent's expectations about what would happen later in the environment. The agent constructs E1 with environmental changes (S3) into current perceptual category C2. This construction involves a validation process in which environmental changes are matched with the agent's expectation. "Valid" means the environmental changes are consistent with the agent's projection of such changes from a previous time.

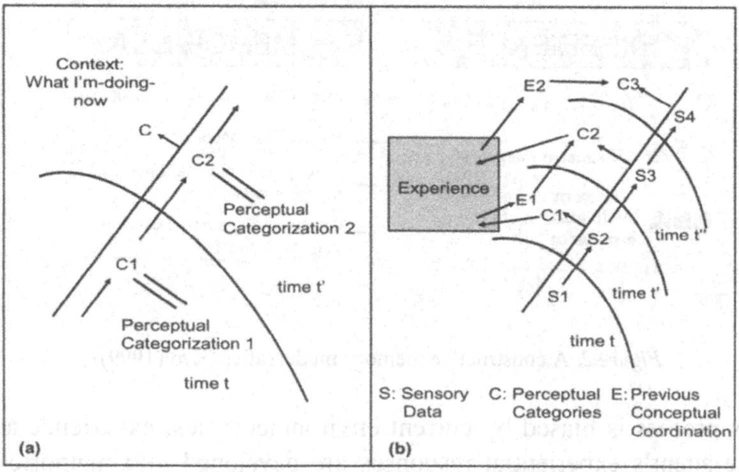

Figure 1. A situated agent learns from its interaction with the environment. (a) shows the conceptual knowledge as a higher order categorization of a sequence (after Fig. 1.6 in Clancey (1999). (b) illustrates a situated learning scenario.

The grounding process then reinforces a valid experience. For invalid expectations, the agent updates its perceptual category (C2) with the latest environmental changes. The concurrence of "situatedness" and "constructive memory" provides the basis for concept formation in a situated manner.

2.2 A Constructive Memory Model

This notion of "constructive memory" contradicts many views of knowledge as being unrelated to either its locus or application (Gero, 1998). A memory can be regarded as a learning process. "Memories are constructed initially from that experience in response to demands for a memory of that experience but the construction of the memory includes the situation pertaining at the time of the demand for the memory" (Gero, 1999). As shown in *Figure 2*, the original experiences (○) are knowledge structures that represent previous memories. Memories (◍, ●, ▲) are constructed from experiential responses to active cues which are demands for memories of current sensory data.

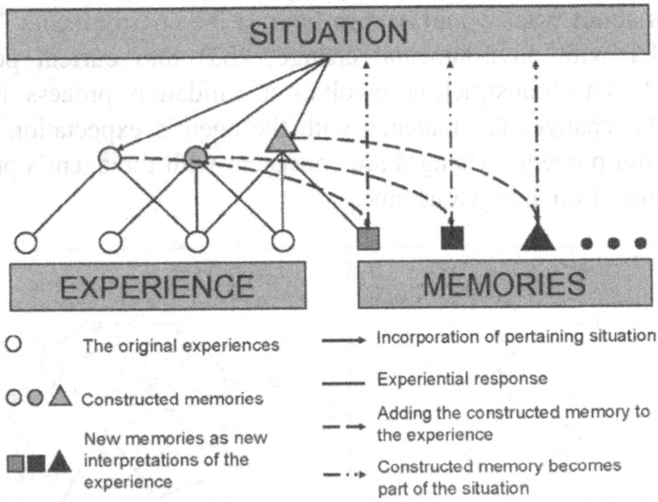

Figure 2. A constructive memory model (after Gero (1999)).

This process is biased by current environment cues, experience and the way the agent's experiential responses are developed into memories from interactions between the agent and its environment. New memories (■, ■, ▲) which are new interpretations of the augmented experience are added to the experience by an experiential grounding process (Gero, 1999; Liew, 2004). In this way, a newly developed experience (■) can be subsequently used to create a newer memory (▲) which in turn grounds itself into another new experience (▲). An agent constructs its memories through its internal representation processes triggered both from external "data-driven" and goal-related "expectation-driven" demands (Gero and Fujii, 2000). A "data-driven" process is triggered by environment changes (i.e. events in the environment). Sensory data are the lowest level of descriptions of environmental changes. They capture the real-time environment stimuli in a format that can be processed by the agent. Sensory data are then mapped into various modalities in producing percepts. Percepts are data structures that can be processed by the agent to generate memory cues. According to its experiential responses to memory cues, an agent commences "expectation-driven" learning processes. In its "expectation-driven" learning processes, the agent validates its hypotheses in interactions. "Data-driven" and "expectation-driven" learning processes are coordinated in this constructive memory model, through which conceptual knowledge can be learned. A memory is activated, reactivated or constructed depending on situations encountered. We now introduce a system architecture via which the agent organizes its experience in time to form concepts.

2.3 A Situated-agent Based Design Optimization Tool

Figure 3 shows the general architecture of a situated agent-based design optimization tool.

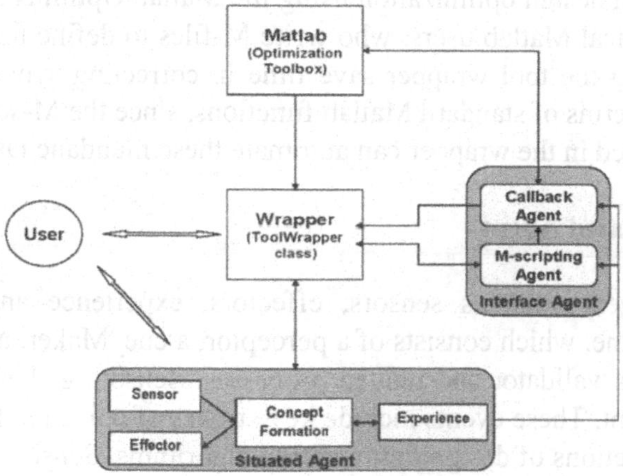

Figure 3. A situated agent-based design optimization tool.

A situated agent wraps around an existing design optimization tool. A user accesses a design tool via a wrapper, where the situated agent senses the events performed by that user. The situated agent uses its experience and concept formation engine to generate a concept, which improves the tool's behaviour in designing. The user can also directly communicate with the agent to obtain additional information. Such a framework provides the means that allows the agent to incrementally learn new experiences. The system consists of two major components: a situated agent and a tool platform which includes a design optimization tool, a tool wrapper and interface agents.

2.3.1 The Tool Platform

In this research, Matlab Optimization Toolbox (version 3.0.1) is chosen as the design optimization platform. It is a collection of functions that extend the capability of the MATLAB numeric computing environment (Release 14). The toolbox includes routines for a variety of optimization classes including unconstrained and constrained nonlinear minimization, quadratic and linear programming, and nonlinear optimization. Via the MATLAB command line, Matlab users use a scripting language called M-file to define and to solve optimization models.

The interface agent, which consists of a Callback agent and a M-scripting agent, enables both users and the situated agent to operate on the engines in the Matlab Optimization Toolbox. A tool wrapper serves as an interface between the user, the tool and agents. It provides a simplified and efficient way to perform design optimization using the Matlab Optimization Toolbox. Unlike the typical Matlab users, who write M-files to define their problems, designers using the tool wrapper save time in correcting syntax errors and searching for terms of standard Matlab functions, since the M-scripting agent that is embedded in the wrapper can automate these mundane tasks.

2.3.2 Situated Agent

A situated agent contains sensors, effectors, experience and a concept formation engine, which consists of a perceptor, a cue_Maker, a conceptor, a hypothesizer, a validator and related processes. Sensors gather events from the environment. These events include key strokes of objective functions and the users' selections of design optimization algorithms. Sensor data takes the form of a sequence of actions and their initial descriptions. For example, some sensor data can be expressed as:

- S (t) {...... "click on objective function text field", key stroke of "x", "(", "1", ")", "+", "x", "(, "2", ")"...}.

The perceptor processes sensor data and groups them into multimodal percepts, which are intermediate data structures of environment states at a particular time. Percepts are structured as triplets:

- E (Object, Property, Values of properties).

For example, perceptual data P1 can be described as (Objective Function Object, Objective_Function, "x(1)+x(2)"). The cue_Maker generates cues that can be used to activate the agent's experience. The conceptor categorizes the agent experience to form concepts. Concepts attach meanings to percepts.

The hypothesizer generates hypotheses from the learned concepts. This is where reinterpretation takes place in allowing the agent to learn in a "trial and error" manner. The validator pulls information from the environment and examines whether the environmental changes are consistent with the agent's responses. An agent needs to validate its hypotheses in interactions to locate a suitable concept for the current situation. An effector is the unit via which the agent brings changes to environments through its actions.

2.3.3 The Agent's Experience and Grounding

The agent's experience is structured as a Constructive Interactive Activation and Competition (CIAC) neural network, which is an extension of a basic IAC network (McClelland 1981, 1995). The CIAC network is implemented in Java using the Repast[2] library. An IAC network consists of two basic nodes: instance nodes and property nodes. The instance node has inhibitory connections to other instance nodes and excitatory connections to the relevant property nodes. The property nodes encode the special characteristics of an individual instance (Medler, 1998). Property nodes are grouped into cohorts of mutually exclusive values. Each property node represents the perceptual level experience which is processed from sensory data. Instance nodes along with the related property nodes describe an instance of a concept. Knowledge is extracted from the network by activating one or more of the nodes and then allowing the excitation and inhibition processes to reach equilibrium (Medler, 1998). Such knowledge is a dynamic construction in the sense that the agent develops adapted experience as environment stimuli change. The experience grounds via weight adaptation and constructive learning.

Experiential grounding is the process that verifies the usefulness of a related experience in current situation (Liew, 2004). A grounding process tests whether a constructed memory correctly predicts environmental changes. Grounding via weight adaptation adjusts the weights of each excitatory connection of the valid concept, so that those nodes that fired together become more strongly connected. Weight adaptation is formulated similar to a Hebbian-like learning mechanism (Medler, 1998).

[2] available from http://repast.sourceforge.net/

Grounding via constructive learning incorporates new instances or reconfigures existing instances. The conceptor performs conceptual labelling in the conception process, in which the agent uses its experience and applies various inductive learners to attach meanings to newly formed memories.

2.4 Concept Formation from Memory Activation, Reactivation and Validation

Memory activation from the agent's reflexive experience is represented by solid arrowed lines in *Figure 4*. An agent reflexes when its experiential response to the current memory cue is sufficiently strong to directly influence the action. Pulling environmental changes and comparing these changes with the activated memory, the agent is able to evaluate the experience of that memory.

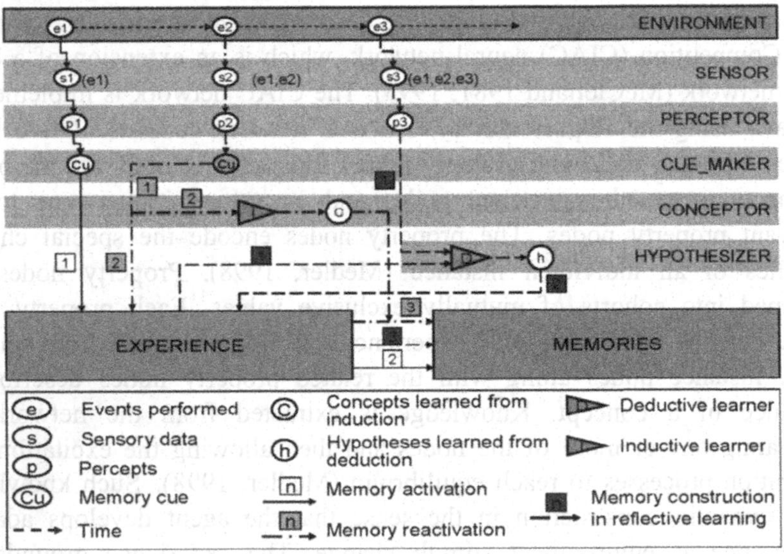

Figure 4. Memory activation, reactivation and construction in a constructive memory model.

Memory reactivation occurs when the agent's initial experiential response to a memory cue fails in the validation process. A memory can be reactivated when a memory cue is able to be subsequently identified in the environment. This process is illustrated as "Long Dash Dot" lines in *Figure 4* where sensor data (e1, e2) can reactivate a memory of an experience. A concept is a hierarchical structure of a design instance, related design features and their generalized descriptions. It can be obtained from this reactivated memory. Validation allows a concept to be evaluated in interactions. An agent grounds the activated or reactivated memories with positive validation value into its experience via weight adaptation ("Dash Dot" lines in *Figure 5*). In this way, a concept is learned from grounded memories that are activated or reactivated from environmental demands of such memories in interactions.

2.5 Memory Construction in Reflective Concept Learning

Reflective learning is activated by discrepancies between the agent's expectation and current environmental changes. In *Figure 4*, "Long Dash" lines show how a memory is formed reflectively. The agent's perceptual experience captures the knowledge of sequencing, in which lists of design events are grouped into a design instance node. The agent's conceptual experience (or label) holds regularities or commonalities that share among a group of design instances. Such abstract structures can be re-usable in more than one context without having to repeat all the details in every context. The conceptual experience (or label) provides domain theories for a deductive learner (i.e. an explanation-based learner). A deductive learner is employed to generate hypotheses for the failure of the agent's experience for environmental changes. The agent draws its goal concept and the explicitly represented domain theories to deduce an explanation for the current sensory data. Backward-chaining reasoning is implemented here.

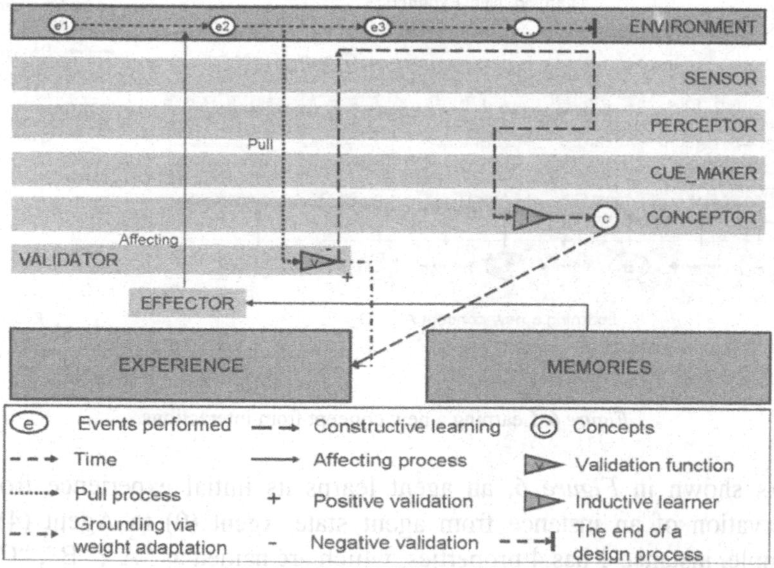

Figure 5. Validation and grounding processes in a constructive memory model.

2.6 Learning Scenarios in Interactions

We use the following five internal states and their changes to illustrate how an agent constructs concepts from its interactions with the environment:

- Conceptual experience is the agent's high-level experience which are domain theories it uses to classify and explain its observations;
- Expectations about environmental changes are generated by the agent's experiential responses to environmental cues;
- Validator status shows whether an agent's expectation is consistent with the environment changes;
- Hypotheses depict the agent's reinterpretation about its failures in creating a valid expectation.
- Experience is structured as a Constructive Interactive Activation and Competition (CIAC) neural network, which is composed of instance nodes connecting to a number of property (or feature) nodes.

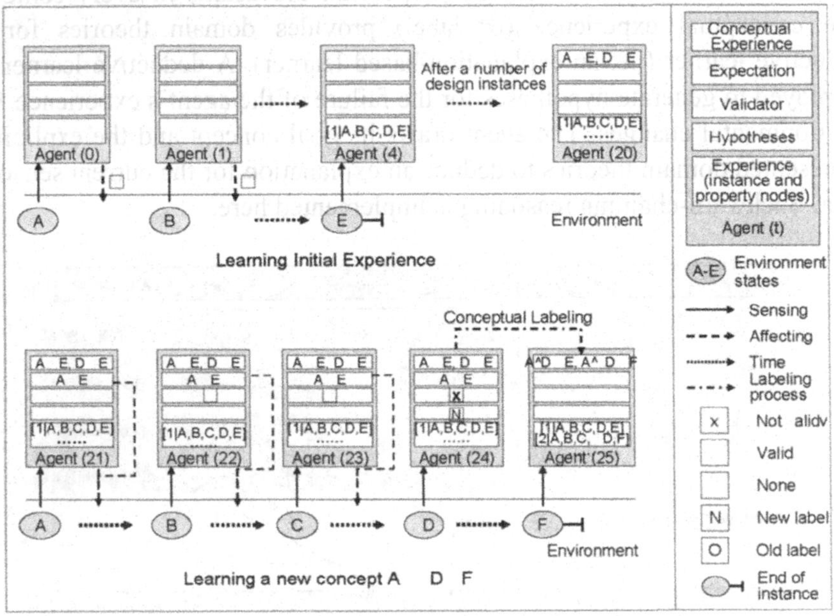

Figure 6. Learning a new concept from interactions.

As shown in *Figure 6*, an agent learns its initial experience from an observation of an instance from agent state Agent (0) to Agent (4). For example, instance 1 has 4 properties, which are named as "A", "B", "C" and "D". Provided a goal concept E, the agent learns its initial experience from a constructive learning process. The learned experience is expressed as

"[Instance No| Properties, Goal Concept]", i.e. "[1|A,B,C,D,E]" for the agent's state Agent (4). After a number of design instances being observed, the agent is able to generalize a conceptual knowledge from its self-labeling process, i.e., "A→E" and "D→E". "A→E" means that a feature of "A" leads to a goal concept "E". These conceptual labels serve as domain theories, upon which the agent is based to create expectations and hypotheses.

The bottom half of *Figure 6* shows the agent constructing new concepts "A^D→E" and "A^¬D→F". In Agent (21), an experiential response "A→E" is created and used to affect the environment. In the process of moving from Agent (22) to Agent (24), the agent validates its expectation. When the agent fails in this validation process at Agent (24), it creates a hypothesis for such a failure. In this example, the hypothesis shows that the agent may face a new concept. As a result, the agent subsequently activates a grounding process, in which two new conceptual labels ("A^D→E, A^¬D→F") and a new instance are incorporated into the agent's experience.

Figure 7 illustrates a detailed example of how new concepts of selecting design optimization algorithms are formed.

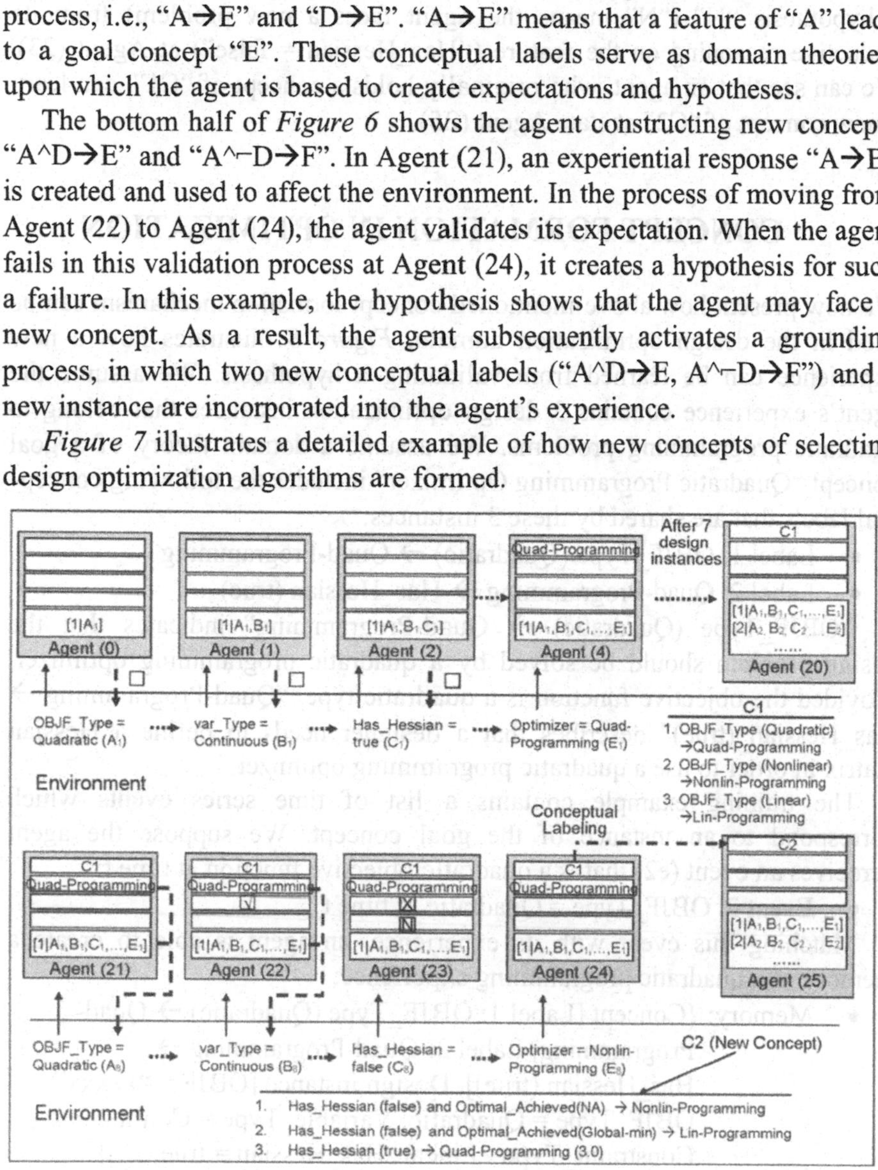

Figure 7. An example of concept formation in design optimization ("OBJF_Type" refers to objective function type, "var_Type" and "cons_Type" denote variables and constraints type, "Has_Hessian" shows if there is a Hessian function defined).

The upper part of the diagram describes a scenario that an agent learns a new concept "C1" from a constructive learning process. As illustrated in the lower part of *Figure 7*, the agent reinterprets its environment and generates a hypothesis "N" ("N" means the agent faces a new problem) from its deductive reasoning on the feature ("Has_Hessian = false") at Agent (23). We can see that an agent, who originally holds a concept of "C1", can learn a new concept of "C2" at state Agent (25).

3. CONCEPT FORMATION IN OPTIMIZATION

We now present how above-mentioned concept formation mechanism can be used in the design optimization domain. *Figure 8* illustrates how a new experience can be learned from validating a hypothesis. We assume the agent's experience contains 3 design optimization instances that belong to quadratic programming problems. We assume a domain theory of a goal concept "Quadratic Programming Optimizer" includes the following conceptual labels that are shared by these 3 instances:

- Label 1: OBJF_Type (Quadratic) → Quad-Programming
- Label 2: Quad-Programming → Has_Hessian (true)

"OBJF_Type (Quadratic) → Quad-Programming" indicates that the design problem should be solved by a quadratic programming optimizer, provided the objective function is a quadratic type. "Quad-Programming → Has_Hessian (true)" describes that a designer needs to define a Hessian matrix in order to use a quadratic programming optimizer.

The training example contains a list of time series events which correspond to an instance of the goal concept. We suppose the agent perceives an event (e2) that is a quadratic objective function at time t:

- Event 2: OBJF_Type = Quadratic at time t

Matching this event with its experience, an agent is able to create a memory of a quadratic programming experience:

- Memory: {Concept [Label 1: OBJF_Type (Quadratic) → Quad-Programming; Label 2: Quad-Programming → Has_Hessian (true)], Design instance [OBJF = "xxxxx", OBJF_Type = Quadratic, Variable_Type = Continuous, Constraint_Type= Linear, Has_Hessian = true, …]}

As shown in *Figure 8*, such a memory fails the validation process since the event e_m – "Has_Hessian (false)" violates the agent's memory of a quadratic programming experience. Based on conceptual label 1 and 2, the hypothesizer creates explanations:

- Hypothesis 1: {Not a linear programming concept because [OBJF_Type = Quadratic]}
- Hypothesis 2: {Not a quadratic programming concept because [Has_Hessian = false]}

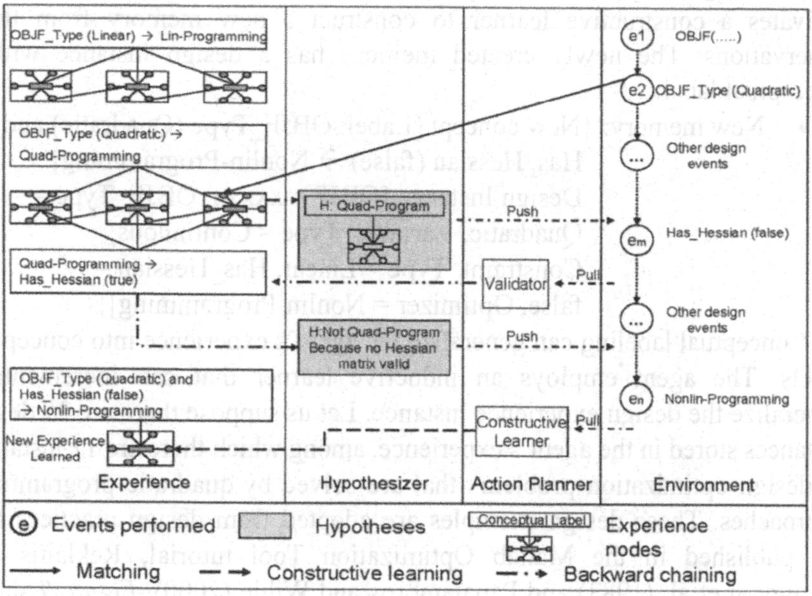

Figure 8. New concept learning from interactions.

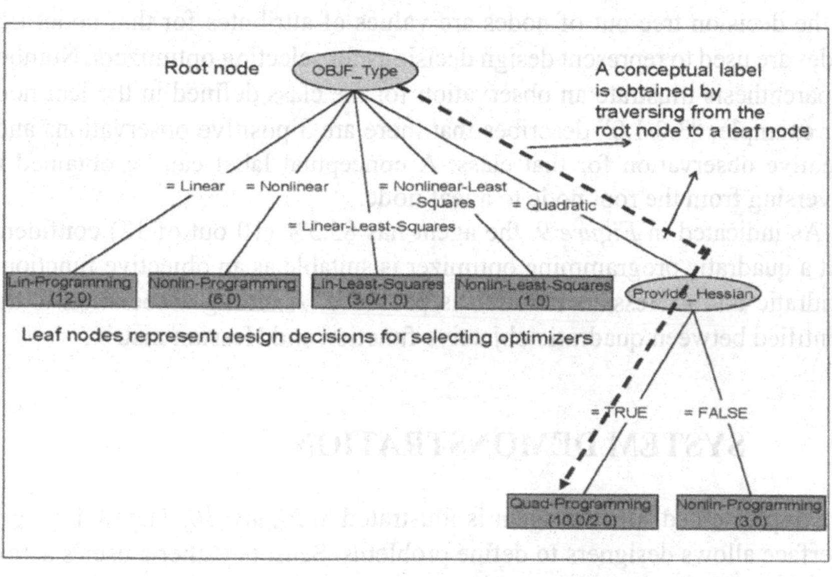

Figure 9. Conceptual labels learned from an inductive learner that uses C4.5 from WEKA[3].

[3] www.cs.waikato.ac.nz/ml/weka/

If no agent experience gives a positive answer, the action planner activates a constructive learner to construct a new memory from these observations. The newly created memory has a design instance with a conceptual label:

- New memory: {New concept [Label: OBJF_Type (Quadratic) and Has_Hessian (false) → Nonlin-Programming]
  Design Instance [OBJF = xxxxx, OBJF_Type = Quadratic, Variable_Type = Continuous, Constraint_Type = Linear, Has_Hessian = false, Optimizer = Nonlin-Programming]}

Conceptual labeling can generalize the agent's experience into conceptual labels. The agent employs an inductive learner that can incrementally generalize the design experience instance. Let us suppose there are 35 design instances stored in the agent's experience, among which there are 12 instances of design optimization problems that are solved by quadratic programming approaches. These design examples are adopted from design practices that are published in the Matlab Optimization Tool tutorial, Reklaitis and Ravindran et al. (1983) and Papalambros and Wilde (2000). *Figure 9* shows the learning results and performance of applying a decision tree learner to the agent's experience. Each non-leaf node stands for a test on an attribute. Edges of the decision tree out of nodes are values of attributes for that node. Leaf nodes are used to represent design decisions for selecting optimizers. Numbers in parenthesis illustrate an observation for the class defined in the leaf node. For example, "3.0/1.0" describes that there are 3 positive observations and 1 negative observation for that class. A conceptual label can be obtained by traversing from the root node to a leaf node.

As indicated in *Figure 9*, the agent has 83.3% (10 out of 12) confidence that a quadratic programming optimizer is suitable as an objective function is quadratic and a Hessian function is provided. A strong association is thus identified between quadratic objective function and Hessian matrix.

4. SYSTEM DEMONSTRATION

The implemented situated agent is illustrated in *Figure 10*. The tool wrapper interface allows designers to define problems. Sensors gather a user's actions that comprise a design optimization process and activate a perceptor to create percepts. A percept cues the agent's initial experience. Activation diagrams output the neurons winning at the equilibrium state, which represent the activated memory. Based on the responses from the CIAC neural net, the agent constructs initial concepts and displays the constructed knowledge in the tool wrapper. The grounding process initiates a validation function which matches the initially constructed concepts with environmental changes. Weight adaptation increases connection weights of the valid concept and grounds experience A to experience B. In the agent's reflective concept learning process, the explanation-based learner is used to form a new concept. A percept at runtime can also be developed as a new concept by a constructive learning process. Experience C is learned from constructive learning and the related self conceptual labeling process.

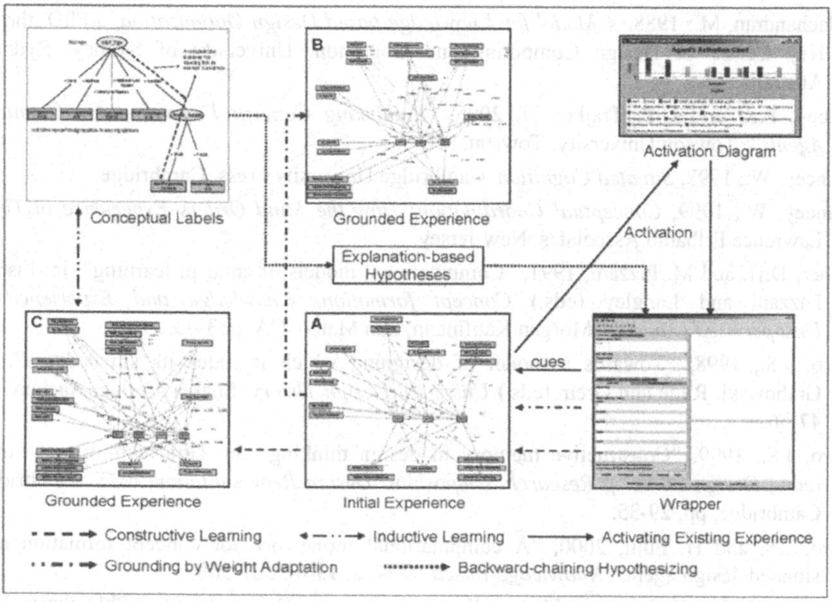

Figure 10. A prototype design optimization system that learns by its use.

5. CONCLUSION AND FUTURE RESEARCH

This paper demonstrates a prototype design optimization system that can learn through its use. From its concept formation processes, the agent develops its structure and behaviour specific to what it is confronted with – its experience. Based on the conceptual knowledge learned, the agent can further improve the behaviour of the tool. As a result, designers can integrate their expertise with the knowledge learned from the agent to develop design solutions. A tool can incrementally learn by their uses. Such a tool has the potential to improve the efficiency of a design optimization tool by reducing the number of design cycles. Future research will focus on training and testing the implemented system.

6. ACKNOWLEDGEMENT

This work is supported by a Cooperative Research Centre for Construction Innovation (CRC-CI) Scholarship and a University of Sydney Sesqui R and D grant.

7. REFERENCES

Balachandran, M.: 1988, *A Model for Knowledge-based Design Optimization*, a PhD. thesis, Key Centre of Design Computing and Cognition, University of Sydney, Sydney, Australia.

Bisbey, P.R. and G.P. Trajkovski, 2005, "*Rethinking Concept Formation for Cognitive Agents*", Towson University, Towson, MD.

Clancey, W., 1997, *Situated Cognition*, Cambridge University Press, Cambridge.

Clancey, W., 1999, *Conceptual Coordination: How the Mind Orders Experience in Time*, Lawrence Erlbaum Associates, New Jersey.

Fisher, D.H. and M. Pizzani, 1991, "Computational models of concept learning", in: Fisher, Pazzani and Langley (eds.) *Concept formation: Knowledge and Experience in Unsupervised Learning*, Morgan Kaufmann, San Mateo, CA, p. 3-43.

Gero, J.S., 1998, "Towards a model of designing which includes its situatedness", in: Grabowski, Rude and Grein (eds.) *Universal Design Theory*, Shaker Verlag, Aachen, pp. 47-56.

Gero, J.S., 1999, "Constructive memory in design thinking", in: Goldschmidt and Porter (eds.) *Design Thinking Research Symposium: Design Representation*, MIT, Cambridge, Cambridge, pp. 29-35.

Gero, J.S. and H. Fujii, 2000, "A computational framework for concept formation in a situated design agent", *Knowledge-Based Systems*, **13**(6): 361-368.

Liew, P., 2004, *A Constructive Memory System for Situated Design Agents*, a PhD. thesis, Key Centre of Design Computing and Cognition, University of Sydney, Sydney, Australia.

McClelland, J.L., 1981, "Retrieving general and specific information from stored knowledge of specifics", in: *Proceedings of the Third Annual Meeting of the Cognitive Science Society*, Erlbaum, Hillsdale, NJ, pp. 170-172.

McClelland, J.L., 1995, "Constructive memory and memory distortion: a parallel distributed processing approach", in: Schacter (eds.) *Memory Distortion: How Minds, Brains, and Societies Reconstruct the Past*, Harvard University Press, Cambridge, Massachusetts, pp. 69-90.

Medler, D.A., 1998, "A brief history of connectionism", *Neural Computing Surveys*, **1**(1): 61-101.

Papalambros, P.Y. and D.J. Wilde, 2000, *Principles of Optimal Design: Modeling and Computation,* Cambridge University Press 2000, Cambridge, UK.

Reklaitis, G.V., A. Ravindran, and K.M. Ragsdell, 1983, *Engineering Optimization Methods and Applications,* John Wiley & Sons, Inc.

Vygotskii, L.S., 1986, *Thought and Language*, MIT Press, Cambridge, Mass.

A Framework for Situated Design Optimization

John S. Gero and Udo Kannengiesser
Key Centre of Design Computing and Cognition
University of Sydney

Keywords: Design optimization, Situatedness

Abstract: This paper presents a framework for situated design optimization that expands the traditional view of design optimization. It is based on the notion of interaction providing the potential for modifications of various aspects of the optimization process: problem formulation, the optimization tool, the designer and ultimately the result. In contrast to other approaches, these modifications can drive further interactions within the same optimization process. We use parts of the situated function-behaviour-structure (FBS) framework as an ontological basis to describe the effects of intertwined interactions and modifications on the state space of ongoing optimization processes.

1. INTRODUCTION

Optimization in designing is a process that aims to find the best design solution with respect to a selected set of performance criteria (Papalambros and Wilde, 2000). Optimization models are often represented in terms of a set of formal mathematical expressions:

$$\text{Given design variables } x \in \chi \subseteq \Re^n \quad (1)$$
$$\text{and constraints } h(x) = 0 \quad (2)$$
$$g(x) \leq 0 \quad (3)$$
$$\text{optimize objective } f(x) \quad (4)$$

Expression (1) defines the set of design variables as a subset of the n-dimensional real space $\Re^n$. The functions h(x) and g(x) specified by equations (2) and (3) represent equality and inequality constraints, respectively. They correspond to a set of specific performances that determine the feasibility or

Jos P. van Leeuwen and Harry J.P. Timmermans (eds.), Innovations in Design & Decision Support Systems in Architecture and Urban Planning, 309-324.

acceptability of the design. Finally, expression (4) specifies an objective function f(x) to be optimized. This function defines the set of performances that determine the optimality of the design.

Table 1 shows how these mathematical descriptions map onto a common ontological representation of design objects, namely Gero's (1990) function-behaviour-structure (FBS) ontology.

Table 1. Mappings between a mathematical and an ontological model of design optimization (S = structure; B = behaviour; B^c = behaviour specified by constraints; B^o = behaviour specified by objective function).

Mathematical model	Ontological (FBS) model	
$x \in \chi \subseteq \Re^n$	S	
$h(x) = 0$	S, B^c	
$g(x) \leq 0$		
f(x)	B^o	$B = B^c \cup B^o$

Structure (S) consists of the components of a design object as well as the relationships among these components. In most design domains, structure (S) comprises the geometrical, topological and material properties of the object. Table 1 shows that the structure state space (S) of the optimum design can be mapped on the given set of design variables. Some of the constraints specify ranges of values for these variables.

Behaviour (B) is derivable from structure (S) and specifies the performance of a design object. In the optimization model, behaviour (B) is embodied in the objective function and in some of the constraint functions. Accordingly, the behaviour state space (B) of the optimum design can be viewed as the union of the set of behaviour variables defined by the objective function (Bo) and the set of behaviour variables defined by some of the constraints (Bc).

Function (F), as the teleology of a design object, is not included in design optimization models.

The FBS ontology has been used as the basis for modelling design processes as a set of transitions between function, behaviour and structure (Gero, 1990). This model is known as the function-behaviour-structure (FBS) framework. Design optimization is a specific class of design process that can be subsumed in the FBS framework, Figure 1. The FBS framework distinguishes between expected behaviour (Be) and behaviour derived from structure (Bs). Be represents a set of performance criteria used as benchmarks for the design structure (S). Bs is the set of performances that are measured or derived from structure (S). Values for Bs must be within the ranges set by Be.

While the FBS framework encompasses eight fundamental processes, the scope of optimization generally comprises only three of them, namely synthesis (labelled 2 in Figure 1), analysis (labelled 3) and evaluation (labelled 4). Synthesis in optimization generates a candidate solution as an instantiation of a point within the structure (S) state space. Analysis derives the values of the relevant behaviours of that solution using the given set of objective functions and constraints. Evaluation then compares the behaviour of different candidate solutions and decides on either continuing or stopping the search for an optimum. Most design optimization problems require iterative procedures involving large numbers of synthesis-analysis-evaluation cycles.

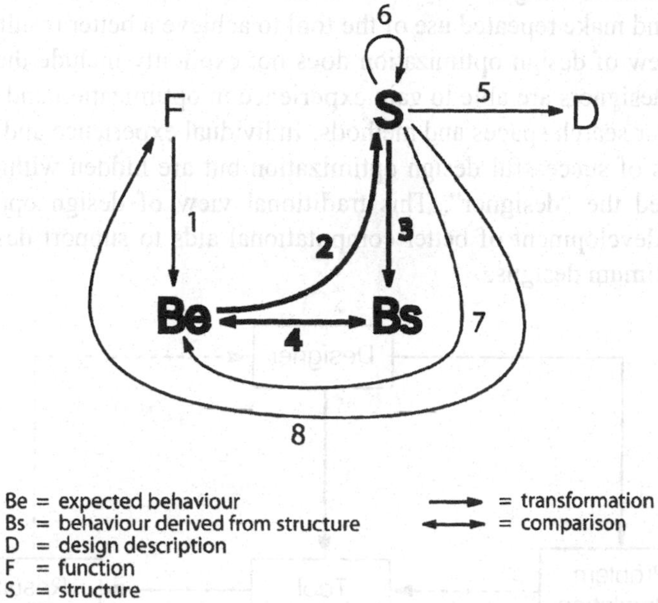

Figure 1. The processes involved in design optimization (highlighted) as a subset of the eight processes in Gero's (1990) framework of designing: (1) formulation, (2) synthesis, (3) analysis, (4) evaluation, (5) documentation, (6) reformulation type 1, (7) reformulation type 2, (8) reformulation type 3.

An extensive body of work exists in the development and application of optimization techniques in design across a large number of design disciplines (Wilde, 1978; Papalambros and Wilde, 2000; Pardalos and Resende, 2002; Parmee and Hajela, 2002). Most research in this area focuses on improving the speed of search for optimal designs and refining the quality of the optimal designs. This has resulted in various new optimization techniques. Recent advances are having only a marginal effect on the efficiency of most design optimization processes. This is mainly due to the following reasons:

- lack of transfer of earlier results as the design changes
- lack of domain knowledge in computational tools

- lack of task knowledge in computational tools
- lack of feedback into process strategies in the tool

These shortcomings are due to a view of design optimization that is too narrow, Figure 2. In this view, the designer starts optimization by selecting a computational tool with appropriate search methods for the particular optimization task and by formulating the problem in a tool-specific form. The tool then produces a result in terms of the structure and behaviour of an optimum design and optionally a set of post-optimality analyses. Based on that result, the designer may then reformulate the problem or tool/method selection and make repeated use of the tool to achieve a better result.

This view of design optimization does not explicitly include the grounds on which designers are able to gain experience in optimization and to decide on particular search spaces and methods. Individual experience and skills are key factors of successful design optimization but are hidden within a black box labelled the "designer". This traditional view of design optimization limits the development of better computational aids to support designers in finding optimum designs.

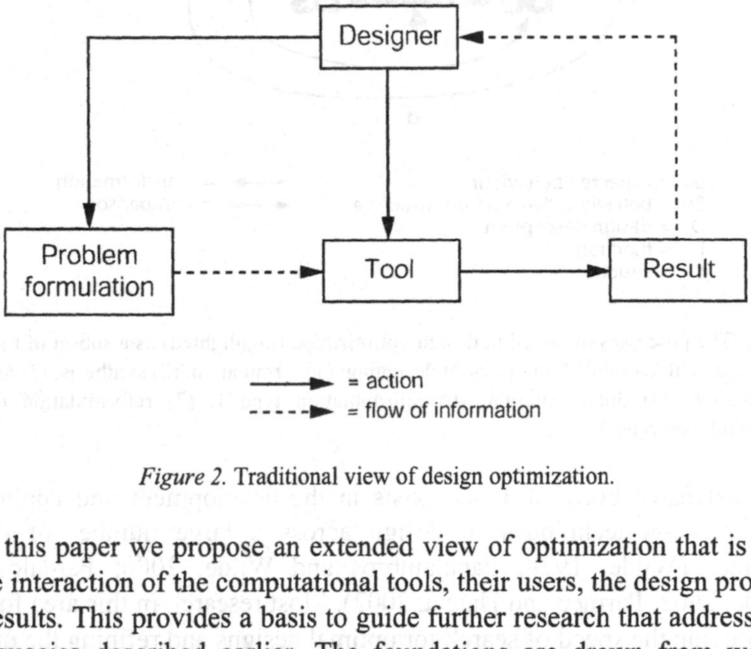

Figure 2. Traditional view of design optimization.

In this paper we propose an extended view of optimization that is based on the interaction of the computational tools, their users, the design problems and results. This provides a basis to guide further research that addresses the inadequacies described earlier. The foundations are drawn from work in situated cognition (Clancey, 1997). We then explore our situated view of design optimization using the FBS schema, which provides a basis for developing new optimization tools that can flexibly and efficiently reason about their interactions and adapt to their use.

2. SITUATEDNESS AND THE NOTION OF INTERACTION

Designing is an activity during which designers perform actions in order to change their environment. By observing and interpreting the results of their actions, they then decide on new actions to be executed on the environment. This means that the designer's concepts may change based on what they are "seeing", which itself is a function of what they have done. One may speak of an "interaction of making and seeing" (Schön and Wiggins, 1992). This interaction between the designer and the environment strongly determines the course of designing. This idea is called situatedness, whose foundational concepts go back to the work of Dewey (1896) and Bartlett (1932).

In experimental studies of designers phenomena related to the use of sketches, which support this idea, have been reported. Schön and Wiggins (1992) found that designers use their sketches not only as an external memory, but also as a means to reinterpret what they have drawn, thus leading the design in a new direction. Suwa et al. (1999) noted, in studying designers, a correlation of unexpected discoveries in sketches with the invention of new issues or requirements during the design process. They concluded that "sketches serve as a physical setting in which design thoughts are constructed on the fly in a situated way".

An idea that fits into the notion of situatedness has been proposed by Dewey in 1896 (Clancey, 1997) and is today called constructive memory. Its relevance in the area of design research has been shown by Gero (1999). Constructive memory is best exemplified by a quote from Dewey via Clancey: "Sequences of acts are composed such that subsequent experiences categorize and hence give meaning to what was experienced before". The implication of this is that memory is not laid down and fixed at the time of the original sensate experience but is a function of what comes later as well. Memories can therefore be viewed as being constructed in response to a specific demand, based on the original experience as well as the situation pertaining at the time of the demand for this memory. Therefore, everything that has happened since the original experience determines the result of memory construction. Each memory, after it has been constructed, is added to the existing knowledge (and becomes part of a new situation) and is now available to be used later, when new demands require the construction of further memories. These new memories can be viewed as new interpretations of the augmented knowledge.

The advantage of constructive memory is that the same external demand for a memory can potentially produce a different result at a later time, as newly acquired experiences may take part in the construction of that memory. Constructive memory can thus be seen as the capability to integrate new experiences by using them in constructing new memories. As a result, knowledge "wires itself up" based on the specific experiences it has had, rather than being fixed, and actions based on that knowledge can be altered in the light of new experiences.

Situated designing uses first-person knowledge grounded in the designer's interactions with the environment (Bickhard and Campbell, 1996; Clancey, 1997; Ziemke, 1999; Smith and Gero, 2005). This is in contrast to static approaches that attempt to encode all relevant design knowledge prior to its use. Evidence in support of first-person knowledge is provided by the fact that different designers are likely to produce different designs for the same set of requirements. The same designer is likely to produce different designs at different points in time even though the same requirements are presented. This is a result of the designer acquiring new knowledge while interacting with their environment.

Gero and Kannengiesser (2004) have modelled situatedness as the interaction of different worlds, including the designer's internal and external world. The internal world contains the designer's experience and goals, while the external world consists of representations of things outside of the designer. Each world can bring about changes in the other world. In addition, constructive memory provides the opportunity to modify the internal world without the need for a changed external environment. The foundation of this ability is established by the designer interacting with their memories.

The notion of interaction has been shown to be central in the concept of situatedness. Interaction may also play a role in non-situated views of the (design) world; here, however, this notion is generally interpreted in terms of a simple feedback loop to inform the actions of a well-defined system that remains itself unchanged. Changes are restricted to take place only in the external environment. In contrast, a situated view allows both entities engaged in an interaction to be affected by change. This creates the potential to emerge new situations for both external and internal environments that could not have been possible with the static, non-situated model.

An important aspect in situated interaction is the notion of interpretation. Rather than being conceived of as a simple flow of information, interpretation is understood as a form of action, originating from both the external and internal environment and resulting in changes in the internal environment. Gero and Fujii's (2000) model of interpretation as intertwined *push-pull* processes can be understood in this sense. A *push process* can be seen as an action driven by the external world aiming to change the internal world according to the data provided by the external world. A *pull process* can be seen as an action driven by the internal world aiming to change itself according to biases provided by the current expectations and goals. The notion of re-interpretation is often used to denote interpretations in which a changed internal world is the major driver of self-directed ("pull") actions leading to substantial changes in the same internal world.

3. AN INTERACTION-BASED VIEW OF DESIGN OPTIMIZATION

Integrating the notion of interaction into a model of design optimization addresses the shortcomings identified in Section 1, as it provides the opportunity for change – both in the internal "knowledge" of an optimization system and in the design it is operating on. This is a condition for future optimization tools to flexibly acquire task knowledge and domain knowledge that is adapted to the classes of optimization problems they are exposed to.

Figure 3 introduces the concept of interaction as the key element of a design optimization system. It establishes the relationships between the designer, the problem formulation, the tool and the result, which were connected previously by unidirectional arrows representing flows of information and action, Figure 2. This concept broadens the scope of optimization as a system in which every component has the potential to induce changes in other components as well as to modify itself. Static flows of information and action may now be viewed as being subsumed in the more general notion of interaction.

The interaction component in Figure 3 stands for three classes of interactions:

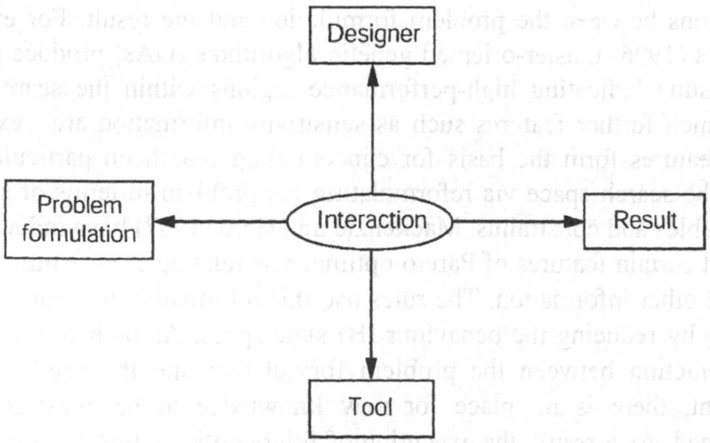

Figure 3. Interaction-based view of design optimization.

1. Interactions involving only one component: These interactions involve the designer or, alternatively, the tool as a component

that interacts with itself. This includes the concept of reflective reasoning about one's own actions.

2. Interactions involving two components: Here all pairwise combinations of individual components can potentially interact. These are:
    a. interactions between the designer and the problem formulation
    b. interactions between the designer and the tool
    c. interactions between the designer and the result
    d. interactions between the problem formulation and the result
    e. interactions between the tool and the problem formulation
    f. interactions between the tool and the result
3. Interactions involving more than two components: These interactions can be viewed as involving composites rather than individual components. Numerous composites and interactions among them are conceivable. An important class of composites represents processes as they can be viewed as input-transformation-output triplets (Gero and Kannengiesser, 2006). A typical triplet consists of the problem formulation (input), the tool (transformation) and the result (output). This triplet represents a search process with which the designer or the tool may interact.

A number of previous models and systems of design optimization can be recast into such an interaction-based approach. Some of them map onto interactions between the problem formulation and the result. For example, Parmee's (1996) cluster-oriented genetic algorithms (GAs) produce preliminary results indicating high-performance regions within the search space from which further features such as sensitivity information are extracted. These features form the basis for concentrating search on particular areas within the search space via reformulating the problem in terms of structure (S) variables and constraints. Mackenzie and Gero (1987) have induced rules to detect certain features of Pareto optimal sets relating to curvature, sensitivity and other information. The rules use this information to reformulate the problem by reducing the behaviour (B) state space. As both components in the interaction between the problem formulation and the result are non-persistent, there is no place for new knowledge to be constructed and maintained. As a result, the execution of future optimization tasks cannot be directly improved without manual intervention.

Other research involves the tool as a persistent component to acquire and reuse new knowledge. A tool developed by Schwabacher et al. (1998) extracts characteristics of optimization results and uses them to formulate new optimization problems. These characteristics include information such as optimal design structure, mappings between structure and behaviour, infeasible behaviour and active constraints. This leads to better problem formulation in terms of reduced search spaces and improved starting points for gradient-based search. Jozwiak's (1987) system learns inactive constraints in order to predict whether or not the computation of particular constraint functions of a future optimization task may be neglected in the tool. Nath and Gero (2004) use machine learning to let a system acquire strategic knowledge as mappings between past design contexts and design decisions that led to useful results. These mappings are then available for the system to achieve solutions to similar design tasks more efficiently. Stahovich (2000) developed a tool that can extract iterative search strategies used by a designer. The tool then reuses these strategies to perform similar design tasks.

In most of this work, any change in the tool takes effect only after the optimization process is completed, i.e. an increase in efficiency can be achieved only in subsequent optimizations. This differentiates the cited work from the concept of situatedness, where changes have an effect on the same instance of interaction that they originate from. Situated design optimization has the potential to modify state spaces and strategies during the process of iterative search. Visualisation systems for process stages have been developed that can be recast to look like this view of design optimization (Ellman et al., 1998). The main driver for interactive change in these systems is still located inside the black box called the "designer".

While the notion of interaction as presented in Figure 3 can substantially shift our understanding of design optimization, it does not distinguish a situated view from the previous work mentioned above. A situated view includes the potential for reformulation of the structure and behaviour during the ongoing process of designing. These two types of reformulation are depicted in Figure 1 as processes 6 and 7. However, Figure 1 does not explicitly include the concept of interaction. Section 4 develops a framework for situated design optimization that is based on both notions, the one of interaction and the one of reformulation.

4. AN ONTOLOGY FOR SITUATED DESIGN OPTIMIZATION

Let us have a closer look at Gero and Kannengiesser's (2004) model of situatedness, Figure 4(a). It subdivides the internal world into an interpreted and an expected world, the latter of which is a subset of the former. These two worlds are connected to each other by a process of focussing on some of the concepts located in the interpreted world and using them as goals that are then located in the expected world. The goals are subsequently used to inform actions changing the external world.

Figure 4(b) presents a specialised form of this view implying a designer or design agent (as the internal world) located within the external world and placing general classes of design representations into the resultant "onion" model. The set of expected design representations (Xe^i) corresponds to the notion of a design state space, i.e. the state space of all possible designs that satisfy the set of requirements. This state space can be modified during the process of designing by transferring new interpreted design representations (X^i) into the expected world and/or transferring some of the expected design representations (Xe^i) out of the expected world. This leads to changes in external design representations (X^e), which may then be used as a basis for re-interpretation changing the interpreted world. Novel interpreted design representations (X^i) may also be the result of constructive memory, which can be viewed as a process of interaction among design representations within the interpreted world rather than across the interpreted and the external world. Both interpretation and constructive memory are viewed as push-pull processes.

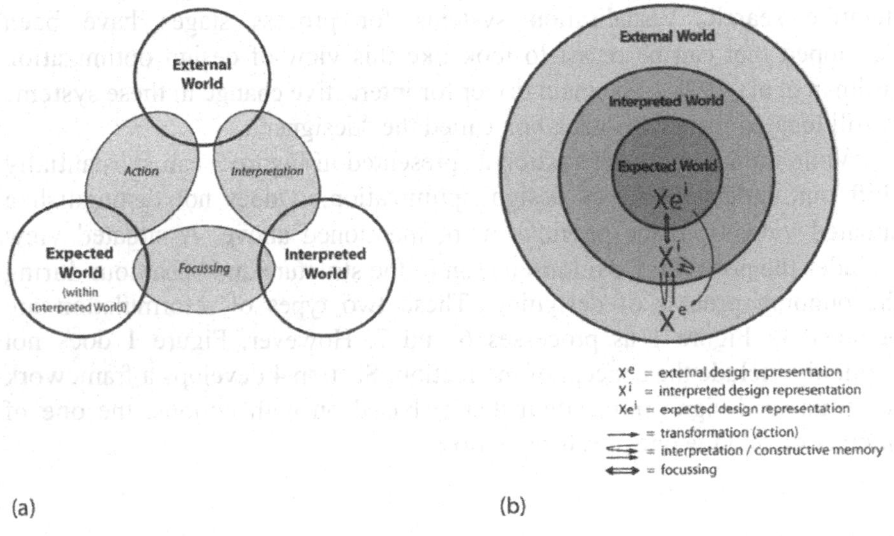

Figure 4. Situatedness as the interaction of three different worlds: (a) general model, (b) specialised model for design representations (after Gero and Kannengiesser (2004)).

The explicit integration of an expected world into a model of interaction accounts for situated designing, as changes in the internal and external world provide the grounds for further changes of the current design process via reformulations of the design state space. It can now be distinguished from other approaches that can be viewed as interaction-based but inconsistent with the idea of reformulating current goals or expectations.

Gero and Kannengiesser (2004) have used this model of situatedness as a basis for a process framework of situated designing derived from Gero's (1990) original FBS framework. Figure 5 presents those parts of their process framework that are relevant in situated design optimization. The three fundamental processes in design optimization, namely synthesis, analysis and evaluation, Figure 1, can now be viewed as consisting of partial processes described as follows:

- *synthesis*: consists of the transformation of expected behaviour (Be^i) into expected structure (Se^i) (process 1 in Figure 5) and the subsequent transformation of that structure into an external structure (S^e) by means of action (process 2).
- *analysis*: consists of the interpretation of the external structure (S^e) to produce an interpreted structure (S^i) (process 3) and the subsequent transformation of that structure into an interpreted behaviour (B^i) (process 4).
- *evaluation*: consists of the comparison of interpreted behaviours (B^i) and expected behaviours (Be^i) (process 5).

These three fundamental processes are no longer viewed as static, as they operate on design representations that are constructed on the basis of situated interaction. As a consequence, there is potential for changes in what is synthesised (i.e., what structure) and in what the design is analysed and evaluated for (i.e., what behaviour). This potential is most significant for optimization tasks that require iterative search procedures involving extensive interaction between expected, interpreted and external structures and behaviours. Situated optimization can be viewed as a process that can modify what it searches as well as what it searches for as it proceeds through the design state space. The design state space relevant for optimization is the union of the structure state space and the behaviour state space, which are represented in Figure 5 in terms of expected structure (Se^i) and expected behaviour (Be^i). Both can be modified at any time during optimization, which is represented by processes 6 and 5 in Figure 5.

Our framework covers opportunities for change in all relevant aspects of situated optimization: change in the current design state space (expected world), change in the designer's experience (interpreted world) and change in the external design representation (external world). In the following, we present a set of instances for each of these changes (denoted by the symbol Δ) relating to the structure and behaviour of the problem formulation and the result. Examples for changes relating to other components or combinations of components proposed in Figure 3 will be included in a forthcoming paper.

Changes in the interpreted world:

- ΔS^i (represented by processes 10 and 3 in Figure 5):
  - ΔS^i of the result:
    The interpreted structure of the result can be generated externally (process 10) or internally (process 3). An example for new S^i is the substitution of a set of design variables by another variable, such as replacing 'length' and 'width' by 'aspect ratio'. These changes are often driven by visual design representations displayed by the tool.
  - ΔS^i of the problem formulation:
    Substitutions of existing structure variables, such as described for the result of an optimization process, may be used to simplify the problem formulation. Additional variables and their ranges of values are often constructed to complement the explicit specifications given to the designer. These variables represent implicit requirements and depend on the designer's subjective experience gathered through previous optimization tasks.

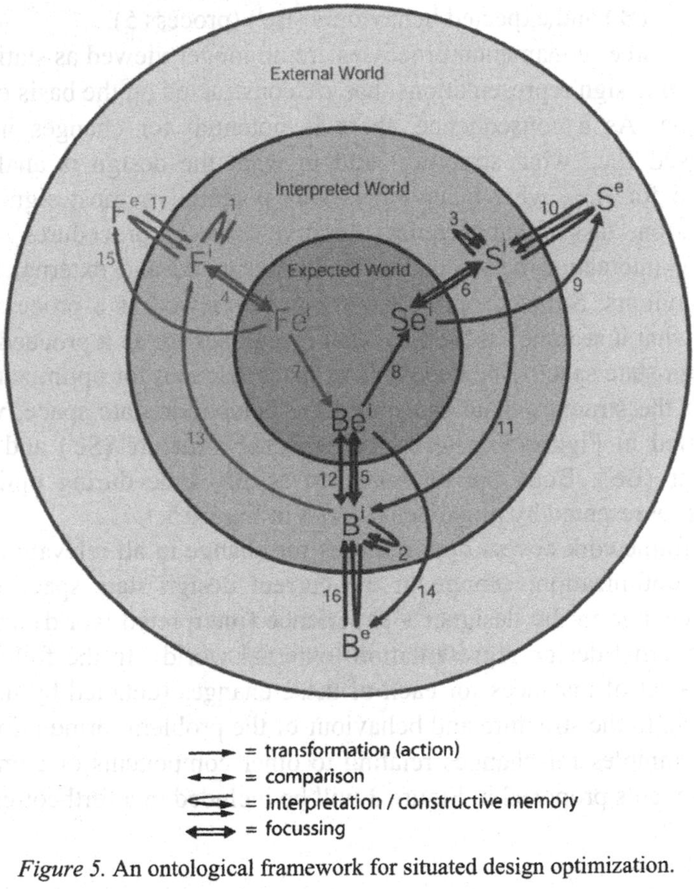

Figure 5. An ontological framework for situated design optimization.

- ΔB^i (represented by processes 16 and 2):
  - ΔB^i of the result:
    New behaviours may be constructed from the results of simulations carried out by the tool that "remind" the designer of additional performances relevant for the particular optimization task. Behaviours may also be constructed internally from design cases retrieved from the designer's memory. In particular, they carry information about the location of optima of similar, past designs.
  - ΔB^i of the problem formulation:
    Initial problem formulations are often represented using natural language expressions, which may give rise to different interpretations regarding the formal set of behaviours required. In addition, a large set of implicit behaviours is often required besides those that are explicitly represented. Implicit behaviours are constructed from existing experience (i.e. the designer's implicit domain knowledge) or from interactions occurring during the current optimization process. Examples are similar to those described for ΔB^i of the result.

Changes in the expected world:

- ΔSe^i (represented by process 6):
  - ΔSe^i of the result:
    This includes the construction of predicted ranges of values for the location of the optimum design solution within the structure state space. They create a subspace representing the designer's implicit assumptions about the result of a routine optimization task that belongs to a well-defined class of similar tasks. These assumptions will be adapted over time to the designer's growing experience and the interpretation of the current optimization process.
  - ΔSe^i of the problem formulation:
    This involves focussing on a set of structure variables and their ranges of values. They represent the state space of all feasible optimum design solutions. Reformulations of the structure state space occur as a result of the designer's increasing domain and task knowledge and the current state of the optimization process.
- ΔBe^i (represented by process 5):
  - ΔBe^i of the result:
    Here the designer assumes an area inside the behaviour state space, which sets predicted bounds for the performance values of the optimum design solution. It forms a subspace that is sensitive to the designer's interaction with previous and current

optimization tasks. Predicted behavioural values are used to evaluate the adequacy of the results produced by the tool and of the search techniques employed.

- ΔBe^i of the problem formulation:
  This involves focussing on a set of behaviour variables and their ranges of values according to all explicit and implicit constraints. Reformulations of the behaviour state space occur as a result of the designer's increasing domain and task knowledge and the current state of the optimization process.

Changes in the external world:

- ΔS^e and ΔB^e (represented by processes 9 and 14):
  - ΔS^e and ΔB^e of the result:
    Although the external representation of the structure and behaviour of optimization results is ultimately produced by the tool, it is part of an action that is informed by the designer's individually constructed expectations and strategies. Results form a basis for new expectations and strategies to be constructed and used to commence new cycles of synthesis, analysis and evaluation.
  - ΔS^e and ΔB^e of the problem formulation:
    This is the direct effect of an action devised by the designer to externally specify a state space of possible designs among which the tool must select the best performing one. Reformulation of the problem can be a consequence of the results of this selection process.

5. DISCUSSION

We have proposed a framework of situated design optimization that represents a departure from previous models as it is based on interaction rather than only on static relationships between the individual components of optimization. Our framework is also distinguished from other approaches by its ability to capture changes in how the optimization process proceeds as a result of its trajectory through the design state space. Using an ontological approach provides a basis for a better understanding and common ground about a situated view of optimization.

The FBS ontology has been shown to be a foundation for representing design knowledge in a general and uniform way (Gero, 1990). This is beneficial for implementing situated design optimization systems that require learning and adaptation to novel design situations. Both knowledge about the object to be optimized and knowledge about optimization strategies can be represented using the FBS schema. Current work in our research centre focuses on the development, implementation and testing of a system using this approach.

6. ACKNOWLEDGEMENTS

This research is supported by a grant from the Australian Research Council, grant no. DP0559885 – Situated Design Computing.

7. REFERENCES

Bartlett, F.C., 1932 reprinted in 1977, *Remembering: A Study in Experimental and Social Psychology*, Cambridge University Press, Cambridge.

Bickhard, M.H. and R.L. Campbell, 1996, "Topologies of learning", *New Ideas in Psychology*, **14**(2): 111-156.

Clancey, W.J., 1997, *Situated Cognition: On Human Knowledge and Computer Representations*, Cambridge University Press, Cambridge.

Dewey, J., 1896 reprinted in 1981, "The reflex arc concept in psychology", *Psychological Review*, **3**: 357-370.

Ellman, T., J. Keane, A. Banerjee and G. Armhold, 1998, "A transformation system for interactive reformulation of design optimization strategies", *Research in Engineering Design*, **10**(1): 30-61.

Gero, J.S., 1990, "Design prototypes: A knowledge representation schema for design", *AI Magazine*, **11**(4): 26-36.

Gero, J.S., 1998, "Conceptual designing as a sequence of situated acts", in: Smith (ed.) *Artificial Intelligence in Structural Engineering*, Springer-Verlag, Berlin, pp. 165-177.

Gero, J.S., 1999, "Constructive memory in design thinking", in: Goldschmidt and Porter (eds.) *Design Thinking Research Symposium: Design Representation,* MIT, Cambridge, MA, pp. 29-35.

Gero, J.S. and H. Fujii, 2000, "A computational framework for concept formation for a situated design agent", *Knowledge-Based Systems*, **13**(6): 361-368.

Gero, J.S. and U. Kannengiesser, 2004, "The situated function-behaviour-structure framework", *Design Studies*, **25**(4): 373-391.

Gero, J.S. and U. Kannengiesser, 2006, "A function-behaviour-structure ontology of processes", *Design Computing and Cognition'06*, Springer-Verlag, Berlin, to appear.

Jozwiak, S.F., 1987, "Improving structural optimization programs using artificial intelligence concepts", *Engineering Optimization*, **12**: 155-162.

Mackenzie, C.A. and J.S. Gero, 1987, "Learning design rules from decisions and performances", *Artificial Intelligence in Engineering*, **2**(1): 2-10.

Nath, G. and J.S. Gero, 2004, "Learning while designing", *Artificial Intelligence for Engineering Design, Analysis and Manufacturing*, **18**(4): 315-341.

Papalambros, P. and D.J. Wilde, 2000, *Principles of Optimal Design: Modeling and Computation*, Cambridge University Press, Cambridge.

Pardalos, P.M. and M.G.C. Resende (eds.), 2002, *Handbook of Applied Optimization*, Oxford University Press, New York.

Parmee, I.C., 1996, "Towards an optimal engineering design process using appropriate adaptive search strategies", *Journal of Engineering Design*, **7**(4): 341-362.

Parmee, I.C. and P. Hajela (eds), 2002, *Optimization in Industry*, Springer-Verlag, London.

Schön, D.A., 1983 *The Reflective Practitioner*, Harper Collins, New York.

Schön, D.A. and G. Wiggins, 1992, "Kinds of seeing and their functions in designing", *Design Studies*, **13**(2): 135-156.

Schwabacher, M., T. Ellman and H. Hirsh, 1998, "Learning to set up numerical optimizations of engineering designs", *Artificial Intelligence for Engineering Design, Analysis and Manufacturing*, **12**: 173-192.

Smith. G.J. and J.S. Gero, 2005, "What does an artificial design agent mean by being 'situated'?", *Design Studies*, **26**(5): 535-561.

Stahovich, T.F., 2000, "LearnIT: An instance-based approach to learning and reusing design strategies", *Journal of Mechanical Design*, **122**(3): 249-256.

Suwa, M., J.S. Gero and T. Purcell, 1999, "Unexpected discoveries and s-inventions of design requirements: A key to creative designs", in: Gero and Maher (eds.) *Computational Models of Creative Design IV,* Key Centre of Design Computing and Cognition, University of Sydney, Sydney, Australia, pp. 297-320.

Wilde, D.J., 1978, *Globally Optimal Design*, Wiley, New York.

Ziemke, T., 1999, "Rethinking grounding", in: Riegler, Peschl and von Stein (eds.) *Understanding Representation in the Cognitive Sciences: Does Representation Need Reality?*, Plenum Press, New York, pp. 177-190.

Learning from Main Streets

A machine learning approach identifying neighborhood commercial districts

Jean Oh[1], Jie-Eun Hwang[2], Stephen F. Smith[1], and Kimberle Koile[3]
1 School of Computer Science, Carnegie Mellon University
2 Graduate School of Design, Harvard University
3 Computer Science and Artificial Intelligence Laboratory, Massachusetts Institute of Technology

Keywords: Main street approach, Community development, Artificial intelligence, Machine learning, Active learning algorithm

Abstract: In this paper we explore possibilities for using Artificial Intelligence techniques to boost the performance of urban design tools by providing large scale data analysis and inference capability. As a proof of concept experiment we showcase a novel application that learns to identify a certain type of urban setting, Main Streets, based on architectural and socioeconomic features of its vicinity. Our preliminary experimental results show the promising potential for the use of machine learning in the solving of urban planning problems.

1. INTRODUCTION

The recent progress in network technologies and computational power has opened up new opportunities for Artificial Intelligence (A.I.), enabling the use of complex data analysis algorithms on a large amount of data available from online information sources. Machine learning, in particular, has been successfully used in various practical problem domains including text categorization (Joachims 1998; Yang 1999; Sebastiani 2002) and computational biology (Baldi and Brunak 1998), demonstrating competitive performance accuracy against human experts. In this paper we explore the possibilities of using A.I. techniques in urban design decision support systems.

Jos P. van Leeuwen and Harry J.P. Timmermans (eds.), Innovations in Design & Decision Support Systems in Architecture and Urban Planning, 325-340.

Levy (1997) summed up the need for urban design[1] in two words: interconnectedness and complexity. Urban design is a complicated decision-making process that involves multiple entities of various interest and constraints. The heterogeneity of urban resources and diversity of involvement raises serious demands for sophisticated communication between government, civic institution, engineers, real estate developers, communities, etc. The communication process, especially with the local community has become more important with the growth of civic rights and public activism (Innes 1998). Moreover, each city has very unique situations and problems. The explicit and insightful examination of the locality is resolved into contemporary major urbanism issues: regional planning, comprehensive planning, and strategic and sustainable development. Therefore, the analysis and identification of existing (and prospective) urban context in a systematic way is the most essential task in the design process.

The Geographic Information System (GIS) is one of the most popular design and decision support tools, facilitating basic data and diverse analysis platforms (Batty, Dodge et al. 2000). Although GIS is a powerful tool it is still limited by the lack of sophisticated data analysis and inference capability. Our research goal is to add intelligence to design support tools such as GIS, and provide support for designers' decision making processes by effectively analyzing relevant collections of information.

A human expert makes use of a pool of accumulated knowledge from the past when solving a new problem. Let us assume that human experts are *rational* users, i.e., experts who make consistent decisions in order to find the optimal solution that would yield the maximum expected future rewards. Given the assumption of rational users we naively hypothesize that a computer system can learn an expert's decision making knowledge.

Representing such knowledge in a structured form, however, is a difficult problem. One of the earliest approaches was a rule-based expert system. A set of rules of thumbs from domain experts defines intelligence in such systems. In practice, a fundamental limitation of an expert system has been the lack of learning capabilities and the fact that knowledge generally must be programmed into them.

A more general approach is to learn a mapping function from an input to an output. In this paradigm a set of descriptive properties called "features" is presented as an input to a decision making problem, and an output is represented as a discrete-valued label, e.g. commonly a binary outcome of "yes" or "no". This paradigm is powerful enough to capture a great many

[1] In this paper, we use the term "urban design" interchangeably with the term "urban planning." Although urban design and urban planning are distinct disciplines, the roles and professions often are commonly overlapped in practice. Since we address the architectural level of space and built form eventually, we consider this study to be in urban design.

real life decision making problems. During the past decade numerous efficient machine learning algorithms have been invented and applied to problems in a wide variety of domains, e.g., ranging from medical decision support systems to spam email filters.

We take a machine learning approach towards solving urban design problems. We hypothesize that there exists a class of urban design decision making problems that can be formulated as a machine learning problem given a set of specific assumptions. Our first assumption is that a human expert has a certain decision making criteria that is represented in domain specific knowledge. The goal is to train the system to learn the expert's decision criteria. The system's model of domain knowledge, e.g., a set of features, is bounded by data availability. So, we further assume that the system has access to domain specific information sources. The growth of the internet has brought easy access to vast amount of domain specific knowledge. For instance, a building' structural information and its land use were publicly available for our experiment.

Although the field of A.I. has been drawing increasing attention in the urban design domain, little has been done to demonstrate true benefits of A.I. techniques in this problem domain. We showcase our proof of concept experiment by building a software program that can identify a certain type of urban setting, Main Streets. Our preliminary results show promising potential for the use of machine learning in this problem domain.

Considering the interdisciplinary nature of this paper we include introductory sections for readers from diverse backgrounds. Section 2 addresses our target research issues in urban design, and Section 3 describes various machine learning paradigms that are used in our research.

2. URBAN DESIGN PROBLEMS

The integrated perspective of form and function in urban studies is not an innovative notion. In fact, it has been the core subject of urban matters for a long time (Sitte 1965; Needham 1977; Rapoport 1990; El-Khoury, Robbins et al. 2003). Previous work, however, has primarily focused on one dominant aspect of either form or function from a particular view point, e.g. architecture, psychology, sociology or economics. Furthermore, the range and definition of form and function varies according to diverse disciplines. For instance, while architects regard form as three dimensional shape of space and building components in the intimate detail, economists rather view it as two dimensional shape of cartographic plane at the regional or national scale. Architects consider function as activities in individual building spaces and the in-betweens, whereas policy makers consider function as performance of parcel or zone in the whole system of the city.

Resolving multiple views has been an important issue in urban design decision making. The urban design profession contributes to shape the city through designing physical structures; however, it has generally been an execution of form-based policy in this respect (Krieger 2004).

Recognizing the importance of considering interdisciplinary aspects of a problem, urban designers have developed methodological frameworks to investigate urban morphology in a manner that combines multidisciplinary aspects. Our research contributes to this effort, by applying AI techniques to develop improved representations and methods for reasoning about urban design issues in an integrated fashion. In this paper we focus on an important methodological framework, *typology*, which represents the understanding of urban settings by classification based on present architectural and socio-econimic elements. Specifically, we aim to develop the urban typology process by formulating it as an A.I. classification problem.

2.1 Main Streets

We have conducted our study on an illustrative example of urban district of *Main Streets*. A Main Street is a typical commercial district in the center of a residential area particularly in the Untied States. Throughout the history of urbanization, Main Street has evolved to serve goods, services, and public activities in a residential neighbourhood. The prosperity of Main Streets, however, declined with the rapid spread of big-box style national chain regional malls along highways during the suburban sprawl era (Isenberg 2004). Along with the widely raised criticism about losing local economic sustainability and cultural identity, revitalization of Main Streets has become the core issue of community development. Many American cities have developed innovative policies intended to restore prosperity and vitality to downtowns. Currently, over 1,000 Main Street communities have adopted "the Main Street approach", which includes a four-point program: organization, promotion, design, and economic restructuring (Robertson 2004).

The design of Main Street is complicated in the both spatial and temporal scale, as many urban design subjects are. In a regional scale, the Main Street is a set of segment of street connected in the whole vessel network of the city. At the same time, the Main Street is a district consisting with buildings and open spaces in local neighbourhood. As historical process of shaping urban landscape, the Main Street approach conveys redefinition and reallocation of architectural and socioeconomic resources over times. Since each Main Street has unique characteristics and problems, it is important to understand and identify the local context of the community.

The process of Main Street development raises an important issue that stems from the complexity of communications among multiple actors. The set of actors involved in Main Street design process includes city officials, local directors, design professionals, communities, developers, investors, etc. The key to a successful Main Street design lies in resolving diverse interests and constraints of multiple actors from architectural, social, economic, and historical perspectives.

2.2 Urban Typology

Urban typology is the study of pieces and cells that generate and change the cityscape (Moudon 1998). Urban typology (Caniggia and Maffei 1979; Krier and Rowe 1991) has long been a study of subject yet little or no attempts have been made to automate this process. In general, urban typology analysis is a time consuming task that requires complex data analysis and field studies.

For instance, the ARTISTS (Arterial Streets Towards Sustainability) project in Europe was developed to identify a set of types of streets in order to provide better insights to urban planners and economists. This 2.2 billion euros budget project involved 17 European countries and took three years to classify five categories of streets (Svensson 2004). The experimental results show how they classified 48 streets into 5 categories based on their decision rules. Our attempt is to carry out a similar classification task on Main Streets but in an automated way using machine learning techniques.

Moreover, a systematic approach can support a designer's typology view by providing a theoretical justification for his/her classification decision criteria which is a perspective envision of larger and more complex urban settings based on her unique design concepts. In practice, urban typology studies have been presented in subjective exhibitions without comprehensive denotations of the decision criteria. A result of typology study is often represented as symbols or specific representative cases.

In this paper we propose the use of machine learning in representing such decision making knowledge. We demonstrate typological process by identifying Main Street as one type of urban districts out of bulky urban landscape data.

3. MACHINE LEARNING

This section provides a high level overview of machine learning techniques that are used in our research. Machine learning refers to a software system that self-improves through autonomous acquisition and integration of knowledge. In other words, machine learning is a system that builds knowledge from past experience to improve expected future performance (Mitchell 1997). For example, historical medical records form a knowledge source that can be used to diagnose new patients. A doctor makes diagnostic decisions based on a patient's symptoms and a careful analysis of historical medical records. Similarly, machine learning can be used to make predictions for new findings based on what has been observed. We focus on two types of learning -- classification and clustering.

A *classification* is a learning task of mapping a set of input variables into a set of outcome variables also referred as "classes". A *classifier* is a system that learns such mapping functions from a set of training examples whose classes are correctly labeled. A classification is a type of *supervised* learning in which the set of output variables are well-defined, e.g., text categorization. It is "supervised" because it requires an expert who can map each training data point to its correct class label. For example, in order to learn how to classify a vicinity of Main Streets the system needs to observe some set of examples of Main Streets and non Main Streets. Further technical details of a classification are discussed in section 3.1.

A *clustering*, on the other hand, is a type of *unsupervised* learning in which there is no distinction between input variables and output variables. Thus, labeling is not necessary in clustering. For instance, a clustering algorithm can group data into a set of clusters each of which contains only those data points that are similar to one another. In our experiment, we use clustering algorithms to pre-process low level data, e.g., grouping buildings into a set of neighborhoods.

3.1 Classifiers

There are two types of classifiers: generative classifiers and discriminative classifiers (Mitchell 1997). Let X be an input vector of n features, $X = \{X_1, X_2, ..., X_n\}$ where X_i denotes a random variable representing the ith feature. Let Y be an output variable. Let us also denote x_k be an instance of X, y_l be an instance of Y. Learning of a classifier is approximating an unknown mapping function $f: X \rightarrow Y$, or equivalently estimating a probability of Y given X, denoted by $P(Y|X)$. Generative classifiers assume that input data is generated from a certain statistical model of input variables, i.e., $P(X)$. Based on this assumption, such classifiers estimate *P(X)* and *P(X|Y)* from training data and apply Bayes rule to compute $P(Y|X=x_k)$.

Equation 1 Bayes rules

$$P(Y \mid X) = \frac{P(X \mid Y)P(Y)}{P(X)}$$

$$P(Y = y_i|X = x_k) = \frac{P(X = x_k|Y = y_i)\,P(Y = y_i)}{\sum_j P(X = x_k|Y = y_j)\,P(Y = y_j)}$$

Discriminative classifiers, on the other hand, make no assumption on underlying distribution model of X. Instead, P(Y|X) is directly estimated from training data. In our experiment we mainly used a Support Vector Machine (SVM) classifier. An SVM is a discriminative classifier that learns an optimal classification boundary that minimizes training set error (Burges 1998). The sets of data points near the separation hyperplane are called support vectors. The *Figure 1* shows a simple noise-free SVM. An optimization objective of the linear SVM in *Figure 1* is maximizing the separation margin width *M*, which is equivalent to minimizing *w.w*.

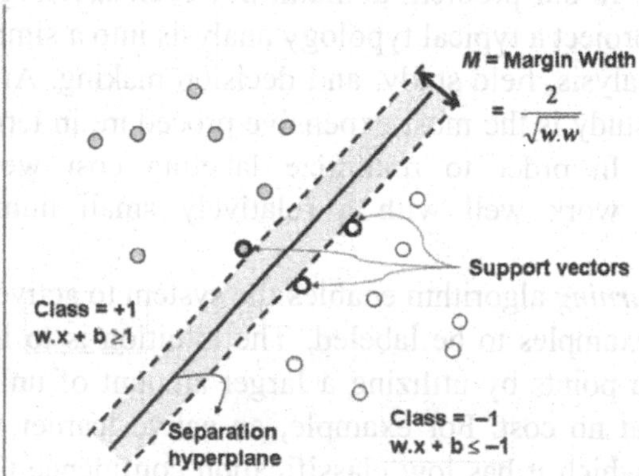

Figure 1. Linear Support Vector Machine.

In our initial experiment, we tried a set of classifiers to determine the best-fitting classifier in our particular problems. Among a set of Decision Trees, a Naïve Bayes classifier, a k-Nearest Neighbors (kNN) classifier, and an SVM classifier, an SVM classifier best performed (Yang 1999)[2]. In

[2] Due to limited space we omit formal definitions of various classifiers and refer to Yang's work (1999) that extensively evaluates various types of classifiers.

general, SVM is considered one of the best performing classifiers in many practical domains. Despite SVM's high quality performance users outside A.I., such as designers, tend to prefer Decision Trees or generative models due to the fact that their results are more comprehensible. Another limitation of the discriminative approach is handling of outliers, i.e., data points far from training set. Since discriminative classifiers do not attempt to model distribution of input variables it is hard to recognize outliers.

In order to overcome general drawbacks of discriminative approach an outcome from an SVM classifier can be presented in alternative ways. For instance, we train a decision tree that is equivalent to the learned SVM classifier in terms of classification results over the test set. That is, after training an SVM classifier using a set of train data, the system labels the remaining set of data with SVM's prediction. Finally we train a decision tree using the original set of train data augmented with the remainder of data labeled by the learned SVM. Similarly, we can train a Bayesian classifier to take advantage of benefits of generative models.

3.2 Active Learning

In some problem domains, cost of labeling is not an issue, e.g., historical medical records. In our problem domain, however, labeling is an expensive procedure. We project a typical typology analysis into a simplified three-step process: data analysis, field study, and decision making. Among these three steps, the field study is the most expensive procedure in terms of both labor cost and time. In order to minimize labeling cost we favor learning algorithms that work well with a relatively small number of training examples.

An *active learning* algorithm enables the system to actively choose a next set of training examples to be labeled. The intuition is to identify the most informative data points by utilizing a larger amount of unlabeled data as a learning guide at no cost. For example, an active learner would select the data points for which it has low classification confidence (Tong and Koller 2000).

A *semi-supervised* method also utilizes distribution of a large amount of inexpensive unlabeled data to guide supervised learning. For example, the co-training method (Blum and Mitchell 1998) learns two classifiers using disjoint sets of features, i.e., two different views over the same set of data, and admits only those predictions upon which both classifiers agree. A more recent approach includes incorporating clustering into active learning (Nguyen and Smeulders 2004). Using prior data distribution their system first clusters data and suggests cluster representatives to the active learner.

Their algorithm selects not only the data points close to classification boundary but also representatives of unlabeled data.

4. MODELING

Modeling an urban typology as a machine learning problem is based on two important assumptions: 1) a set of relevant features that define an input to a learning algorithm are known in advance, and 2) data that describe the features are a well-structured set of vectors. Applying machine learning algorithms to a well defined set of data is a straightforward task. However, a major difficulty of formulating urban typology into a machine learning problem resides in feature space modeling and compiling a set of relevant data.

The human experts' elicitation of relevant features is often vague and incomplete. We exemplify a modeling of feature space in *Figure 2*. This example depicts the feature dependency graph that represents a perception of *public-ness*. Public-ness is a meaningful concept in urban design and relates to how people perceive whether a given urban component is public or private. We modeled this example based on a survey that was conducted on both domain experts and non-experts. Although this example does not directly address the problem of Main Streets the features in the graph, such as Massing, are commonly used as urban decision criteria, thus they are relevant to our discussion.

Among these features the entries that are drawn in boldface in *Figure 2* are the set of features that users considered important in decision making. Because the system can only recognize well-structured data, e.g., features stored in databases, only the features shown in grey are included in our model. This example illustrates our modeling assumption that domain experts' model of relevant features are often abstract semantic concepts that *depend* on descriptive features that are available in low level databases.

Massing, for instance, is a feature that differentiates buildings by their structural size information. In our information sources Massing is represented as multiple features, height, area, periphery, distance to nearest neighbor, etc. Our survey result also reveals the existence of hidden features that are completely isolated from what is available in low level database. These hidden features were denoted by *intangible* features in the picture, e.g., features related to "Use Patterns".

We learn from this example that a majority of features in a human user's model are abstract concepts, whereas the system only has access to low level databases. We make a specific assumption that abstract concepts that human experts consider relevant in fact depend on low level features in databases.

We also assume that the system has access to such domain specific information sources. The challenge then is to infer the mapping from low level features to abstract concepts.

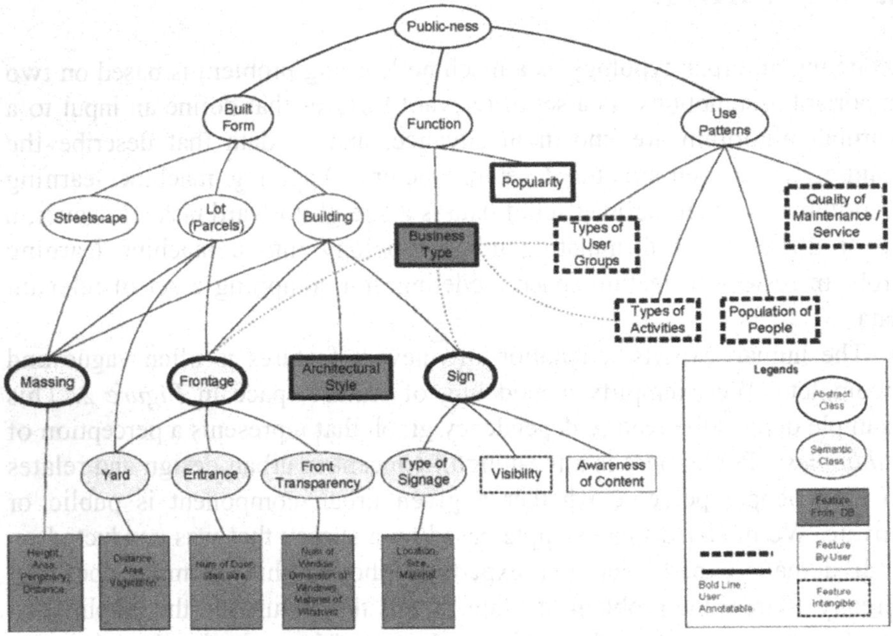

Figure 2. Features for determining public-ness of an urban component.

5. EXPERIMENT

This section describes a preliminary experiment carried out to verify our hypothesis. We chose the specific problem of identifying a certain type of urban setting, Main Streets, based on architectural and socioeconomic features of its vicinity. The criteria of classifying a commercial district varies from city to city, thus it is hard to find a generalized set of rules to distinguish Main Streets from rest of districts. Therefore, we aim to establish the customized decision criteria of identifying Main Streets over the proposed data model for a particular city.

Most machine learning algorithms expect data to be a well defined set of tuples, but in reality this is rarely the case. For example, if data is stored in a relational database with multiple tables the data must be merged into a giant single table. Building an inference network from a relational database is an interesting area of research (Getoor 2001) and we also anticipate that our future work may be in this area. For the sake of simplicity we assumed that we already have the data formatted into a set of tuples in this experiment.

5.1 Data Pre-processing

We investigated Main Streets in the city of Boston for this study (*Figure 3*). Boston provides an ideal testbed for evaluation because a complete set of districts were already identified as Main Streets by field experts. We used relational database tables exported from GIS information sources that are available from the city of Boston. The data was then pre-processed to get it into a format suitable for general classifiers.

Initially we started with two database tables: buildings and parcels. Note that a data entry in these tables represents a building and a parcel, respectively, whereas our target concept, Main Streets, is defined as a district which is usually composed of hundreds of buildings and parcels.

First, we applied unsupervised learning methods to group buildings and parcels into a set of candidate districts. We used a single-linkage clustering algorithm in which every data point starts with a separate cluster and merges with the closest neighboring cluster until a given proximity threshold is satisfied. The proximity threshold was chosen empirically to generate reasonable size clusters.

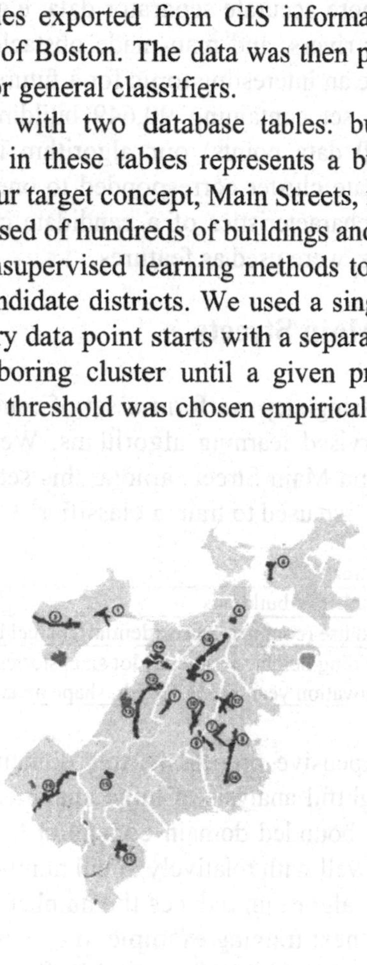

Figure 3. Main Streets in Boston, Massachusetts.

Our algorithm for identifying district candidates consists of two clustering steps. Since the backbone of Main Streets is a strip of commercial buildings we first clustered buildings that are associated with commercial land use code in order to retrieve strips of commercial buildings. For this research study small clusters that contained less than 5 commercial buildings were filtered out. In the second step, the commercial strips identified in the first step were treated as a single cluster when the second round of clustering started, i.e., the set of initial clusters in the second round was the union of commercial strips, non-commercial buildings, and all of parcels. The number of buildings and parcels in the resulting district candidates were in the range of hundreds.

For simplicity, we used Euclidean distance between the two centers of buildings as the distance measure. In order to refine cluster boundaries we need to incorporate more accurate separator data, e.g., geographic obstacles such as mountains or rivers, and man-made obstacles such as bridges and highways. This will be an interesting topic for a future work.

Using a raw data set containing 90,649 buildings and 99,897 parcels (total around 190,000 data points) our algorithm identified 76 candidate districts. Each candidate cluster corresponded to one data row for a classifier, and aggregated characteristics of a candidate cluster, such as average height of the buildings, were used as features.

5.2 Finding Main Streets

In the data pre-processing step we form a set of candidate districts of Main Streets using unsupervised learning algorithms. We then use a supervised learning method to find Main Streets among this set of candidates. Table 1 lists the set of features we used to train a classifier[3].

Table 1. Main Streets features.

Aggregated features	Number of buildings
Ratio	land use (commercial/residential), parcel business type
Average	building height, perimeter, lot size, stories, built year, renovation year, shape length, shape area, gross area, living area

Labeling is an expensive process in this domain because labeling one district requires thoughtful analysis of huge amounts of data and it involves field study. This cost-bounded domain constraint leads us to favor learning algorithms that work well with relatively small number of training examples.

An active learning algorithm reduces the number of training examples by actively choosing the next training example to be labeled. We selected Tong and Koller's approach over SVM (Tong and Koller 2000). The basic idea is to suggest data points that are near the separation boundary, which is quite intuitive and has also proven to be very effective in other practical domains such as text classification. We extended the SVM-KM Toolbox (Canu et al. 2003) to implement the active learning strategy.

[3] We used all numeric features in the buildings and parcels database, some of which are not included in Table 1.

We also tried incorporating pre-clustering (Nguyen and Smeulders 2004) to find the initial samples to be labeled. This technique, however, didn't have a significant impact on performance in our experiment mainly because the size of unlabeled data was not large enough (After pre-processing we had only 76 district candidates). We would expect higher impact on performance if we had a larger set of data.

5.3 Experimental Results

We have built an intelligent urban design decision support system that aims to solve a complicated decision making problem of urban typology, Main Street. First, we evaluate a general classification performance of our system to prove that the typology of Main Street can be efficiently developed by a machine learning approach.

We used precision, recall, and their harmonic mean as evaluation measures. In our example, precision p is the ratio of the number of correctly predicted Main Streets to the total number of positive predictions. On the other hand, recall r is the ratio of the number of correctly identified Main Streets to the total number of Main Streets in Boston. Because the two measures are in inverse relation their harmonic mean is often used as a compromising measure. F1 measure, which is a harmonic mean of precision p and recall r, is defined below.

Equation 2 F1 measure

$$F1 = \frac{2pr}{p+r}$$

Since we had relatively small sized data points after pre-processing we used Leave-One-Out-Cross-Validation (LOOCV) to evaluate the general performance of Main Streets classifier. LOOCV is a cross validation technique where one data point is left for testing while a classifier is trained using the rest of data points. The LOOCV results in Table 2 shows promisingly good performance by achieving high F1 measure of 0.8. The results read that the system made 6 correct predictions out of every 7 trials, identifying 76% of Main Streets.

Table 2. Leave-One-Out-Cross-Validation Result.

	Precision	Recall	F1 measure
LOOCV	0.842	0.762	0.800

We also compared the performance of the active learning strategy to the performance of the random learning strategy. Under the random learning strategy the system also learns an SVM classifier by incrementally taking more training examples. Whereas the active learning strategy takes advantage of distribution of unlabeled data in selecting a next data point, the random learning strategy chooses an arbitrary data point. We evaluated the performance of the two approaches in terms of their learning speed.

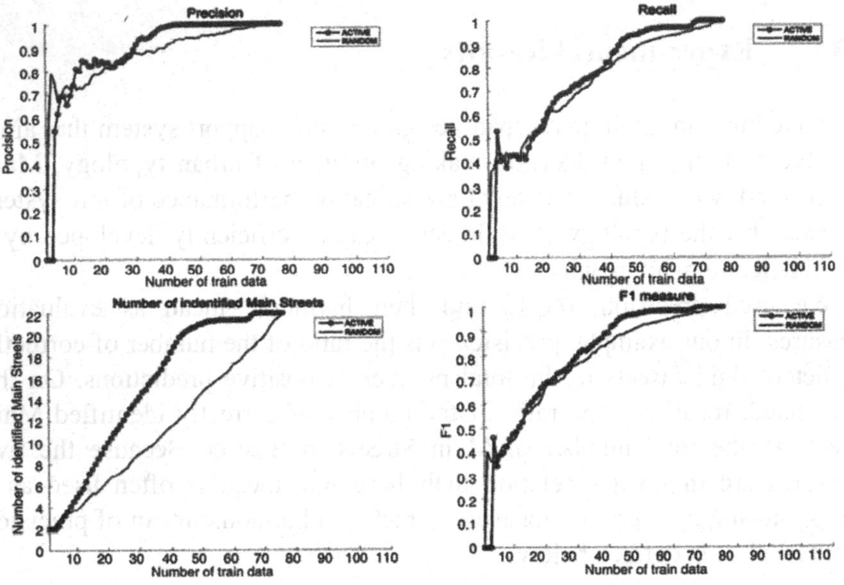

Figure 4. Active learning strategy vs. Random learning strategy.

Figure 4 shows the performance of active learning strategy and random learning strategy. We included the number of identified Main Streets (the lower left in *Figure 4*) which also clearly demonstrates superior performance of the active learning algorithm. The experimental results in *Figure 4* are average performance of a set of 20 independent trials.

As shown in *Figure 4*, the results illustrate the benefit of learning, i.e., a steady progress in performance proportional to the amount of experience, such as a number of training examples (X axis). First, we notice that the precision of our system under both active and random strategies reach nearly .7 which is a reasonably high precision level after observing approximately 10 training examples. The level of recall also reaches .5, i.e., a half of Main Streets have already been identified, given 20 training examples. The experimental results first indicate that finding Main Streets is a class of urban design decision making problems that can be solved by a machine learning approach. The results also show that the active learning algorithm significantly[4] outperforms the random learning algorithm, achieving high classification accuracy after given a relatively small number of examples.

Our initial hypothesis was that there exists a class of urban typology problems that can be modeled as a classification problem. Our current study shows that classifiers perform very well on identifying Main Streets even with an incomplete set of features. Especially using an active learning algorithm our system can cleverly choose better samples to be labeled, outperforming a random selection model significantly.

6. CONCLUSION AND FUTURE WORK

Urban design professionals need to collect and thoroughly analyze large amounts of data in order to make robust plans towards long-term goals and to communicate with diverse stakeholders. This is normally a careful and time-consuming task, due in part to limited financial resources but also because design decisions often generate cascading effects contingent on both pre-existing physical urban structures and future design decisions.

Although there is growing interest in using A.I. in urban design and planning this community remains a field dominated by human experts. Recent catastrophic disasters such as hurricane Katrina, however, have underscored the need for increased automation and more efficient urban design processes. For example, finding good locations for temporary housing is one of the most urgent decision making tasks in post-disaster urban planning that would benefit from A.I. approach.

In this paper we described our efforts to apply A.I. techniques to urban design problems. We conveyed a methodological framework of urban design studies: *typology*. As a proof of concept experiment we showcased an application of machine learning that actively learns to identify a certain type of urban settings, Main Streets. Our preliminary experimental results show promising potential for utilizing A.I. techniques in this problem domain.

Our ongoing research focuses on 1) improving the models and heuristics; 2) developing the human-comprehensible explanation; and eventually 3) embracing morphological approach of urban network (Hillier and Hanson 1984).

7. ACKNOWLEDGEMENTS

The authors thank Yiming Yang for fruitful discussions. This research was sponsored in part by the Department of Defense Advanced Research Projects Agency (DARPA) under contract #NBCHD030010.

[4] This is statistically significant with a strong evidence of p-value = 0.01.

8. REFERENCES

Baldi, P and Brunak, S., 1998, *Bioinformatics: The Machine Learning Approach*, MIT Press.

Batty, M., M. Dodge, et al., 2000, *New Technologies For Urban Designers: The VENUE Project*. Center for Advanced Spatial Analysis working Paper Series.

Berges, J.C., 1998, "A Tutorial on Support Vector Machines for Pattern Recognition", *Data Mining and Knowledge Discovery*, **2**(2): 21-167.

Blum, A. and Mitchell, T, 1998, "Combining labeled and unlabeled data with co-training", *Proceedings of the Workshop on Computational Learning Theory*, pp. 92-100.

Caniggia, G. and G. Maffei., 1979, *Architectural Composition and Building Typology: Integrating Basic Building*. Firenze, Italy, Alinea Editrice.

Canu, S.; Grandvalet, Y.; and Rakotomamonjy, A. 2003. *Svm and kernel methods matlab toolbox*. Perception Systems et Information, INSA de Rouen, Rouen, France.

El-Khoury, R., E. Robbins, et al., 2003, *Shaping the city: studies in history, theory and urban design*. New York, NY 10001, Routledge.

Getoor, L., 2001, *Learning Statistical Models from RelationalData*. PhD thesis, Stanford University.

Hillier, B. and Hanson, J., 1984, *The social logic of space*. Cambridge; New York, Cambridge University Press.

Innes, E.J., 1998, "Information in Communicative Planning", *Journal of the American Planning Association*, **64**(1): 52-63.

Isenberg, A., 2004, *Downtown America: a history of the place and the people who made it*. Chicago, University of Chicago Press: xviii, 441 p., [2] p. of plates (col).

Joachims, T., 1998, "Text Categorization with Support Vector Machines: Learning with Many Relevant Features", *Lecture Notes in Computer Science*, 1398, Springer-Verlag.

Krieger, A., 2004, *Territories of Urban Design*, Harvard Design School.

Krier, R. and Rowe C., 1991, *Urban space*. London, Academy Editions.

Lopilato, L., 2003, *Main Street: some lessons in revitalization*, University Press of America, Inc., New York.

Mitchell, T.M., 1997, *Machine Learning*, McGraw Hill, New York.

Moudon, A.V., 1998, "The changing Morphology of Suburban Neighborhoods", in *Typological process and design theory, Aga Khan Program for Islamic Architecture at Harvard University and Massachusetts Institute of Technology*.

Needham, B., 1977, *How cities work: an introduction*. Oxford; N.Y., Pergamon Press.

Nguyen, H.T., Smeulders A., 2004, "Active learning using pre-clustering", *Proceedings of International Conference on Machine Learning*.

Rapoport, A., 1990, *The meaning of the built environment: a nonverbal communication approach*. Tucson, University of Arizona Press.

Robertson, K.A., 2004, "The Main Street Approach to Downtown Development: An Examination of the Four-point Program." *Journal of Architectural and Planning Research* **21**(1): 55-72.

Sitte, C., 1965, *City planning according to artistic principles*. New York, Random House.

Sebastiani, F. 2002, "Machine learning in automated text categorization" *ACM Computing Surveys*, **34**(1): 1-47.

Svensson A., 2004, *Arterial Streets for People*, Technical report, Lund University, Department Department of Technology and Society, Sweden.

Tong S. and Koller D., 2000, "Support Vector Machine Active Learning with Applications to Text Classification", *Proceedings of 17th ICML*, pp. 999-1006.

Yang, Y. 1999, "An Evaluation of Statistical Approaches to Text Categorization", *Information Retrieval*, Springer, **1**(1-2): 69-90.

Culturally Accepted Green Architecture Toolbox

Pre-design helping tool and rating system for new built environment in Egypt

Usama El Fiky and Mark Cox
Eindhoven University of Technology

Keywords: Software program, Building culture, Green architecture

Abstract: This paper describes and analyses the process of developing computer software to incorporate both green architecture design strategies and their cultural indicators in one easy tool to help a designer to match his / her design with green architectural principles in the pre-design stage. The three resources of architectural identity in Egyptian, current and past green building practices with the up to date foreign knowledge of green architecture were combined in one toolbox with their building cultural indicators in Egypt. Using the toolbox in primary design stage helps the designer by providing more information about each green design strategy, how to use it effectively. Finally, the toolbox provides indicators about how much the whole project will be accepted in Egyptian society and to what extend it applies green architectural principles. For verification reason, the toolbox was tested with two groups of students in Egypt and The Netherlands as well as professional architects from both countries.

1. INTRODUCTION

There is wide knowledge and experience on green architecture principles, but the choice of other priorities or values than environment and climate in the building construction affects the degree of adaptation to climatic comfort and environmental sensitivity. There are many examples around the world of how people do not respond primarily to climate, environment, or economic factors but to their culture (Rapoport, 1986). Indeed, cultural values form a great gap between theoretical principles and the implementation processes of green architecture.

Jos P. van Leeuwen and Harry J.P. Timmermans (eds.), Innovations in Design & Decision Support Systems in Architecture and Urban Planning, 341-356.

This research theory focuses on the identification and incorporation of cultural aspects of current and past green building practices in Egypt with the current foreign knowledge of green architecture. Avoiding cultural obstacles, the green design strategies generated from previous three resources of architectural identity of Egyptian community are culturally tested by non professionals in Egypt thus each green design strategy is labeled with a building cultural indicator which means the acceptance level. User oriented technology is used for verification reason where two separated student design workshops were organized at the Technical University of Eindhoven, The Netherlands and Alexandria University in Egypt. The final version of the toolbox will be tested again by professional architects from The Netherlands and Egypt.

This research product is an easy primary design toolbox to be used by the designer in order to incorporate green architecture principles in new urban settlements generally in hot arid zones and particularly in Egypt's vast desert. Delphi Computer language is used to produce a software version of the toolbox.

Since this paper presents the final results of a PhD research, it will briefly describe how the contents of the toolbox (green design strategies) are collected and how they are culturally tested by questioning non-professionals in Egypt then the procedure of utilizing the toolbox through the student's design workshops for verification reasons and finally the programming process of the final software.

2. DATA COLLECTION FOR THE TOOLBOX

2.1 Rating Systems

More efforts have been done to benchmark the built environment from a green architecture point of view. Thus a large number of systems have been developed to analyse building sites and buildings. Rating systems are organized in form of checklist which gives credit points to existing buildings.

The production of environmental programmes and building codes is, of course, not entirely a matter of science. Rather, it is highly social and contentious processes in which some interests are suppressed and others are reinforced. Commercial construction certification schemes like LEED and BREEAM are just a few examples. (Smith, 2005)

This research studies both the contents and the format of those rating systems. Studying the content helped to understand different green architectural design strategies concerning housing in hot arid regions around the world. Studying the format helped to understand the order and mechanism of such rating systems in order to develop another tool for Egypt. Most of the existing rating systems have been investigated. The following chapter will briefly survey the characteristics of some of them.

2.1.1 Existing Rating Systems

2.1.1.1 EcoHomes

EcoHomes is the homes version of BREEAM (Environmental Assessment Method of Building research establishment). It covers all standard housing developments in England, Scotland, Wales and Northern Ireland. It provides an authoritative rating for new, converted or renovated homes, for both houses and apartments. EcoHomes balances environmental performance with the need for a high quality of life and a safe and healthy internal environment.

The issues assessed are grouped into seven categories: ecology and land use, water, energy, pollution, materials, health and well-being and transport. A project judged by EcoHomes can achieve 'pass', 'good', 'very good' or 'excellent' rating based on the number of points achieved in these seven environmental categories (EcoHomes, 2003).

2.1.1.2 LEED rating system

The Leadership in Energy and Environmental Design (LEED™) Green Building Rating System initiated by the U.S. Green Building Council's provides a national standard for developing high-performance, sustainable buildings.

Based on well-founded scientific standards, LEED emphasizes state of the art strategies for sustainable site development, water savings, energy efficiency, materials selection, indoor environmental quality and innovation. A project judged by LEED can achieve 'Certified', 'Silver', 'Gold', or 'Platinum' rating based on the number of points achieved in these six environmental categories.

Members of the U.S. Green Building Council (USGBC), representing all segments of the building industry, developed LEED by consensus and continue to contribute to its evolution. (The U.S. Green Building Council, 2002)

2.1.1.3 Built Green

Built Green is an environmentally-friendly, non-profit, residential building program of the Master Builders Association of King and Snohomish Counties, developed in partnership with King County, Snohomish County, and other agencies in Washington State.

Built Green provides a framework for assessing building performance and meeting sustainability goals in the areas of site and water, energy efficiency, materials selection, indoor environmental quality, homeowner education and innovation. A building project earns points in each of these six environmental areas to achieve a Built Green™ one, two or three star rating (Washington State and Home Builders Association of Metro Denver, 2003).

2.1.2 Analysis of Existing Rating Systems and Guides

1 - Existing rating systems need other programs, codes and regulations to assess building performance to be an evidence of rating the building from the green architecture point of view. Other codes such as; International Energy Conservation Code 2000 (IECC), International Residential Code (IRC), E-Star Colorado, American Lung Association Health House Standards and programs such as; RESCheck software and more.

2 - Existing rating systems cover only green buildings design aspects. They don't address green urban design aspects. For example they concern about the close distance of building to the mass transportation and provision of bicycle storage but they don't address the design of the city to be walkable, cycle able and all houses are within short distances to mass transportation.

3 - Existing rating systems do not differentiate between design features and design targets. They mix them in the checklist and each design target or design feature achieve some credit points. For example, LEED does not differentiate between certified wood as a design feature and rapidly renewable materials as a design target. Actually, it gives credit points for both. EcoHomes does not differentiate between provision of drying space and cycle storage as design features and improving the performance of the building envelope as a design target actually it gives credit points for all.

4 - The order of most existing rating systems does not meet the requirements needed to develop a culturally accepted pre-design tool (the research target) which covers all phases of the design process. The new tool needs to address the design features instead of the design targets to be easily culturally tested by non professional people.

5 - The weight balance for different categories of green architecture is culturally sensitive and varied depending on the country of production. For example, energy use gets 21% in EcoHomes and 25% in LEED and 42% in Built Green Colorado. Health gets 17% in EcoHomes and 22% in LEED and 13% in Built Green Colorado. The new toolbox for Egypt will give fixed point to each green design strategy and the cultural acceptance will give weight balance for each strategy.

6 - Some green design strategies (features) serve in different elements of natural environment. For example, using local materials serves in both

categories material efficiency and energy efficiency. Mass transportation serves in both energy and sitting. This confusion leads to differences in existing rating systems.

7 - The method of Built Green rating system is suitable for the new toolbox where it demonstrates the green architecture design strategies which covers whole live cycle of the buildings (design, construction, operation and demolish phases). The new toolbox will cover only the design phase but for both fields architecture and urban design.

8 - The green design strategies which cover all aspects of housing design and how they are linked to and promote some aspects of natural environment or human comfort will be used as one of the resources to develop the new toolbox.

2.1.3 Green Architecture Categories

From previous green architecture principles and rating systems 15 categories of green architecture are generated. Each category includes some theoretical green architecture design strategies.

Regarding the elements of natural environment (energy, water, materials, atmosphere, land fauna and flora in addition to human being), five main categories could be generated as the main green architecture categories from the previous 15 categories demonstrated by existing green architecture rating systems and checklists. Indeed, green architecture principles come to make a balance among the elements of the natural environment.

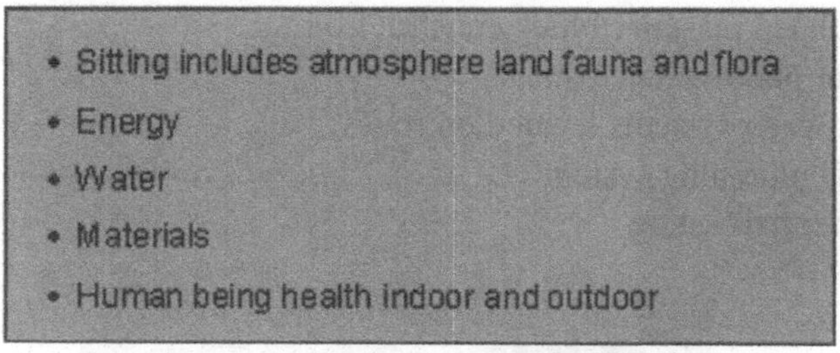

Figure 1. Main Categories of Green Architecture.

2.2 Projects Applying Green Architecture Principles

To understand practically the different green architectural design strategies concerning housing in hot arid regions, there are clear lessons to be learned by considering historical precedents and current examples around the world which claim applying green architecture principles.

2.2.1 Projects Examples

By demonstrating the following projects the research will address just the proven green design strategies for both architecture and urban design. Currently, the research is not going to judge the green design strategies from Egyptian building culture view. The only judgement will come later in chapter 4 by questioning non professional people in Egypt.

In the following examples, the green design strategies (features) will be arranged according to the design process phases instead of previous green architecture categories.

2.2.1.1 Civano: Tucson's solar Village (Corbett and Corbett, 2000)

The master plan for Civano Solar Village is distinctive by the following green design strategies:

- Compactness - High residential densities
- Mixed land use (integrates residential with shopping, workplace, school, and civic facilities)
- Using native, drought-tolerant plants
- Internal circulation system is designed to encourage both bicycle and pedestrian traffic
- The project has the latest fiber optic telecommunications
- Thermal mass technique is used for the walls
- Straw bales, wood frame, adobe, fly ash concrete and Rastra construction
- Insulated masonry block construction
- Solar photovoltaic panels
- Hot water systems located on roofs
- Dual plumping system
- Water harvesting

Figure 2. Straw bale construction.

2.2.1.2 Findhorn ECO Village

The master plan for ECO village Findhorn is distinctive by the following green design strategies:

- Shared facilities (laundry, kitchens, lounges) avoiding unnecessary duplication
- Use of passive solar features where possible through orientation and window layout
- Cellulose insulation (made from recycled paper)
- Non-toxic organic paints and wood preservatives throughout
- Boarding manufactured without the use of toxic glues or resins
- Locally grown and harvested timber from managed forests
- Local stone for skirting, patios and pathways
- Roofing with natural clay tiles

Figure 3. Roof garden house (Eco Village Findorn).

Figure 4. Dome and vaults in New Gourna village – Egypt.

2.2.1.3 New Gourna Village (West Luxor, Egypt) (Fathy, 1986)

The master plan for New Gourna village Findhorn is distinctive by the following green design strategies:

- Courtyard house
- Domes and Vaults
- Mud brick construction system
- Fountain
- Wind catcher (MALGAF)
- Wooden Screen with a lattice-grill work (MASHRABIYA)
- Clerestory

2.3 Field Survey

The field survey is to figure out the influence of building culture on architecture performance. It also contributes to give insight and outline a clear image of Egyptian building culture and to figure out the potentials of green architecture practices in. Finally, it aims to produce a building culture indicator for each green design strategy (to what extent each green design strategy is accepted in Egypt).

2.3.1 The Survey Procedures

2.3.1.1 Focused Issues of the Field Survey

The field survey addressed selected aspects of green architecture design strategies of the following main topics:

- Urban scale: land use, building density street design and transportation
- Building management: Ownership, Design and construction management
- Architectural scale: Construction system, Building materials and Passive climatic design features
- Building facilities and installations: Artificial lighting, Electric appliances, Potable water fixture and Energy supply
- Landscape: Vegetation, Water bodies

2.3.1.2 Sampling Frame and Method used for Data Analysis

Samples of the Egyptian inhabitants 'Non-professionals' will act as informants for the purpose of collecting data. One member of each household has been asked to fill out the questionnaire.

In order to outline a real building culture image for Egypt, it would be impractical to execute a detailed survey for all Egyptians; therefore a random stratified sampling method is used to get right samples where four different main areas were identified to participate, distinguished by both climatic and cultural characteristics. One city and one village are chosen to represent each area. Both genders of different level of education and age were targeted by the survey. (Bernard, 1995)

The four regions are: (Hemdan, 1980) (CAPMAS, 2004)

- Lower Egypt (all delta region including Cairo and Alexandria cities)
- Desert (all oasis in western desert)
- Upper Egypt (all cities and village along The Nile River Valley)
- Nubia (Nubians who live around the aria of Lake Nassir from Aswan to Abu sembel cities)

Minimum of 50 surveys were distributed in each region for a minimum total of 200 distributed for each questionnaire. The response rat was 80% (160 respondents), which helped to conduct a statistical analysis.

An SPSS 12.0.1 database was created to record participant's individual answers and analysis participant's answer.

2.3.1.3 Field Survey Steps

The field survey started with the fact mission trip followed by two different questionnaires.

A trip of ten days around Egypt with the supervisor of the research has been done on December 2003 to get acquaint with the general building cultural phenomenon in Egypt.

A survey of four different regions in Egypt (questionnaire I) was carried out on January 2004 in order to figure out the potentials of green architecture practices in Egypt. Three questions were asked for both current and past building practice then the preference in the future. A fourth question was asked to know the reasons behind this preference.

After the detailed analysis of the first questionnaire's findings, the current and traditional green practices in Egypt both with world green experience were presented in questionnaire II which was carried out on June 2004.

2.3.2 Field Survey Findings

Some of current traditional building practices which were considered as green architecture design strategies were included in the toolbox and were labelled with building culture indicator (the level of acceptance).

All green architecture design strategies which were collected around the world and presented to non-professionals in Egypt were also included in the toolbox and labelled with building culture indicator as well.

The findings of the field survey, together with accumulated experiences of green architecture around the world were used to develop the culturally accepted green toolbox.

3. THE TOOLBOX AND VERIFICATION PROCESS

3.1 Text Version of the Toolbox

In the text version of the toolbox, a list of green architecture design strategies for hot arid region around the world are organized in two main groups urban design strategies and Architecture design strategies. Each main group is divided into seven sub-fields.

The seven fields are organised to follow the logic phases of the design process to help the designer to corporate the green design features in the project easily during the first design stages. Each item is technically described in details. The following is the order of the main sub-fields in each main group of the toolbox.

A. Urban	**B. Architecture**
A.1 Urban fabric	E.1 site selection
A.2 Land use	E.2 Form, zoning and orientation
A.3 Public Landscape	E.3 Building Envelop
A.4 Streets	E.4 Construction Systems
A.5 Open spaces	E.5 Building Materials partially used
A.6 Transportation and accessibility	E.6 Building facilities and installations
A.7 Infrastructure	E.7 Private Landscape

Figure 5. The seven main fields of design process for urban and architecture design.

The same list of green design strategies is attached to the first list which contains the same green design strategies but with acceptance level generally for whole Egypt as well as for the different four regions around Egypt. The acceptance level was presented by percentage where more than 50 % means accepted and less than 50% means unaccepted.

3.2 Student's Design Workshops

The student's design workshops have been one of the elements of the research methodology that has been applied for the verification of the usability of the toolbox. The student's design workshops, presents the analytical comparison of the utilization of the developed culturally accepted green toolbox by the students during the two workshops in the Netherlands and Egypt. It also presents the critical feedback of the two groups of students about the toolbox. The findings of the two workshops, together with student's critical feedback were used to refine the toolbox.

3.2.1 The Workshop's Process

The main goal of the workshops was to test the developed culturally accepted green toolbox (text version) by utilizing it as a primary design toolbox to design a sustainable prototype design for the new region of Toshka in the southwest desert of Egypt, which has to comply with both green architecture principles and Egyptian building culture. As user oriented technology, the students have evaluated the green toolbox afterwards.

A workshop with third year students from different departments (Building Technology, Architecture, Urban Design and Technology Management) – Faculty of Architecture, Building and Planning at the Technical University of Eindhoven in the Netherlands was carried out on September 2004. The second workshop was carried out with third year architectural students in Alexandria University, Egypt on February 2005. The toolbox was presented to the students as a print out text version.

3.2.2 Toolbox Refinements

Regarding the findings of the two workshops, a list of refinements has been generated by student to improve the toolbox.

Refinement 1 some groups in both workshops did not choose from listed items within green toolbox but they add new items for some categories of design phases. The toolbox needs an easy technique to help the user add new items with his responsibility for greenness and cultural acceptance. This technique (adding new items if any) improves innovation and creativity of the user.

Refinement 2 both groups of students misunderstand the use of different items within the toolbox where they use the items that could not be used with each other. The toolbox needs some constraints to be added to notify the user with such conflicted items.

Refinement 3 changing the method of showing acceptance level for green design strategies from percentages to level of acceptance is recommended and very readable for architects and urban designers for example; strongly accept, accept and somewhat accept.

Refinement 4 showing total culture indicator and green certificate for the whole project is very important to help the designer to figure out to what extend the project has achieved green points and to what extend it will be culturally accepted.

Refinement 5 in addition to technical details of each green design strategy, more information is needed to be added to the toolbox explaining why such design strategies are green and culturally accepted. Other information like preconditions for such design strategies to achieve high level of greenness also is needed. Relevant references like books, project examples and web site links are very important.

Refinement 6 considering that both groups of student do not follow specific sequence of design process, the toolbox has to be designed with free movement among its chapters.

Refinement 7 since not all items in the toolbox have the same weight in the design process, different green values (points) have to be given to each green design strategy in the toolbox.

Refinement 8 since most features of the toolbox are used to some extend, so a gradation should be included at every feature (low, average, high) or to be branched into more detailed items.

Refinement 9 A software version of the toolbox must be designed in order to contain all previous refinements. The software also must be flexible to accept addition of new green architecture design strategies in the future.

3.3 Professionals Test

After the refinement process of the toolbox according to the student's feedback a final version for the toolbox was prepared and a final verification is planed with some professional architects from Egypt and Netherlands to test the final version of the toolbox.

4. PROGRAMMING PROCESS

4.1 Delphi Computer Language (Cantu, 2003)

Green Architecture Design strategies; GADS Toolbox is not an interface for an ordinary database. It consists mainly of irregular structures, especially the constraints and the counting of the Green points and Culture indicator makes it very difficult to make such a design tool (stable and fast) in an ordinary database program. Delphi is chosen but also Visual Basic or C++ could have been chosen to develop the calculations routines (core) of which the majority are programmed as a library file (dll). This means that most windows applications (presentation) can read such files. Even Microsoft Excel and Microsoft Access can use the dll's functions without problems.

Actually the program was designed as follows:

1. input files (in txt format for simplicity)
2. core (calculations and constraint check)
3. presentation (screen and user interaction)

Therefore with minor changes, the GADS Toolbox software can be translated or converted to a different programming environment. The program is stable and can be independently extended easily on all three levels. This is due to the fact that all three levels are independent of each other but interacting in a well defined way.

The Delphi solution has been chosen, mainly because the final user does not need to possess or buy Excel or Access because the GADS Toolbox is an independent program. Consequently, updating, maintaining and protection of the program is rather easy now.

4.2 Green Architecture Design Strategies (GADS)

4.2.1 Definition of Toolbox Software

GADS Toolbox is a primary design tool. It is a collection of green architecture design strategies organised in good order to be used by architects and urban designers with different concepts among three levels.

Firstly, it could be used for Toshka region Southwest desert of Egypt with average building culture indicator of different regions of Egypt. Secondly, it could be used particularly in different four regions around Egypt with building culture indicator for each region. Thirdly, it could be used generally for Hot Arid zone around the world as green design strategies without any culture indicator.

GADS toolbox provides green certificates for achieving greenness in the design as well as levels of acceptance indicator in Egypt for whole project during the primary design phase.

4.2.2 The Toolbox Software Main Features

1 - Through out GADS Toolbox, architects and urban designers can choose appropriate items for their design from a pool of green design strategies and print them out in two different forms; one with the titles of selected green design strategies and another with the titles in addition to their technical description.

2 - There are seven main fields in both branches of the Toolbox – urban and architecture design. Each main item contains more detailed items. The designer can start from any field he would like.

3 - Once the item is clicked, the information window at the bottom of the screen containing more technical information for each green design strategy will appear. A separate box shows some photos for each detailed item.

4 - There are two radio buttons at the upper right corner of the screen for the choosing process; 'Select' and 'Not initialized' buttons. 'Not initialized' button will be the default. The designer must click the select radio button if he decides to use specific item in the project. It is also possible to unselect

the item by clicking the ‘Not initialized’ button whenever the designer wants even if he moved to another sub-field.

5 - Some of the green design strategies cannot be used in parallel or will add just one green point for both items so the program will notify the designer about these constraints. Still he is able to select the item after acceptance of the notification.

6 - Each design strategy is labelled with building cultural indicator in Egyptian society and accumulated points are collected when selecting every item in order to be calculated for whole project. The total cultural indicator for whole project will be shown during the navigation through the program and reported at the print out form. There are six levels of acceptance which can be achieved varying from ‘strongly accepted’ to ‘strongly un-accepted’.

7 - Some green design strategies have no cultural indicator. Either because they were not tested or not relevant to culture practice – people never considered them.

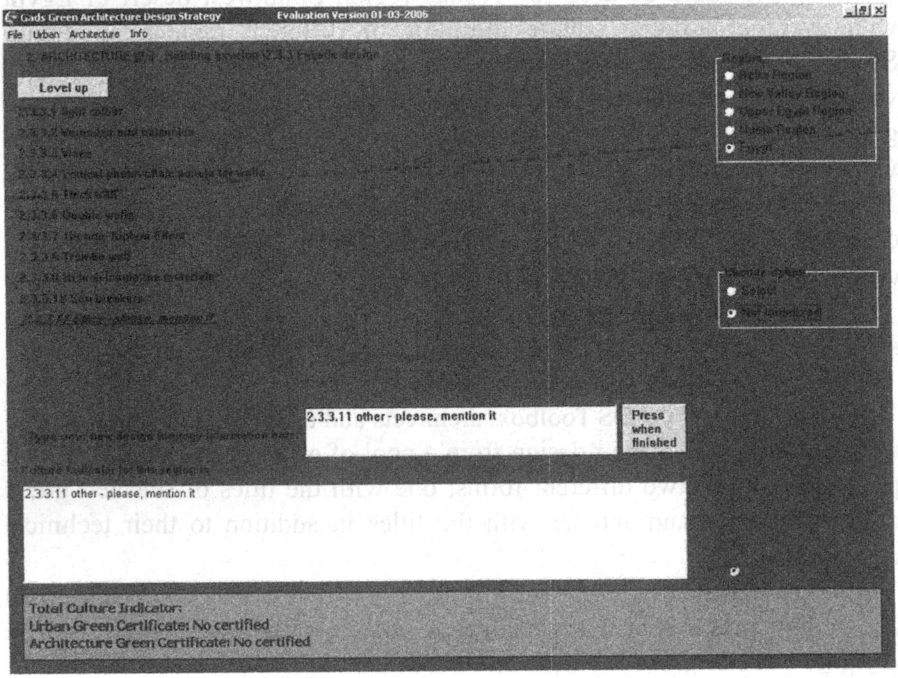

Figure 6. GADS interface.

8 - Each design strategy is labelled with scored point for green architecture and accumulated points are collected when selecting every item. Green certificate for whole project in both main branches Urban and architecture will be shown during the navigation through the program and

reported at print out form. Each design has to achieve specific amount of points to score different level of green certificate which starts from not certified, bronze, silver, golden and platinum.

9 - Instead of selecting from the listed green design strategies, the designer can choose to add his own design strategy at the end of each field and write down a brief description of such design strategy in the information box, which will appear in the print out form without building cultural indicator or green point (on his responsibility).

10 - The software is designed to read plain text files which easily could be altered or redefined. Separated text files are dedicated to different kind of data for example; toolbox text file which contains all green design strategies and their technical description other contains building cultural indicators and other contains constraints.

5. CONCLUSIONS

1 - With this culturally accepted green toolbox GADS, a minimum level of applying green architecture principles in Egypt is insured. In order to achieve high level of applying green architecture in Egypt the building culture of Egyptian has to be changed through:

- Teaching upcoming architects (the university students) using the GADS toolbox;
- Apply GADS with Governmental, regional and local authorities building;
- Both pioneer architects and owners to utilize GADS toolbox.

2 - Applying sustainable settlements in the new region of Toshka south west desert of Egypt is possible by using the developed software program GADS toolbox.

3 - By testing the three resources of Architecture identity of Egyptian community (current and past green building practices in Egypt with the current foreign knowledge of green architecture) through questioning non-professionals, Egyptian building culture has great potentials to implement green architecture principles in Egypt.

4 - One of the effective approaches to apply green architecture in Egypt is the cost of green architecture practices. The more green architecture is cheaper the more green architecture will be applied. More concern is needed about affordable green architecture practices such as existing mud brick house as well as the current situation of less ownership of private car.

6. REFERENCES

Cantu, M., 2003, Mastering *Delphi 7*, Sybex, London.

Corbett, M. and Corbett, J., 2000, *Designing Sustainable Communities: Learning from Village Homes*, Island Press, Washington.

EcoHomes, BRE, 2003, *Rating Prediction Checklist*, Building Research Establishment Ltd., Watford UK – Viewed in website http://www.breeam.org/ecohomes.html

El Fiky, U., 2002, "Arid Desert Colonization: Toshka Region – Egypt as a Case Study", *Housing Construction: an International Task: Proceedings of XXX IAHS World Congress on Housing, Coimbra, Portugal*, Wide Dreams – Projectos Multimedia Lda, Coimbra, pp. 303-312.

El Fiky, U., Hamdy, I. and Van Dansik, D., 2004, "The Potential Green Architecture Design Strategies in Egyptian Building Culture", passive *and Low Energy Architecture: proceeding of Plea2004 – The 21st Conference, Eindhoven, the Netherlands*, Technical University Eindhoven, Eindhoven, pp. 373-378.

Fathy, H., 1969, *Gourna: A tale of two villages*, Dar el Kateb el Arabi Press, Cairo.

Fathy, H., 1986, *Natural Energy and Vernacular Architecture, Principles and Examples with Reference to Hot Arid Climates*, the University of Chicago Press, Chicago.

Rapoport, A., 1986, *Settlement and energy: historical precedents, Environment-behaviour research*, University of Arizona, Tucson.

Smith P. F., 2005, *Architecture in a Climate of change: A guide to sustainable design*, Architecture Press, Oxford.

The U.S. Green Building Council, LEED 2002, *Leadership in Energy and Environmental Design*, Green Building Rating System www.usgbc.org/DisplayPage.aspx?CategoryID=19

Washington State Built Green programs and Home Builders Association of Metro Denver, 2003, *Built Green handbook – HOME BUILDERS Self-Certification Checklist 2003 version* Colorado – Viewed in website http://www.builtgreen.net/

Bernard, H. R., 1995, *Research Methodology in Anthropology Qualitative and Quantitative Approaches*, AltaMira press, Walnut Creek.

Hemdan, G., 1980, *The Character of Egypt: The place also is genius*, Alam el kotob, Cairo.

CAPMAS, Central agency for public mobilization and statistics, 2004, Egypt, *CAPMAS Statistical Year book 2004*, Cairo.

Urban Decision-Making

An Urban Decision Room Based on Mathematical Optimisation

A pilot study supporting complex urban decision questions

P.P. van Loon and E. Wilms[1]

Faculty of Architecture, Delft University of Technology, The Netherlands
[1]Adecs BV, Delft, The Netherlands

Keywords: Decision support system, Urban planning, Design optimisation

Abstract: In general the Urban Decision Room is an interactive computer system based around a digital model for the simulation of complex urban decision questions. Such questions involve various parties with often differing interests. The UDR can assist in finding collective solutions. The UDR is a useful instrument for making the great variety in interests and ideas of the participants manageable. Furthermore, insight is quickly and clearly provided into the results. This enhances the efficiency and effectiveness of processes of urban development. A pilot of the Urban Decision Room based on mathematical optimisation has been made for Schieoevers, an industrial area on the bank of the river Schie to the south of Delft. This pilot is based on a feasibility study for a new urban development in this area which has been carried out by a consultancy firm (Adecs BV) under commission from the municipality of Delft.

1. INTRODUCTION

In the Urban Decision Room Schieoevers, the various parties are each seated behind a computer. In a series of rounds, a collective solution is constructed on a central computer. Each party enters proposals in the search for a collective solution. The parties provide sub-solutions, based on their professional knowledge and their role in the development process, for the problems that arise. A group decision is reached in a repeating series of these interactive actions and decisions. During the process, the intermediate solutions, possibilities, and (then) unsolvable issues are presented to the participants (for instance on projection screens). This provides the parties

Jos P. van Leeuwen and Harry J.P. Timmermans (eds.), Innovations in Design & Decision Support Systems in Architecture and Urban Planning, 359-374.

with the information they need to collectively work towards a solution. Thus, the input from the various parties, each with differing disciplines and stakes, leads via an iterative process to feasible planning options.

The UDR is intended to be a methodological answer to one of the most radical changes which have taken place in the fields of urban development and urban decision making over the past few decades. Namely the replacement of hierarchic by decentralised urban planning and decision making. During this replacement four essential shifts took place:

1. The shift from superior-subordinate positions of authority to parallel positions of authority between equal partners, each with their own goals and means to achieve them (Figure 1.).
2. The shift from hierarchically ranked decision-making areas to individually positioned decision-making areas in which each stakeholder in the process is responsible for decision-making in his own particular area (Figure 2.).
3. The shift from decision making instructions to decision making negotation relationships between decision makers, which involve a special form of cooperation (Figure 3.).
4. The shift from a hierarchically structured to a participatory structured relationship, each stakeholder has his own relationships with the decision making environment and therefore his own conception of (a part of) the decision making problem and decision making solution (Figure 4.).

The OPM Research Group of Delft University of Technology has developed a new methodologically consistent framework for decentralised urban decision making in 'inter organisational' urban planning and decision making situations (Van Loon, Micheels, Wilms, 1987; Van Loon, 1998; Van Gunsteren, Van Loon, 2000). This framework is an operational extension of earlier frameworks in the field of 'urban and regional planning models' (Freind and Jessop, 1969; Catanese, 1972; Lee, 1973). In addition, the new framework has a systematic connection with a 'reflective group decision making environment' (Schön, 1983; Schön and Rein, 1994). Based on this framework ADECS, a consultancy firm for advanced urban planning and decision making, has made a pilot for an Urban Decision Room for Schieoevers, an industrial area on the river Schie to the South of Delft.

2. THE BASIC PREMISE OF THE URBAN DECISION ROOM

In the practice of decentralised, multi-actor and multi-stakeholder urban planning and decision making, the number of alternatives and related components might be so great that the issue of choice becomes too large and

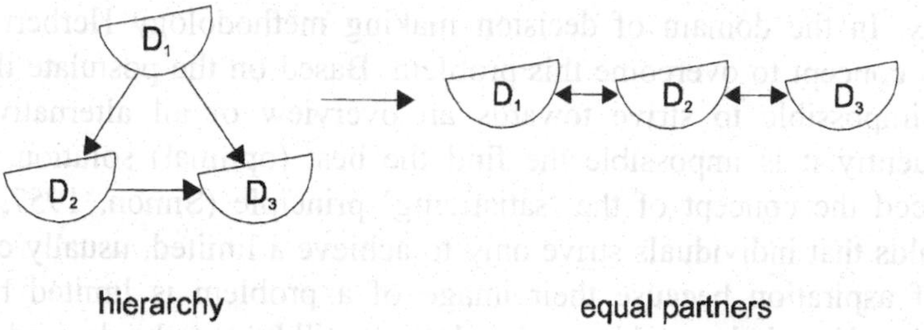

Figure 1. The shift to parallel positions of authority between decision makers (Dn).

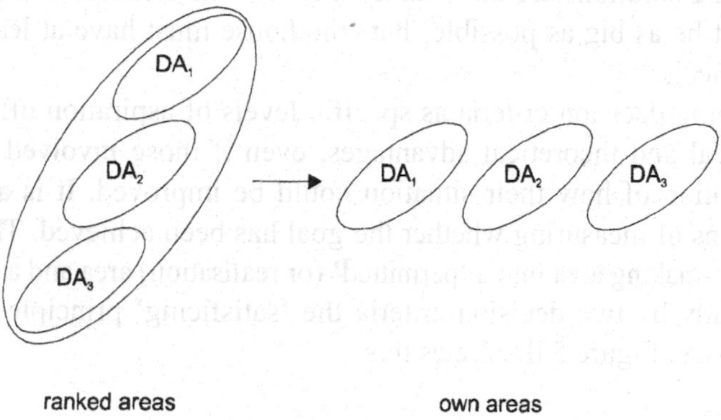

Figure 2. The shift to individually positioned decision-making areas (DAn).

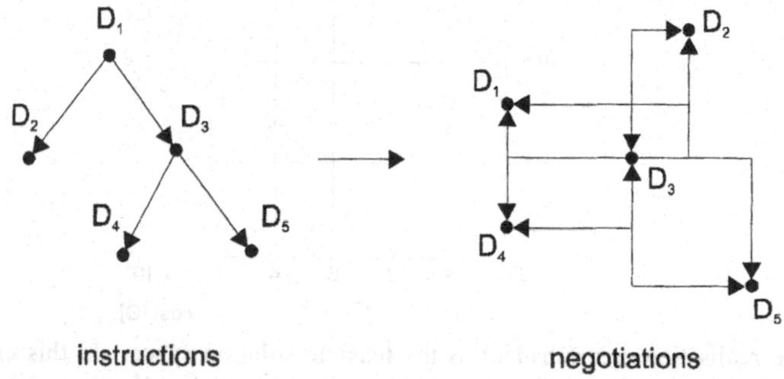

Figure 3. The shift to decision making negotiation relationships.

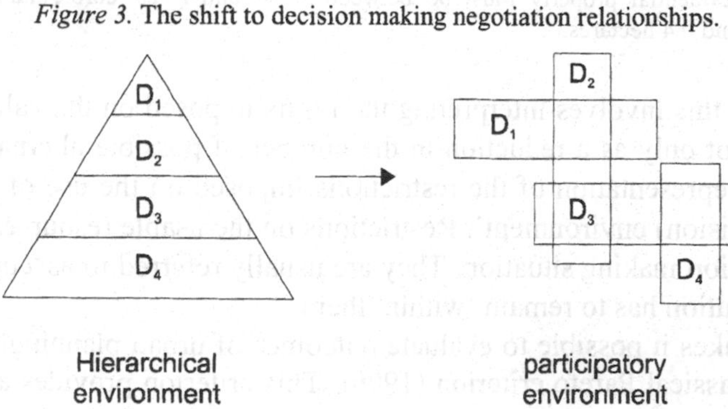

Figure 4. The shift to a participatory relationship.

complex. In the domain of decision making methodology Herbert Simon offers a concept to overcome this problem. Based on the postulate that it is nearly impossible to strive towards an overview of all alternatives and consequently it is impossible the find the best (optimal) solution, he has introduced the concept of the 'satisficing' principle (Simon, 1957, 1969). This holds that individuals strive only to achieve a limited, usually concrete level of aspiration because their image of a problem is limited by their incomplete knowledge and because solutions still have to be devised and the effects of the solutions are not entirely known. The criterion is not then 'the house must be as big as possible' but 'the house must have at least 200 m2 of floor space'.

Describing decision criteria as specific levels of aspiration offers important practical and theoretical advantages, even if those involved have only a vague notion of how their situation could be improved. It is an unambiguous means of measuring whether the goal has been achieved. This divides the decision-making area into a 'permitted' (or realisation) area and a 'forbidden' area. Already by two decision criteria the 'satisficing' principle leads to a solution space. Figure 5 illustrates this.

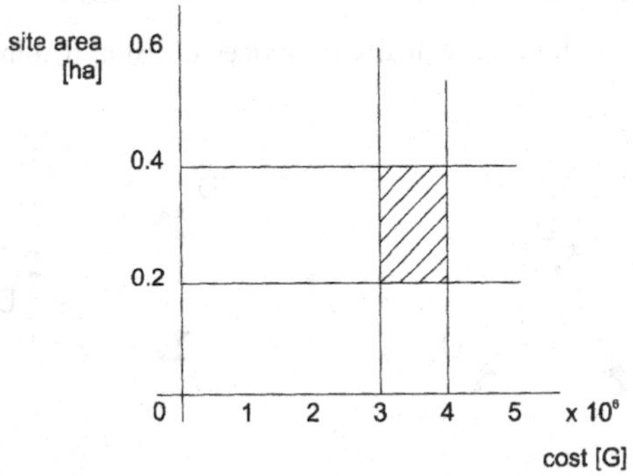

Figure 5. The realisation area (shaded) as the feasible solution space. In this example, the cost of the residential property must be between 3×10^6 and 4×10^6 euros, and the site area between 0.2 and 0.4 hectares.

Briefly, this involves interpreting the limits imposed on the values of the variables not only as a reduction in the number of possible alternatives, but also as 'a representation of the restrictions imposed on the use of resources by the (decision) environment'. Restrictions on the usable resources occur in every decision making situation. They are usually referred to as 'constraints', and the solution has to remain 'within' them.

This makes it possible to evaluate outcomes of urban planning processes with the classical Pareto criterion (1906). This criterion provides a scale for measuring increase in the collective welfare of a group. Collective welfare is deemed to have increased if the welfare of one or more members of the group increases without diminishing the welfare (the constraints) of the other members. The criterion not only comprises a measure of the direction of change, but also its end point. According to this criterion, collective welfare is optimum as soon as it is no longer possible to increase the welfare of one or more individuals without decreasing (violating the constraints) that of one or more of the others.

3. CONSTRAINTS AS THE BASIS OF THE URBAN DECISION ROOM

For using constraints in a decentralised multi-actor urban decision making process, a distinction between the quantitative and qualitative aspects of a constraint occurred to be very important.

The quantitative aspect can be expressed as a maximum (maximum resources to be used), a minimum (minimum resources to be used), or a fixed amount (resources available). For example, municipality M has made available a maximum of 14 hectares of land for the construction of a number of dwellings in area A. It might argue that the site may not cover more than 14 hectares, or area A will become too full and the natural environment in the area will suffer. An example of a minimum is a local park that must cover at least 2 hectares. This is the minimum amount of space required at the site to ensure that there are enough green areas distributed throughout the urban zone of which area A is a part. An example of a fixed amount is 700 dwellings that have to be built in area A.

The qualitative aspect ties in with this, and can be expressed in the qualities of the object being designed. These qualities must be reflected in the final solution. For example, the 2 hectare local park is to be a botanical garden, and the 14 hectare residential area is to be environmentally friendly.

In general, constraints can be seen as expressions of (sub)solutions. For instance, the constraint 'no more than 14 hectares in area, to be developed in an environmentally friendly way' is obviously a good solution for the municipality, and an important part of its plan in and around area A. As soon as the constraints are open to discussion (14.5 hectares might be acceptable, and a 1-hectare park might also suit the plan for the green areas) their representations in a decision making model become part of the process. This allows the underlying (quantitative and qualitative) considerations to be brought into the open.

This approach to maximum and minimum values – which can easily be expressed as constraints in a mathematical decision model in the form of upper and lower limits – and this treatment of their qualities – which can be expressed mathematically as types of variables as shown below – provide the extra instruments for the mathematical structuring of an urban planning problem which brings together devising (designing) possibilities, indicating limits, etc., and choosing from possibilities. This is represented in Figure 6.

limits	residential area ≤ 14 [ha] park ≥ 2 [ha]
variables	ESRA ≤ 14 (ESRA: Environmentally Sound Residential Area) BG ≥ 2 (BG: Botanical Garden)

Using this approach of constraints, two dimensions recognisable in a constraint are very relevant for the practical use in the Urban Decision Room: the goal-oriented dimension and the resource dimension. These will be explained next.

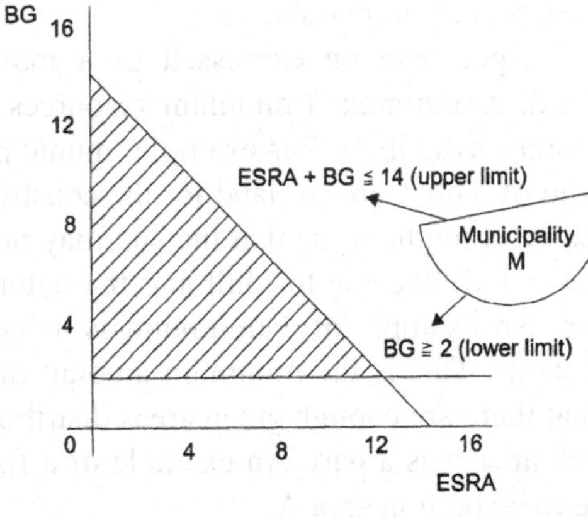

Figure 6. The upper and lower limits of municipality M.

4. CONSTRAINTS IN THE URBAN DECISION ROOM AS A REFLECTION OF GOALS

A constraint is literally a restriction, determined in advance, which the solution must in any event observe. But in an urban planning process, not all constraints are seen in such fixed terms. In fact, a distinction is drawn between 'hard' and 'soft' constraints. Only the hard ones are fixed in advance. The soft constraints are flexible and negotiable, and are determined more fully during the urban planning process. The planners can experiment with these flexible constraints, in the search for the best 'form' and 'position' for those constraints in the solution space or, in other words, the best form and position for the final outcome.

In a multi-party urban planning process, these experiments can soon become complicated. Each party has its own 'flexible' constraints, which they fix only after they have conducted experiments. At that moment, these constraints also become fixed for the other parties. However, if they pose insurmountable obstacles to the other parties' experiments, they will try to render them free again, to allow them to experiment more freely, and set their constraints at a point more convenient for them. This is brought before all the parties, who will respond, and the process may start all over again.

The motives underlying the constraints are very important when it comes to the structuring of the multi-party urban planning brief. In urban planning methodology, these motives are said to belong to the field of goals. After all, a constraint is an expression of a goal, as we have seen in the example of the municipality wanting to create a local park: to ensure a good spread of green areas, a local park of at least 2 hectares must be created in area A.

4.1 Definitions of Goals

The urban planning methodology literature is unanimous about two points when it comes to the function of goals in the urban planning process: they must be set at the beginning of the process, and at the end of the process the outcome must conform to them as far as possible. Incidentally, planning goals are largely referred to as the 'bill of requirements' when it comes to concrete objects such as buildings etc. The term 'project goals' is often used for complex projects, such as neighbourhoods and infrastructure.

However, these two agreed principles are put into practice in wide-ranging, and often confusing, ways. At the beginning of a process goals are seen as a formulation of the brief: 'plan a building B that conforms to requirements 1 to 100'; 'draw up a plan for a neighbourhood N which indicates how project goals 1 to 10 can be achieved'. At the end of the process, goals are used as evaluation criteria: to what extent does the plan fulfil the goals, and where can improvements be made? Between these two meanings lies a vast area in which goals are also used to derive sub-solutions, select from alternatives, guide negotiations in a certain direction, and support particular choices. In urban planning situations where there are several specialists, goals are also seen as a basis for the allocation of tasks: each specialist is responsible for the goals connected with his own speciality. Goals are also seen as an instrument the project manager can use to steer and regulate the process. If someone comes up with a sub-solution that deviates too widely from the goals, the work must be corrected.

This confusing multiplicity of uses and definitions makes it all the more difficult to represent goals in urban decision models in a way that is understandable to all. The term 'goal' is first systematically analysed below, and a description of a goal that can be used in urban decision models is then distilled (Berkhout, Van Loon, Micheels, 1982; Van Loon 1998).

A goal is a primary mental fact. The goal of a particular act is what a person hopes to achieve by his action, or non-action. It is thus a conceived state of affairs, not a description of a state of affairs. The term 'goal' refers to things one is seeking to achieve. Goals can be regarded as more or less standardised conceptions of what is desirable. The goal (of an actor) is a future situation, which may or may not involve the actor, that he aims at in a given starting situation, or at least accepts as desirable, and whose advent he believes he can promote or bring about.

A number of examples: 'efforts should be made to ensure that the land-use planning brings about a regional hierarchy of urban shopping centres, which are situated as favourably as possible with respect to the existing transport system'; 'efforts must be made to attract a mixed residential population to the inner city'; 'specific parts of the organisation must be recognisable in the building'; 'the construction costs must be financially sound and may not exceed a given budget'; 'the architecture of the building must blend in with its surroundings'.

It is possible to distinguish between a 'goal', an 'end', and an 'objective'. The first refers to something specific the individual in question has in mind, whereas an end is often further away, and is only vague in outline. The term objective is used for a formally described and established aim or end: "an objective is a formulation pertaining to a goal; it is a sentence that constitutes a statement about a goal". The term 'goal' will be used below only to refer to the general concept of a 'goal'.

'Goal' is a collective term for a number of specific, closely inter-connected concepts such as value judgments, value goals, object goals, tasks, etc. The whole can be referred to as a 'goal complex' whose elements – the actual goals – are linked in various ways. Below I shall summarise the categories that can be distinguished within such a goal complex, in ascending order of 'concreteness'.

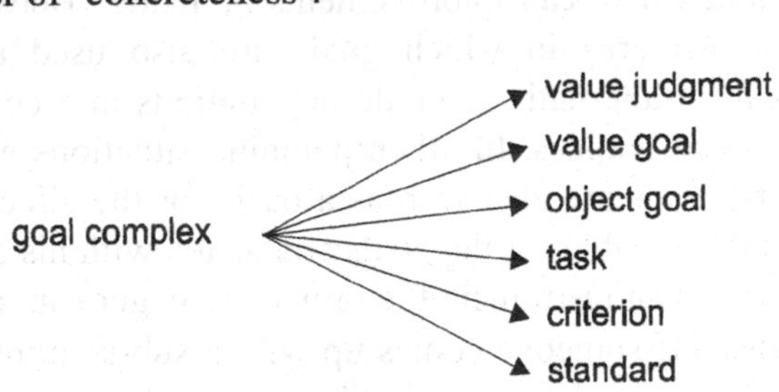

Figure 7. Categories within a goal complex.

– A value judgment is a statement involving a value, such as 'enhance individual freedom', or 'guarantee individual safety', whereby individual freedom and safety are values.

- A value goal is a statement in which the goal is expressed in terms of more general and often more abstract values, such as 'strive for optimum self-development', or 'improve freedom of movement'.
- An object goal refers to more specific and concrete variables of the object under consideration, such as 'create a safe living environment' or 'improve access to public transport'. Whereas value goals can be seen as general statements about the direction which the solution of problems should take, object goals set the requirements for the solutions themselves. The goal 'improve access to public transport' is thus a more concrete expression of 'improve freedom of movement'.
- A task is a fully measurable goal, such as 'there must be at least a 5% increase in the number of dwellings in area A'.
- A criterion indicates that something should be achieved as well – or as much or as little – as possible. For example, 'the maximum possible number of dwellings must be built in area A' or 'the number of dwellings to be built should meet demand as far as possible'.
- A standard indicates that something should be bigger or smaller, or must lie between two values. For example, 'there must be more than 1000 dwellings in area A' or 'there must be between 300 and 500 parking spaces in area A'.

4.2 Mathematical Analysis of Goals

In mathematically linear terms there are five types of goal:

$A = B$	(something must equal something else)
$A \geq B$	(something must be greater than something else)
$A \leq B$	(something must be smaller than something else)
A max!	(something must be as big as possible)
A min!	(something must be as small as possible)

These five types are subject to one definition, composed of three elements. A goal is an imperative, normative, relational statement:

- imperative because the described goal must be achieved as fully as possible, or to a certain level;
- normative because the desired situation specified in the goal functions as a desired 'norm';
- relational because it describes a relationship between two (or more) variables.

This last characteristic deserves further explanation. It appears that many goals are statements about one variable, such as 'improve the living environment' (which can be classified as a value goal). But what is in fact being said is that the future new living environment (first variable) must be better (relationship) than the current living environment (second variable). A statement such as 'there should be enough dwellings' (an object goal) also actually concerns two variables (available dwellings and demand for dwellings) that have to be attuned to each other. A statement such as 'the dwelling density must be 40 dwellings per hectare' (a task) is also relational. It states that the ratio between the number of dwellings and the total site area must (imperative) be such that there are an average of 40 dwellings per hectare (norm).

A goal statement can thus be divided into two (or more) variables and the relationships between them. For example:

'the living environment in area A must be (equal to) a varied living environment'

variables	Living Environment in area A (LE_a) Varied living environment (Vle)
relationship	$LE_a = Vle$

Another example:

'the percentage of subsidised dwellings in area A must be at least 40% of the total number of dwellings in area A'

variables	number of Subsidised Dwellings (SD) Total number of Dwellings (TD)
relationship	$SD \geq 0.4 \times TD$

Finally, an example with more than two variables:

'the housing supply, by building type, size and cost category must match housing demand as well as possible'

(two groups of) variables	Housing Supply ($HS_{b,s,c}$) (by building type, size and cost category) Housing Demand ($HD_{b,s,c}$)
relationship	$HS_{b,s,c} \geq\leq HD_{b,s,c}$ (match as well as possible)

In this way, entire collections of goals can be analysed and represented clearly and consistently.

5. CONSTRAINTS IN THE URBAN DECISION ROOM AS A STATEMENT ABOUT RESOURCES

The other dimension (other than the imperative, goal-oriented dimension discussed in the previous section) of constraints is the resource dimension. After all, a constraint indicates what resource(s) may be employed to achieve the goal. For instance, the constraint:

'the park in area A must be at least 2 hectares'
$PA \geq 2$ (meaning: a park of at least 2 hectares would provide sufficient greenery for this residential area)

implies that the resource 'park' must be used to ensure there is sufficient greenery. This might also be achieved using 'smaller green areas' or on the basis of 'existing landscape elements'. In the constraints: 'the living environment in area A must be (equal to) a varied living environment', and 'the percentage of subsidised dwellings in area A must be at least 40% of the total number of dwellings in area A', the resources 'living environment' and 'dwelling' are the resources for the realisation of the goals.

In the discussion of the origin of constraints in the previous section it was stated that each constraint is linked to its own unique (individually-based) conception of how it is to be realised. Since each constraint also involves a resource, an individual conception of this resource is linked to every constraint.

5.1 Definitions of Resources

Generally speaking, a resource is that which one employs to achieve a goal. Resources might be available in different forms: as a quantity of something (ordered or otherwise), such as an area of land to build on, a sum of money to invest, a stock of building materials for a new building, or a certain space for an activity; a ready-made solution, such as a dwelling in which to live, a park for recreation, a room to work in; a combination of these two, i.e. a certain number of solutions, such as a stock of dwellings, a green area for parks and playgrounds, a building with rooms for offices.

These examples are often referred to collectively as primary resources, to distinguish them from secondary resources. These are resources found more in the 'background', such as specialist knowledge, experience, aids (instruments, machines, energy, etc.). Continuing along the same lines, we may regard procedures, planning, working arrangements and organisational forms as tertiary resources, while management science, methodology and political science would be our quaternary resources. These resources feature in every planning process. This does not, however, mean that each resource has to be identified, considered and weighed up every time. Many resources will be 'fixed', and given.

These classes of resource can be described more fully by relating them to the design-decision process. In this case, I shall take the architectural design-decision process as an example (Berkhout, Van Loon, Micheels, 1982).

In the architectural process, primary resources are used by do-it-yourself enthusiasts and small contractors. In such situations, the solution to an architectural problem depends heavily on the amount, quality, and characteristics of the building materials, and on the number of working hours and the properties of the tools used. The problem-formulating actor both commissions and performs the work. In both capacities he exercises a direct influence on the choice and use of these primary resources.

The use of primary resources in this pure form is rare these days, because do-it-yourself enthusiasts and small contractors have to take into account statutory regulations laid down in building ordinances and land development plans, and by the building inspectorate. They are forced to put their ideas into effect via a more circuitous route, using secondary and tertiary resources – plans and applications for building permits.

Secondary resources come into play in the architectural process whenever the solution to a problem is sought first on the basis of experience with similar problems. The next shed, home, or interior layout will be based on a past design. This serves not only constructive and functional goals and interests but also social and cultural interests. Consider, for example, the many increasingly popular traditional structural solutions and architectural forms. Our appreciation of things built in the past is well known.

Secondary resources appear to make it easier to assimilate the effect of standardised solution models, where the influence of one solution on another is regulated in advance. This is particularly so if these resources are clearly based on previous experience, on what is generally accepted. There will also be less resistance to rules and regulations.

Tertiary resources begin to play a role in the architectural process from the moment professional designers, administrators, and community organisations begin to direct and regulate the problem-solving process. In general, they will set up a number of standard problem-solving routines based on a number of standard problems. If the problem in question can be fitted to one of the standard problems, then the solution will flow more or less automatically from the relevant standard problem-solving routine. Tertiary resources are used mainly to comply with legal and administrative rules. Sometimes the problems do not fit, but have to be forced into the standard, in which case the solution will not be appropriate to the original problem.

Quaternary resources are important in the architectural process at the point where one reflects on actions and processes. This leads to alternative problem-solving routines, the creation of theories about problems and methodologies for routines. This process involves exchange between general reflections on human existence, philosophy and epistemology.

Using these definitions and schematic representations, one can consider resources separately but still in relation to the goals. This is very important when it comes to multi-party design and decision making.

6. A PILOT FOR AN URBAN DECISION ROOM BASED ON A MULTI-ACTOR PROBLEM STRUCTURE

The new method of structuring the multi-actor and multi-stakeholder planning and decision making problem is illustrated in brief by the following pilot study.

Schieoevers, an industrial area on the banks of the river Schie to the South of Delft, covers 130 ha. In the area more then 100 companies, big and small, are located. Parts of the area are deteriorating. The Delft municipality started an urban master plan study three years ago to develop alternatives for improving the area. From the start it was clear that the actual redevelopment of the area would take 10 to 15 years. So a final master plan (blue print plan) would not be appropriate. A flexible planning structure was needed to adapt new insights and an actual solution during the development process. Therefore the study should also result in a decision-making structure which enabled the stakeholders to take actual decisions during the whole development period.

The following stakeholders took part in the planning and decision-making process:

- Delft Municipality (departments of Transport, Environment, and others)
- Delft Town Councillors (Urban Development and Economic Affairs)
- Province of South Holland (Housing Planning Department)
- Regional Government Haaglanden
- Chamber of Commerce
- Business Association Schieoevers

Each stakeholder defined his own objectives and criteria and put forwards his possibilities in the availability of land, types of buildings (houses, offices, factories, etc), and investments. All this information is represented in one mathematical optimisation model (Linear Programming) which gives one common solution space for the planners and decision makers involved. This model as the core of an Urban Decision Room has been used to find out, by means of simulation, what urban redevelopments for the area are feasible.

This example illustrates the two main features of the new structuring of an urban decision making problem.

The first feature concerns the relationship between goals and resources. In the new structure, these are seen as being inextricably linked. In technical terms resources are variables in stated goals: building land is a resource with which to realise a new residential area; dwellings are a resource with which to fulfil a need for housing. Thus stated goals directly determine the plan in the sense of being a proposal for the use of available resources.

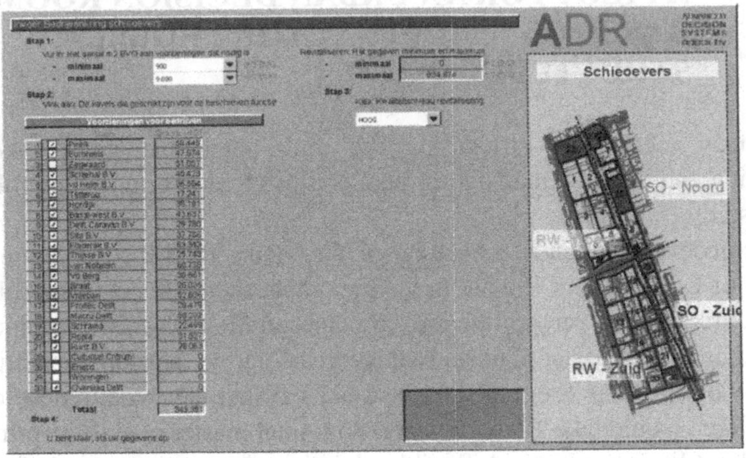

Figure 8. Input sheet Business Association Schieoevers.

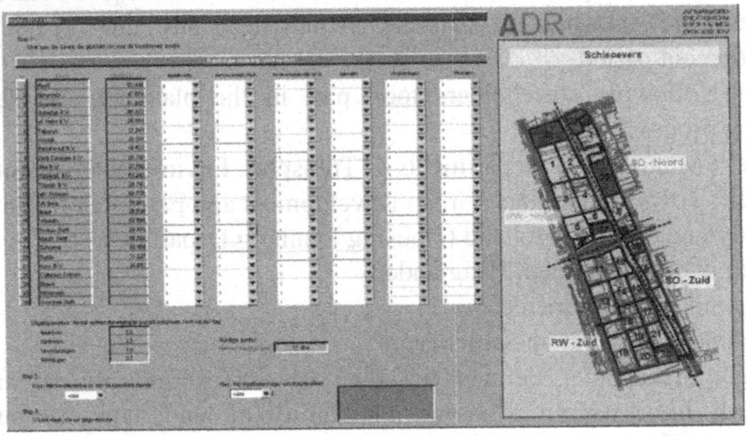

Figure 9. Input sheet Delft Municipality.

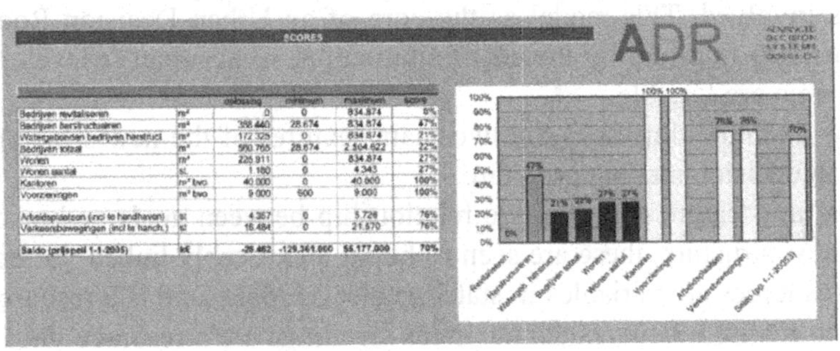

Figure 10. Output of a group decision.

The second feature concerns the concept of goal realisation. The new structure is geared to a plan at the level of the normal situation. For this purpose the goals are formulated in such a way that they relate to the current situation, to what is customary in practice: in a residential area it is customary to have a neighbourhood park; a family home with garden normally requires an area of 120 m2. This enables the process to be geared to the allocation of the resources available and accepted at that particular moment.

7. CONCLUSIONS

This structuring of a multi-actor design decision making problem can be used under the following conditions:

First, everyone involved must accept the goals of each party as the point of departure. While this appears logical enough in a multi-party process, it is frequently not the case. In such a situation the parties adopt diametrically opposed positions and tend to withhold their resources. The result of this is that the process grinds to a halt.

Second, each party must have their own decision-making area, as this enables them to seek upper and lower limits for the realisation of their goals both together and from an individual, autonomous negotiating standpoint. This too would appear to be logical in a multi-party process, but what in fact usually happens is that parties tend to avoid setting limits, because in practice they will either be set too rigidly, too high, which is often unrealistic, or too conservatively, which blocks innovation.

Third, it must be acknowledged that every quantitative statement has a qualitative implication. In practice this often proves a stumbling block, as all too often the assumption is that quantitative and qualitative aspects are separate and, more especially, opposite poles. The parties who present only quantitative goals and constraints are accused of not being concerned about the quality of the plan and qualitative goals are often labelled as being too idealistic and impractical.

The new structuring of the design decision making problem allows the gap between the decision-making model and the decision-making environment to be solved in the first place by the direct linking of the unpredictable decision-making area to all the components of the design commission. This removes the distinction often made in design and planning methodology (Van Loon, 1998) and also in Operations Research between: the 'goal design', a conceptual model of the desired new situation, and the 'resources design', a concrete proposal for a situation which can be achieved in reality. This provides the opportunity during the process of working both with various images of the goal realisation and with a wide variety of proposals for the deployment of resources. This forms the methodological basis of the multi-party process.

In the second place, the problem of the gap is solved as all ideas, wishes, requirements, constraints, etc. are 'translated' into goal statements which cover the specific elements of the situation. This creates a neutral benchmark for all parties: the existing reality in which elements can be identified and named and to which all ideas, no matter how abstract, vague, or idealistic, can be linked. This enables direct communication between the parties.

The authors have recently started as series of real-life experiments with their UDR approach. Urban planners and other professionals in the field of urban decision making are participating. These experiments should provide insight into how the UDR can and will be used in practice.

8. REFERENCES

Berkhout, E.E., P.P v. Loon, and S. Micheels, 1982, *Ontwerp en Planning Methodologie*, Delftse Universitaire Pers, Delft.

Catanese, A.J., 1972, *Scientific Methods of Urban Analysis*, Leonard Hill Books, Aylesbury.

Friend, J.K., and Jessop, W.N., 1969. *Local Government and Strategic Choice, An Operational Research Approach to the Process of Public Planning*, Tavistock Publ. Ltd., London.

Lee, C., 1973, *Models in Planning, an Introduction to the Use of Quantitative Models in Planning*, Pergamon Press, Oxford.

Loon, P.P. v., 1998, *Interorganisational Design, a new approach to team design in architecture and urban planning*, Ph.D. Thesis, Delft University of Technology, Faculty of Architecture, Delft.

Loon, P.P. v. and L.A. v. Gunsteren, 2000, *Open Design, a collaborative approach to architecture*, Eburon Publishers, Delft.

Loon, P.P. v., S. Micheels, and E. Wilms, 1987, *Planninginformatica voor Bouwprogramming, 5 Delen*, TU Delft, Faculteit der Bouwkunde, OPM-Groep, Delft.

Micheels, S., and Wilms, E., 2004, *Schieovers*, Adecs BV, Delft.

Schön, D.A., 1983, *The Reflective Practitioner, How Professionals Think in Action*, Basic Books Inc, New York.

Schön, D.A., and Rein, M., 1994, *Frame Reflection: Toward the Resolution of Intractable Policy Controversies*, Basic Books, New York.

Simon, H, 1957, *Administrative Behavior*, The Macmillon Company, New York.

Simon, H.A., 1969, *The Sciences of the Artificial*, Massachusetts Inst. of Techn.

Forms of Participation in Urban Redevelopment Projects

The differing roles of public and stakeholder contributions to design decision making processes

John G. Hunt
School of Architecture, University of Auckland, New Zealand

Keywords: Urban redevelopment, Public participation, Stakeholder participation, Design negotiation, Design decision support

Abstract: This paper examines how political commitment to participatory design within the context of a major urban redevelopment project was translated into a strategy and a course of action for achieving effective participation within a demanding project timeframe. The project in question involves a new transport interchange for the city of Auckland (New Zealand), the redevelopment of a number of heritage buildings, and the introduction of new buildings to create a mixed use precinct covering three city blocks.

The project, currently being implemented, has involved extensive public consultation and stakeholder participation as it has proceeded through the stages of project visioning, an open public design competition, and the development of the competition winning design. The paper draws a distinction between the contributions of stakeholders versus the public at large to the decision-making process, outlines the different kinds of participatory processes adopted by the local authority (Auckland City Council) to effectively engage and involve these two different groups and the stages in the evolution of the project at which these different contributions were introduced.

The model of 'open design' proposed by van Gunsteren and van Loon is used as a basis for explaining the success of multi-stakeholder inputs at a crucial stage in project development. The paper concludes by examining the limits of applicability of the 'open design' model in the context of urban redevelopment projects in which there is broad public interest, and by suggesting a number of design decision support guidelines for the management of participatory processes.

Jos P. van Leeuwen and Harry J.P. Timmermans (eds.), Innovations in Design & Decision Support Systems in Architecture and Urban Planning, 375-390.

1. AN OVERVIEW OF THE BRITOMART PROJECT

In her seminal contribution to the issue of citizen participation, Arnstein (1969) offers a typology of eight levels of participation, arranged as rungs in a 'ladder' of participation. These range from forms of non-participation in which citizens are merely advised of project intentions and their assumed benefits, to forms of participation based on significant degrees of citizen involvement in decision-making. Midway on this ladder is 'consultation', in which citizens both hear and are heard, but without the power to ensure that their views will be heeded by decision-makers.

The history of the Britomart redevelopment project reveals significant shifts in the kinds of participation included in the design decision-making process. A proposal prepared in 1994 included forms of participation located in the bottom half of Arnstein's ladder, while a subsequent proposal (currently being implemented) has been based upon forms of participation occupying the top half of Arnstein's ladder. These dramatic shifts in both design direction and in the accompanying forms of participation are briefly outlined below, as background to a fuller discussion of participatory initiatives in the current project.

The Britomart precinct comprises a strategically significant 5.2 hectare downtown site with an interesting and problematic history as part of Auckland's historic waterfront port, as the site of the city's original railway station, and subsequently as a bus terminal. In 1994 the Auckland City Council completed its purchase of the land and heritage buildings within the precinct, and through its own property division the Council prepared a comprehensive redevelopment proposal based on a number of tower blocks rising from a large pedestrian plaza, with a five level transport interchange below. Of the sixteen heritage buildings only two were retained, together with the facades of a further six. Widespread public opposition to this loss of heritage buildings, and the lack of effective public consultation on the proposal as a whole, sparked a public reaction that brought the project to a halt and was a significant factor in precipitating the demise of the mayor and key city councillors in local government elections in 1998.

The incoming mayor and council, elected in part on their promise to rethink the project, initiated an extensive process of public consultation, and in 2000 initiated a public design competition as the basis for establishing the form that future development of the precinct should take. From the 153 designs submitted in the initial stage, seven were selected for development in a second stage. The outcome of this two-stage process was a unanimously selected proposal that differed radically from the initial 1994 design. In the winning design the historic pattern of city streets was reinstated, and all the heritage buildings were retained. A number of functions of the transport interchange were woven into this pattern of streets, rather than being located entirely below ground. New buildings were to be of a scale and character compatible with the heritage buildings. Positive urban open spaces replaced the amorphous character of spaces between the high rise towers of the earlier design. While in one sense the competition-winning proposal relied upon traditional patterns of city making, in another sense it was innovative in its integration of these traditional patterns with the requirements of a multi-modal transport interchange

2. FORMS OF PARTICIPATION IN THE BRITOMART PROJECT

In analysing the participatory processes that underpin the current design proposal, a clear distinction may be made between the roles of the public at large and project stakeholders in the design decision-making process. In this section of the paper the different ways in which these two groups were drawn into the decision making process are outlined and diagrammatically summarized. These are traced through a typical project development sequence, involving project visioning and the formulation of project objectives, preparation of the design brief, and an iterative process of design generation and evaluation at increasing levels of detail.

2.1 Public Participation

For several years prior to the commencement of the Britomart project the City Council had conducted a number of public participation exercises, focussing on broader waterfront redevelopment issues. A significant aspect of this process was the use of focus groups, in which participants imaginatively envisioned the future waterfront, and on this basis developed a number of broad goals and principles.

Immediately prior to the inception of the Britomart design competition, these principles, together with the working assumptions that the City Council had adopted for the project, were tested via a public questionnaire. This questionnaire was supported with an exhibition of information regarding broader decisions already taken for waterfront redevelopment, and regional and city transport strategies (of which the Britomart project would become a critical element.) More than 1400 questionnaires were completed and analysed and the results of this analysis were included in the design brief for the open first stage of the design competition. In addition to focus groups and a public questionnaire, members of the public were also invited to submit entries for the first stage of the competition, in which a single A1 drawing was required. In order to maximize opportunities for public participation in what was, in fact, a complex and challenging urban design problem, entrants were invited to submit proposals for the whole project, for the building sites only, or for public open spaces within the precinct. Of the 153 design submissions received, relatively few were from entrants who were not qualified design professionals, suggesting that this form of public participation lacked broad appeal. Nevertheless, considerable public exposure was given through the media to the range of ideas included in the Stage 1 designs.

At the conclusion of this first stage the public at large had an opportunity to contribute further to the decision-making process, by visiting the exhibition of Stage 1 designs, selecting up to 5 'preferred' designs and recording their reasons for this selection. Approximately 11,000 people visited the exhibition and over 600 questionnaires were completed. These responses were analysed in terms of the reasons given for the preferences (rather than the preferred designs themselves), and responses were grouped into a number broad themes. This analysis was conveyed to the Stage 1 judging panel and included in the Design Brief for the second stage of the competition. These themes could be summarized as development that is 'distinctively of our place', a people-friendly environment with a strong waterfront presence while also providing an inviting and efficient interchange, and with existing heritage buildings an integral part of the redevelopment.

In this way a broad ranging set of public responses became a significant source of design direction for the seven teams selected to take part in the second stage of the design competition. The final step in the process of public participation was the opportunity for the public to select their three most preferred Stage 2 designs and to rate each on a 1 to 5 scale, for each of twelve key project objectives. Approximately 400 responses were received and a summary of these responses was conveyed to the judging panel. Interestingly, the panel had difficulty in reconciling a number of these public evaluations with its own, perhaps as a result of the increasingly complex and technical nature of the Stage 2 designs.

2.2 Stakeholder Participation

This second strand of the participatory process ran concurrently with, and continued beyond, the processes of public participation. The extensive and varied involvements of stakeholders throughout the early phases of the Britomart project reflects the diverse set of interests that intersect in such a major urban redevelopment project, and the recognition that each of these interests might potentially have a significant impact upon the viability of the design outcome. For this reason each of the stakeholder groups needs to be involved in the decision-making process at those point where it's interests can be most effectively accommodated.

The first step in the process of stakeholder involvement was a series of presentations by the project sponsor, outlining the principles and working assumptions for the project. Thirty six stakeholder groups were consulted in this way, including land owners and property interests that might be affected by the redevelopment of the precinct, design and development professional institutes, civic and heritage groups, and organisations representing each of the public transport modes (rail, buses, ferries, taxis, and tourist coaches) that would use the interchange. By involving each stakeholder group at the outset of the process it became possible to establish the most effective way to involve each of these groups in the design decision-making process. Broadly speaking, these groups would assume either an advisory role (in which they would they would be kept informed of design developments and have the opportunity to comment - in other words, to 'hear and be heard'), or a decision-making role (in which they would have direct access to the design consultants and exert influence in the design decision-making process.) Within this second category a core of key stakeholders whose interests were critical to the success of the project were identified, and invited onto the design competition judging panel. Included were organisations representing the city's bus and coach operators, property development interests within the city (the Britomart project being the single largest property development in the city's history), and the local Maori iwi (the indigenous people of the central Auckland area with whom the City Council had a collaborative agreement.) The judging panel also included the Mayor and several senior city councillors. The remaining members of the nine person panel were three independent architects and urban designers, including the panel chair.

A second tier of stakeholders assisted the judging panel with a variety of technical evaluations of the design submissions. These included representatives from the Historic Places Trust (a national organisation responsible for overseeing the protection and redevelopment of historic buildings), the city retailers association, experts in the provision of heavy and light rail transport modes, ferry service providers, and representatives of departments within the City Council's own organisation that would have project delivery responsibilities. Each of these stakeholder representatives was responsible for ensuring that appropriate others whom they represented made their own assessment of design submissions at both stages of the competition, and discussed their views with their stakeholder representative. This key stakeholder role in the competition judging process represents the second step in the process of stakeholder involvement.

The outcome of the competition was the unanimous selection of a single submission at the end of the second stage. In order to facilitate the implementation of this winning design the City Council established a key stakeholder Reference Group that would meet on a monthly basis with the winning design team during the one year technical design development phase that followed the second stage of the competition. This was the third step in the process of stakeholder involvement. This Reference Group comprised members of the judging panel, together with key City Council staff responsible for various aspects of the statutory approvals process that would precede construction work. Several stakeholder groups that had been represented at the second tier during the design competition process also joined this Reference Group – namely the Historic Places Trust and the city retailers association. This reflected the growing importance of these interests at the technical design development stage. The Reference Group was charged with ensuring that the vision embodied in the competition winning design was retained during its development, and that the requirements of each stakeholder group were also met. Significantly, the meetings of this group provided a forum in which competing stakeholder interests could also be debated and resolved.

These roles indicate that the members of this Reference Group were now working in partnership with the City Council and its design consultants – a step on the ladder of participation that Arnstein (1969, p217) characterises in terms of opportunities to negotiate and to engage in trade-offs. At these Reference Group meetings the design consultants would present their design development work, followed by discussion and the preparation of a number of recommendations. A review of the recorded outcomes of these meetings indicates a high level of agreement amongst the various stakeholder representatives in respect of all issues except one. In fact as a consensus-building exercise the process appeared to be very successful. It is also possible to trace the way in which this group shaped the direction of design development, particularly via the search for alternative design solutions when pragmatic and technical demands began to undermine the very clear vision on which the competition-winning design had been based. In this way these stakeholder representatives contributed directly to the creative process by which the innovation embodied in the competition-winning design was maintained and enhanced.

2.3 Public versus Stakeholder Participation

From the above review it is apparent that a clear distinction can be drawn between public and stakeholder contributions to the design decision-making process. While the public were given a number of opportunities to influence the direction and content of the decision-making process they were not involved directly in the processes of negotiation by which design decisions were made. This would seem to be a reasonable and realistic strategy – reasonable because such voluntary public involvement cannot be assumed to be truly representative of the public at large, and realistic because of the limitations to the number of individuals that might be effectively included in the decision-making process, and the technical nature of many of these decisions[1].

In respect of stakeholder participation, the levels of participation varied considerably amongst stakeholder groups, as noted above. However, even where key stakeholders were included directly in the decision-making process, their involvement was not as design collaborators – as full participants in the creative processes by which multi-disciplinary design teams typically develop and refine their design proposals[2]. Given the natural limits to the size of such a core design team, and the diverse range of design professionals comprising design teams for major urban redevelopment projects, this would also seem to be a reasonable and realistic position. Rather than direct creative involvement, the role of key stakeholders becomes primarily evaluative, but with the opportunity to suggest design elements or configurations that would fulfil their individual stakeholder expectations or requirements.

From the perspective of participatory processes in urban projects, Arnstein's ladder of participation offers some useful distinctions (although Arnstein herself favours citizen participation at the upper levels of decision making power.) The Britomart experience suggests that in order to be effective, public participation should take a number of forms, but that in each case the focus should be on developing insight and understanding regarding the aspirations and expectations of the public at large, for the project in question. This requires that the public 'hear and be heard' – the distinguishing characteristic of the mid-level of participation that Arnstein labels 'consultation'. In respect of key stakeholder participation, this ranged from Arnstein's 'consultation' to 'partnership' within design decision

[1] For example, public responses favoured declamation of part of the site, and the introduction of canals, boat basins and such like. However, this presented a number of significant technical and cost problems.

[2] In regard to design collaboration, Cys and Ward (2003) suggest that true collaboration occurs when practitioners consciously step beyond their boundaries and engage in a new process of design that is informed by their collaborators from other professional areas, while Papastergiadis (2004, p. 160) suggests that 'collaboration is a way of receiving others, involving both the recognition of where they are coming from, and the projection of a new horizon line towards which the combined practice will head'.

making, with key stakeholders involved in negotiations and trade-offs around key design decisions.

A diagrammatic summary of these differing kinds of participation over the duration of the project process is provided in *Figure 1*. The one-directional arrows represent processes of consultation, while two-directional arrows indicate an involvement based on negotiation. The diagram provides a basis for project sponsor decision in respect of participatory processes for projects of this kind.

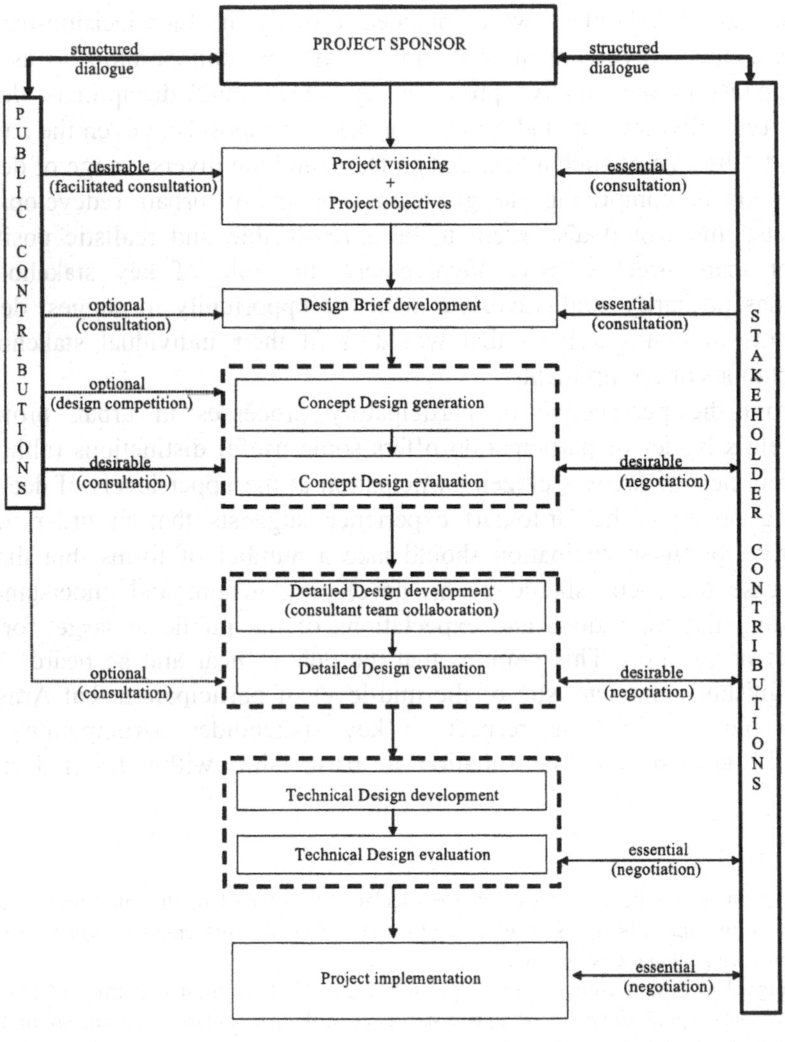

Figure 1. Participatory processes for the Britomart project.

3. NEGOTIATION, CONSENSUS AND COMPROMISE: PARTICIPATION IN PRACTICE

3.1 Perspectives on Multi-stakeholder Negotiation Processes

If negotiation is the hallmark of stakeholder involvement in design decision-making then arguably the most interesting and critical aspect of the Britomart participatory process was the role of stakeholder members of both the competition judging panel and the Reference Group established to guide the development of the competition-winning design. Both processes served as a forum for debate amongst competing stakeholder interests, allowing each of the stakeholders to become aware of the needs and interests of the others, and to temper their individual positions and expectations accordingly. For example, during the judging process transport-related stakeholders came to accept the urban design merits of an on-street arrangement for buses, rather than a transport terminal facility occupying development sites within the precinct. Property development representatives came to realise the urban design merits of a development that was compatible with the scale of existing heritage buildings, rather than maximising development floor area. During the ten half-day workshops in which key stakeholders engaged with the design consultants as part of the Reference Group process, heritage stakeholders came to accept the need for significant change to the principal historic building within the precinct, in order to achieve the larger goal of a pleasant and efficient transport interchange. How might this crucial aspect of effective participatory processes be understood?

The literature in planning includes an extensive examination of participatory processes, although there is little on the particulars of participatory processes within urban design projects. In the context of planning and and urban design Forester depicts the process as one of negotiation, in which each of the stakeholders "take advantage of their differing priorities in order to realise joint gain" (Forester, 1998, p. 8). He observes that this requires stakeholders to come to understand their different priorities in the first place, and to then negotiate well rather that reaching what he refers to as "lousy compromise". Bishop and Bonner suggest that such compromises typically result from unresolved conflicts, in which "the lowest common denominator emerges because nobody has taken hold of and properly managed the partnership process in a proactive, coherent and principled manner" (Bishop and Bonner, 1995, p. 210). They argue that a key feature of consensus building is providing a framework within which people with different views can come together interactively, working towards a mutually satisfactory solution.

Van Gunsteren and van Loon offer a more detailed explanation of multi-stakeholder interactions, pointing out that each individual shapes his or her order of preferred outcomes at the moment of decision. They note that "this implies that where individuals have to take a decision together, something which on paper could be considered a dilemma between them will not necessarily turn out to be so. Conversely, what appears to be a problem-free issue may well prove to be a dilemma in practice" (van Gunsteren and van Loon, 2000, p.16). To explain what guides the process of negotiation and decision-making they invoke Pareto's Criterion: namely that collective welfare is at optimum as soon as it is no longer possible to increase the welfare of one or more individuals without decreasing that of one or more of the others. In the context of multi-stakeholder design decision-making, van Gunsteren and van Loon (2000, p.17) observe that any decision is at an optimum when it can no longer be improved to the benefit of one or more individuals without diminishing the benefits enjoyed by one or more of the others. It is this goal that shapes the direction of the collective decision-making process.

It remains to ask in arriving at such an optimum outcome whether differences of views amongst individuals need to be relinquished. Hillier suggests that this is not necessary, noting Mouffe's (1992) position that "decisions taken in a conflicted field generally imply the repression of some representations" (Hillier, 2002, p. 225). On this basis Hillier argues for a meaning of consensus as con-sensus: as feeling and sensing together, but not necessarily implying agreement.

3.2 The Experience of Multi-stakeholder Negotiations within the Britomart Project Reference Group Process

The above noted perspectives offer an explanation of why the diversity of stakeholder interests comprising the Britomart project Reference Group managed to reach a consensus view on so many of the issues that confronted it. However, as previously noted, there was one issue on which some members of the Panel remained divided, and for which a solution was unable to be reached without separate mediation. This issue went to the heart of the future success of the development, and its resolution provides an example of how innovative design ideas are at risk during design development in multi-stakeholder projects.

The issue in question was the extent to which the former Chief Post Office (CPO) building, being the most important of the heritage buildings within the precinct and the most strategically located, might become an integral part of the experience and operation of the transport interchange. The competition-winning design had proposed that the ground floor of this building would become a giant public vestibule to the interchange, providing ticketing and waiting areas for commuters, and with direct linkages to bus stops, ferry services, and the below-ground rail station. (This feature of the competition winning design is illustrated in *Figure 2*.)

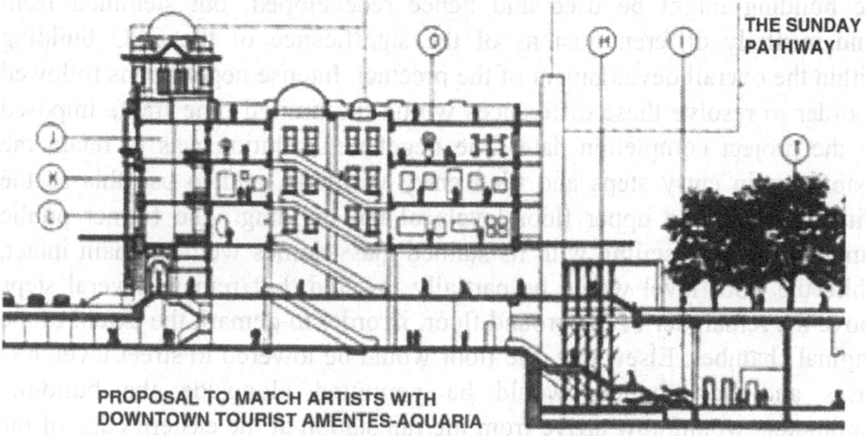

Figure 2. Competition winning design proposal.

To provide a focal public space, it was proposed that the stained glass domed ceiling above the ground floor would be raised three floors to the top of the building, thus transforming the existing central light-well above these domes into a public atrium. Escalators would rise though this space to provide access to upper floors, and descend to the first basement level, where a public concourse would lead to the nearby ferry terminal and descend a further level, to the rail station. In this way commuters and visitors to the city would arrive into this central atrium space. This innovative idea was extended in early meetings with the Reference Group, by proposing to lower the original ground floor level of the building to street level, in order to strengthen connections between the interior and surrounding streets.

A key objective in achieving an efficient interchange was to concentrate bus stops around the former CPO building, thus strengthening its role as a nodal point in the interchange. A loop-based bus circulation system within the precinct was proposed by the design consultants, allowing easy access for buses to three sides of the building. Glass-roofed bus shelters would be introduced alongside the building, in order to provide minimum visual interruption to its heritage features.

At this point in the design development process it became apparent that these proposals were in direct conflict with the interests of the Historic Places Trust, and their desire to have the former CPO building remain as little changed as possible. In particular, their representatives opposed any changes to the ground floor level, and the raising of the domed ceiling of the former banking chamber. Additionally, they indicated a preference to keep buses and bus stops away from the streets immediately alongside the building. These differences of view were not merely in respect of how the building might be used and hence redeveloped, but stemmed from fundamentally different visions of the significance of the CPO building within the overall development of the precinct. Intense negotiations followed in order to resolve these differences within the limited time frame imposed by the project completion date. The negotiated solution was to retain the historic main entry steps and tiled entry vestibule, and to use this as the principal access to upper floor levels of the building. The former public banking chamber ceiling with its stained glass domes would remain intact, while the floor level would be partially lowered, but remain several steps above the remainder of the ground floor, in order to demark the extent of the original chamber. Elsewhere, the floor would be lowered to street level, and buses and bus shelters would be permitted alongside the building. Commuters would now arrive from the rail station at the eastern edge of the building rather than at its centre (as proposed in the competition-winning scheme). These design changes are indicated in *Figure 3* below.

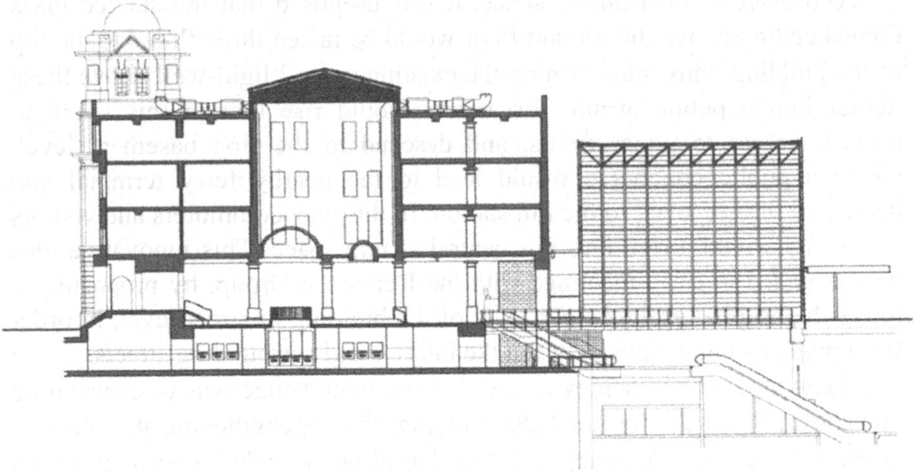

Figure 3. Stakeholder-negotiated design solution.

This instance of negotiated consensus, while experienced by the affected stakeholders as less than ideal, was nevertheless a form of agreement based upon Hillier's (2002) particular interpretation as 'con-sensus' noted above, and ensured that the former CPO building would be an integral part of the transport interchange. However, this con-sensus solution also had wider ramifications, with the need to reposition the below-ground station eastwards, and to introduce a much larger entry structure (the so-called 'glasshouse') to the interchange, alongside the eastern face of the CPO building. This in turn has significantly reduced the size of the public open space (Station Square) that formed an important ground level link between the interchange and the remainder of the development, and which was a distinctive feature of the competition-winning design. *Figure 4* illustrates the design arrangement achieved through negotiated agreement amongst the stakeholders, including the 'glasshouse' structure alongside the Station Square frontage of the CPO building.

Figure 4. As-built relationship between below-ground station and former CPO building.

This example illustrates not only the high degree of interdependence amongst design decisions that is typical of urban design projects, but in addition the need for the consequences of decisions taken by stakeholder groups to be fully explored at the time that they are first promulgated, and any necessary trade-offs agreed. In this sense these key stakeholders become active participants in the creative process by which design innovation may be either achieved, or retained during the design development phase.

4. CONCLUSIONS: DECISION SUPPORT GUIDELINES FOR THE MANAGEMENT OF PARTICIPATORY PROCESSES

Barlow (1995, p.1) identifies three key factors that shape the process of participation: the types of techniques employed, the types of participants, and the stage in the planning/design process in which these participants become involved. In examining the politics of urban development he notes that participation may have a number of objectives: to provide information, to test opinions or assumptions, to invite new ideas, and to help to achieve participant aims. He also observes that in developing participatory strategies planners will need to both gain public support and retain sufficient control in order to deliver on the promises of the public agencies that they represent (Barlow, 1995, p. 59).

The Britomart project illustrates these points quite clearly. Participation techniques were selected to suit both the expertise of the participants and the points in the process at which this expertise could be most effectively used. Each of the techniques sought to inform, to test assumptions, and/or to invite new ideas, while all techniques sought to achieve an outcome that would meet participant aims. Control of the process remained with the planners and other City Council staff, in order to meet project timeframes and other aspects of the political commitments made by the new City Council to the public at large.

The experience of the Britomart project suggests that in deciding on participatory processes within urban redevelopment projects, a key decision becomes the balance between consultation and partnership (as these terms are used by Arnstein, 1969), and the extent to which stakeholders are given decision-making powers. Stakeholders may have an advisory role, be empowered to participate in negotiations around design decisions, or have full design collaboration roles. The character or 'profile' of the participatory process will be determined to a large extent by which stakeholder groups are accorded what level of decision-making involvement.

Arnstein favours forms of participation based on partnership, delegated power and citizen control (being the top three rungs on her 'ladder of participation'). Levels of participation below these she disparagingly refers to as 'tokenism' and 'non-participation'. However, this paper questions whether such levels of public involvement are achievable in urban redevelopment projects. In particular the complex and often technical nature of design decision-making would seem to preclude the close involvement of lay persons. Key stakeholders may need to participate in these decision-making processes, but in many instances these stakeholders will not have the skills to engage directly in the processes of design manipulation that characterize the involvement of design collaborators[3]. So is the 'open design' model outlined by van Gunsteren and van Loon fully applicable in urban development projects that impact upon the public realm and therefore upon public interests at large? Interestingly, these authors provide an example of 'open design' in an urban planning context, but it would appear to be atypical of urban development projects, since it involved the selection between two competing design proposals by two stakeholder groups – future residents of the proposal, and the local housing association. The potential exists for such stakeholders to be involved in a truly representative way, in contrast to public interests at large, which cannot be so represented in the decision making process. This in turn suggests the need for forms of participation that lie outside of those on which the 'open design' model is based, if the public at large are to have an effective involvement.

The Britomart project clearly demonstrates that participatory processes can and should be an integral part of design decision support strategies for urban redevelopment projects. Specifically this project suggests the following guidelines for the management of participatory design decision making:

1. That public participation should be focussed on early project visioning stages, and may need the facilitated involvement of members of the public at large, in order to avoid superficial consultation outcomes.
2. That open invitations for the public to evaluate project objectives and design proposals are likely to be more effective than direct involvement in design-generative activities.
3. Public involvement in design-generative activities raises questions of the representativeness of participants and of skills needed to make decisions in the face of competing interests. However, designing as an envisioning activity is a potentially effective way for members of the public to articulate aspirations and visions, without the need to address project- specific technical issues.
4. The distinction between public and stakeholder interests is fundamental in developing a strategy for participation.
5. Stakeholders will represent diverse and sometimes competing interests. Those interests which are critical to the success to of the project need to be distinguished from those that are merely important for the project.

[3] Achten (2002, p. 7) identifies the opportunity for all participants to manipulate the design at certain moments as one of the defining characteristics of collaborative design.

6. The involvement of key stakeholders in processes of negotiation and trade-offs during design decision making is both highly desirable and achievable.
7. The need for the project sponsor or commissioning agency to retain control over the participatory process is a realistic constraint that need not undermine the integrity of participatory processes.

5. REFERENCES

Achten, H.H., 2002, "Requirements for Collaborative Design in Architecture", in: *Design and Decision Support Systems in Architecture: Proceedings of the 6th International Conference*, Ellecom, The Netherlands, Eindhoven University of Technology, pp. 1-13.

Arnstein, S.R., 1969, "A Ladder of Citizen Participation", *Journal of the American Institute of Planners*, July 1969, pp. 216-224.

Barlow, J., 1995, *Public Participation in Urban Development: the European Experience*, Policy Studies Institute, London.

Bishop, J. and A. Bonner, 1995, "Participation, partnership and consensus – making 'parts' into 'wholes'", *Town and Country Planning*, August 1995, pp. 209-211.

Cys, J. and S. Ward, 2003, "Using the c-word:collaboration in pedagogy and practice", *Design and Research: Second international conference of the Association of Architecture Schools of Australasia*, published at http://www.arbld.unimelb.edu.au/events/aasa

Forester, J., 1998, "Creating public value in planning and urban design: the three abiding problems of negotiation, participation and deliberation", *Urban Design International*, **3**(1): 5-12.

Gunsteren, L.A.v, and P. v Loon, 2000, *Open Design: A Collaborative Approach to Architecture*, Eburon Publishers, Delft.

Hillier, J., 2002, *Shadows of Power: An Allegory of Prudence in Land-Use Planning*. Routledge, London.

Mouffe, C., 1992, *Dimensions of Radical Democracy, Pluralism, Citizenship and Community*. Verso Publishers, London.

Papastergiadis, N., 2004, "Creative Practice and Critical Thinking", in: Wissler, R., B. Haseman, S.Wallace, and M. Keane, *Innovation in Australian Arts, Media, Design: fresh challenges for the tertiary sector*, Post Pressed, Australia, pp. 159-170.

The Neighbourhood Wizard

Cause and effect of changes in urban neighbourhoods

Jos P. van Leeuwen and Léon A.H.M. van Berlo

Eindhoven University of Technology, The Netherlands

Keywords: Participatory Design, Bayesian Networks

Abstract: The Neighbourhood Wizard is a website that makes citizens aware of the consequences of the changes that they would like to realise in their neighbourhood. Users of the website can suggest changes to their neighbourhood. A Bayesian Belief Network is used to predict the effects of the changes on several indicators of liveability as experienced by the community. The Neighbourhood Wizard also shows what would be the optimal experience of liveability for different sections of the population.

1. INTRODUCTION

In the list of criteria that people use when buying or renting a house, the quality of the house itself plays a dominant role. However, the quality of the neighbourhood of the house, both physical and social, plays an increasingly important role as well when people are selecting their future home. It can often be noted that the inhabitants of neighbourhoods make an effort to keep up the quality of their surroundings and even try to improve it, as they realise that the 'liveability' of their environment is strongly determined by the social and physical quality of the neighbourhood. In many neighbourhoods in the Netherlands local initiatives for neighbourhood improvement are taken by groups of inhabitants or neighbourhood associations.

Municipalities generally have the policy to support and promote these initiatives. A good approach is to start by initiating dialogues on the issues involved and nurturing these dialogues until they mature, until they lead to new, shared, moral understandings (Etzioni 2004). Many municipalities

Jos P. van Leeuwen and Harry J.P. Timmermans (eds.), Innovations in Design & Decision Support Systems in Architecture and Urban Planning, 391-406.

therefore actively seek the participation of inhabitants in the development and (re-)design of neighbourhoods.

Two issues are commonly encountered in the process of citizen participation in urban development. Firstly, citizens are not generally educated to acknowledge the complexity and range of problems in their neighbourhood, but rather tend to focus on the problems they encounter in their daily activities. Citizens are not always able to acknowledge the viewpoints and needs of all members of society and their suggestions for improvement therefore tend to be too constricted.

Secondly, citizens tend to express themselves in terms of solutions when asked to describe the problems they encounter and the wishes they have for the improvement of their neighbourhoods.

Both issues are addressed in the research project that is reported in this paper. The paper first introduces the objective of the research project. Section 3 outlines the general approach and research method that was followed for the development of the project. Section 4 introduces the term liveability, which plays a key-role in this work. Section 5 discusses how people experience liveability and how this can be modelled. Sections 6 and 7 explain how we built a knowledge representation from data that was collected regarding the experienced liveability of neighbourhoods in the city of 's-Hertogenbosch. Sections 8 and 9 discuss the development and evaluation of the prototype system. Finally, in section 10 we draw conclusions regarding the work done and future work.

2. OBJECTIVE

The research project presented in this paper aims to support the process of participation by neighbourhood inhabitants in (re-)designing their neighbourhood. In this project we have focused on making citizens realise what the consequences are of their ideas for changes in the neighbourhood. These consequences are often more complex than citizens can oversee and have to do with many different aspects of the quality of the neighbourhood. Changes in the neighbourhood may have a positive influence on one aspect, but work out negatively for other aspects. Furthermore, proposed changes may have a positive effect for one group of inhabitants, but be assessed negatively by inhabitants who have different requirements.

The objective of the project was to develop a tool that allows citizens to propose changes to their neighbourhood and assess the quality of these changes. The assessment is done in the context of how the community will experience the various aspects that determine the liveability of the neighbourhood.

This tool can be used by citizens to gain a deeper insight in their own desires and in the multifaceted qualities of the changes they propose. In the process of participatory planning and design, this kind of tools can have an important educational and motivating function (Frissen 2003).

3. APPROACH

The main problem that needed to be addressed in this research project was how to assess the proposed changes in the context of multiple aspects of liveability, aspects that are appreciated differently by different sections of the population, e.g. teenagers or elderly people. Having to deal with uncertainty, we selected Bayesian statistics as the methodology to model the causal relationships between neighbourhood characteristics and inhabitants' experiences.

The development of the prototype was limited by narrowing its scope from 'any' neighbourhood to the plaza type of habitat. The prototype was tested in the Dutch city of 's-Hertogenbosch. Recent research by the so-called Bosch Architecture Initiative (BAI) has delivered a comprehensive set of data regarding how people experience a considerable number of physical and qualitative aspects of plazas in the city. This data collection played a central role in constructing the knowledge representation for the system.

4. LIVEABILITY

The term *liveability of the built environment* is often seen in policy reports on both national and local levels. However, the usage of the term is rarely univocal; no clear and commonly accepted definition is available (Michalos 1997). In these documents, many other terms are used to differentiate the term liveability, such as 'welfare in the habitat' and 'quality of living'. Although most of these terms have a certain overlap in their meaning, we must conclude that liveability is not univocally defined, let alone in a measurable way. What is commonly understood, however, is that liveability is influenced by the personal appreciation of inhabitants and the relative importance of a range of aspects. The experience of liveability is personal, differs for every inhabitant, and is expressed using a varying number of other terms (Jirón and Fadda 2000).

5. EXPERIENCING LIVEABILITY

To be able to predict the liveability of a (changed) neighbourhood, we need to model the way inhabitants experience the liveability of their neighbourhood. *Figure 1* shows the model by Leidelmeijer and Marsman (1999) that we use for this purpose.

The model in *Figure 1* shows that an inhabitant experiences the liveability of an environment by evaluating his satisfaction of a number of characteristics of the environment. The satisfaction of each characteristic is determined by his appreciation of the state of that characteristic and is weighed by the importance that this characteristic has to the inhabitant.

Each inhabitant has a personal interpretation of the state of characteristics (e.g. the state of the characteristic 'busyness' might be 'I think it is quite busy here'), a personal appreciation of that state (e.g. 'I like it busy!'), and the importance of that characteristic for the experienced liveability (e.g. 'busy or not, it does not matter to me' or 'the busyness here matters a lot to me'). Every inhabitant of a neighbourhood will have a different, personal experience of how liveable the environment is.

The values that inhabitants give to the characteristics in the model are influenced by their personal preferences in the context of the habitat of the environment. The characteristics also influence each other, which is indicated in the model by the dotted lines (see section 5.1 and *Figure 1*).

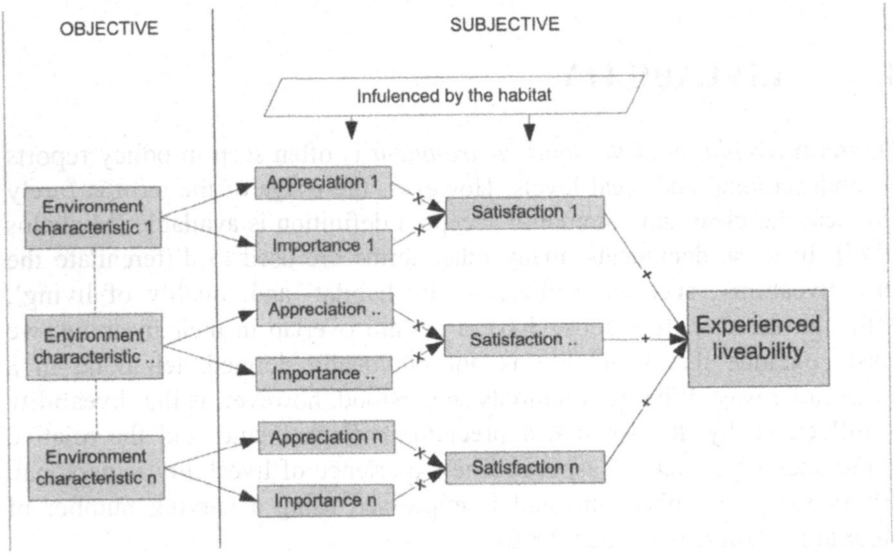

Figure 1. Experiencing liveability (Leidelmeijer and Marsman 1999).

Figure 2 shows an example of how a particular individual experiences the liveability of a particular habitat with respect to three of its characteristics.

The person in this example finds the composition of the environment 'simple'; its status is perceived as 'popular'; and the security of the environment is perceived as 'safe'. The person's appraisal of these characteristics is found in the second column. This appraisal leads to a satisfaction per characteristic and the aggregation into the overall experienced liveability.

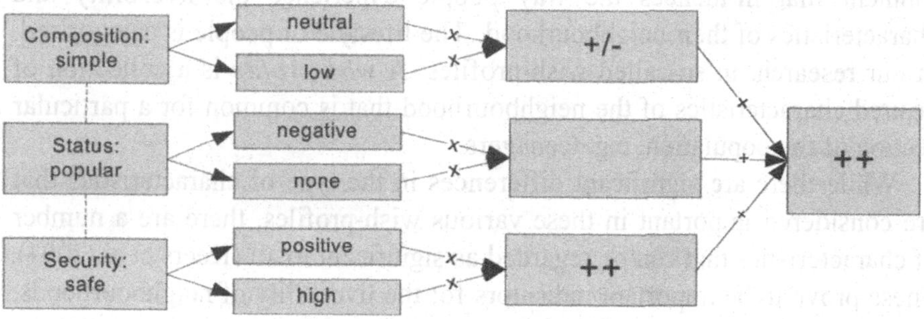

Figure 2. Example of the liveability of a habitat as experienced by an individual.

5.1 Mutual Influence of Characteristics

In the given example only three characteristics of the environment were taken into consideration. Actually, the number of characteristics of the environment is theoretically unlimited. In the example, the characteristic 'Status' is not given any importance in this context. There are many such characteristics that appear not to have a significant contribution to the overall experience of liveability. These characteristics we have termed *elements*. The characteristics that do have a significant influence on the overall experience of liveability are called *aspects*.

Although the elements do not have a direct influence on the experience of liveability, they do influence the state and appraisal of other characteristics. Therefore we cannot prune them from the network. The state of the aspects also influences the perception of other characteristics (both elements and aspects). Changing the state of a characteristic will influence the perception of other characteristics: if we reduce the number of cars, the plaza may appear more green.

These interrelationships are modelled in the network, which forms the basis for the Neighbourhood Wizard application. Changing the state of elements will influence the state of aspects, or even trigger a chain of changes in other aspects or elements, and will eventually lead to a differently perceived liveability.

5.2 Habitat and Wish-Profiles

The term *habitat* can be defined as the physical/functional and social environment in which people live (de Leeuw-Hartog 1988). *Lifestyle* is the behavioural pattern that can be identified for people with similar demographic, social-structural, and cultural characteristics (Stoppelenburg 1982). Research has shown that the lifestyle of people and their habitat are related, albeit not one to one (Anderiesen and Reijndorp 1990). In our research, we define the habitat as the present physical and social environment that influences the way people experience the liveability and characteristics of their neighbourhood. The lifestyle of people is categorised, in our research, in so-called wish-profiles. A *wish-profile* is a collection of desired characteristics of the neighbourhood that is common for a particular section of the population, e.g. teenagers.

While there are significant differences in the type of characteristics that are considered important in these various wish-profiles, there are a number of characteristics that can be regarded as significant to all (Keers et al. 2004). These prove to be important indicators for the liveability of neighbourhoods, as perceived by the majority of a population. However, the choice of indicators to be involved in the evaluation of a particular neighbourhood can be made dependent of the objectives of the evaluation. In the Neighbourhood Wizard, this means that for each application of the system, the moderator (normally the municipality) can decide which elements and aspects of the neighbourhood should be included in the user-sessions.

6. DATA COLLECTION

At the start of the research project, data was made available from a recent survey of people's experiences of liveability regarding the city of 's-Hertogenbosch in The Netherlands. In this survey, people were invited to participate in several walks through the city and were asked to fill out a questionnaire. The questionnaire consisted of differentials that helped people to indicate their experiences of characteristics such as 'public furnishing', 'available facilities', 'public accessibility', 'status', 'appearance', 'ambiance', etc.

For plazas, over 40 characteristics were included. For each of these characteristics, the subjects were asked how they experienced the characteristic on the plaza they were visiting and to indicate this on a differential with a scale of seven possible values ranging from deficient, through moderate and neutral, to ample and excessive. Per characteristic, specific terms were used to indicate these values. The classification and categorisation of the terms was helping people to really understand the meaning of the elements.

Over 250 subjects have participated in this enquiry, 15 plazas were visited. Within the scope and constraints of the project, the data was an adequate starting point for the development of the knowledge representation and the first phases of prototyping.

7. KNOWLEDGE REPRESENTATION

The causal relationships between the elements and aspects of a plaza can be modelled on the basis of the data retrieved from the BAI enquiry. This data can be used as input for the construction of a Bayesian Belief Network (or Bayesian Network, BN in short). In short, a BN can be described as a directed graph where the nodes represent variables (here, characteristics of the environment) and the connections between the nodes represent the causal relationships between them. The variable represented in a node has a state that is determined by a conditional probability table in which the possible states of related nodes are the attributes.

Determining the structure of a BN is the first important step in modelling the knowledge domain. This can be done by the knowledge expert who constructs a network from the knowledge that is acquired through, e.g., communication with domain experts. Another way to construct the network is by examining significant amounts of data from the particular domain. In this project, the latter approach is used to come to a base network which was refined by domain experts.

7.1 Structural Learning

There are several methods to learn the structure of a Bayesian-network model from data. To learn the structure for the prototype system the NPC algorithm (necessary Path Condition) was used. This algorithm is a constraint-based learning algorithm that derives conditional independence and dependence statements (CIDs) by performing statistical tests on pairs of variables in the data set.

7.2 Structure of the Bayesian Network

The first network that was found is shown in *Figure 3*. In this first structure, the significance level of the dependency test was set to 0.05. We see that there are some characteristics that do not have a significant relation with any other characteristic (see *Figure 3*). Even when the dependency test significance level is set to 0.3, these relations still do not appear in the learned structure. This means that the data does not show sufficient evidence for the existence of these relations. For some relations this intuitively seems strange. For example, the missing influence of the playground elements seems incorrect. The municipality of 's-Hertogenbosch noted that the placement of playground elements on a plaza always leads to a lot of criticisms from inhabitants.

There are several possible explanations why we did not find such relations:

- The collected data is incorrect;
- The influence of the playground elements is processed by one or more other characteristics. This means that the effects of changing playground elements will be the same as changing other elements;
- The criticism from inhabitants is not founded.

It is very difficult to point out the most likely of these possibilities. Probably all three reasons play a role. However, the second possibility can be checked with corresponding statistical tests, such as a Chi-Square test. The Chi-Square test did not find any relation between the playground elements and any other characteristic in the collection of data.

Therefore we must conclude that, given the situation and data, the state of playground elements does not have a significant influence on other characteristics of the built environment, in the context of how inhabitants

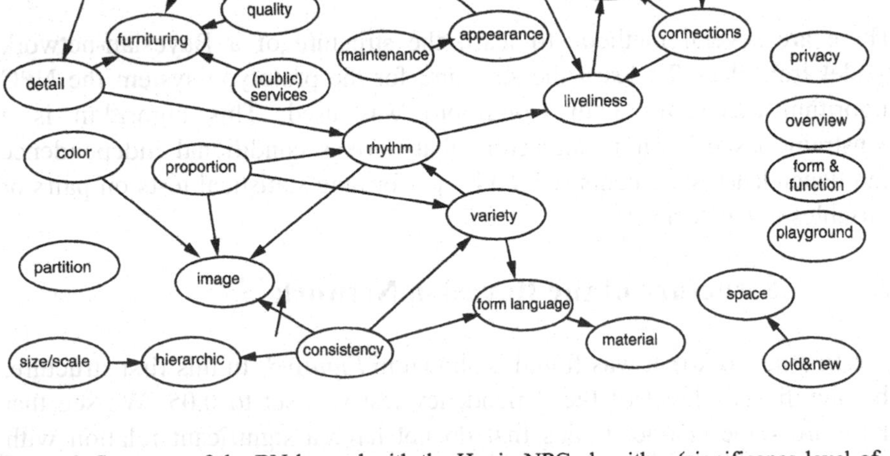

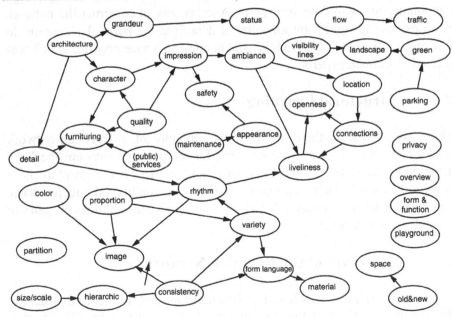

Figure 3. Structure of the BN learned with the Hugin NPC algorithm (significance level of dependency test = 0.05).

7.3 Resulting Network

Besides expected relations that were not found, there were also unexpected relations that were found. The reliability of these relations could also be confirmed with a Chi-Square test. This does not mean that these relations really exist in reality; it only means that they are present in the given data set. In reality, the apparent dependency between variables may in fact be a coincidence. The network structure was verified and refined with the aid of a number of professional city designers. Some additional relations were found using the NPC algorithm with a different dependency test significance level (0.2 and 0.3). The structure of the BN that was used for testing the prototype is shown in *Figure 4*.

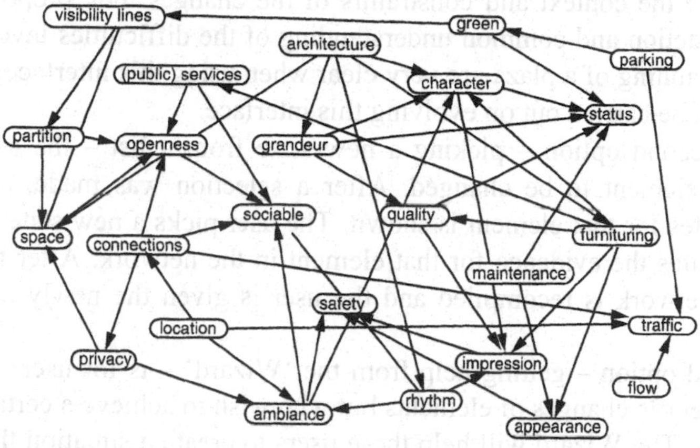

Figure 4. The Bayesian Network that was used for testing the prototype.

8. PROTOTYPE DEVELOPMENT

A prototype web application was developed to utilise the Bayesian Network. The development of the prototype was based on the following principles:

- User-interaction focused on a task assigned to the user. Users can experience this like a game;
- Representing the effects of changes to elements of the neighbourhood that users propose on the various aspects, or indicators of liveability;
- Representing the desired states of the aspects for different sections of the population. This way users can evaluate how changes will be appreciated differently by the different types of people;

- Availability of the system on Internet with no unnecessary threshold for usage by a large public;
- Easy to use interface and obvious navigation.

8.1 Changing Elements of the Neighbourhood

Changing the state of an element can be done in three different ways:

- Drawing;
- Picking a new state from a list of possible states;
- Getting help from the 'Wizard'.

The 'drawing' option gives the users the ability to draw modifications on a photograph of the plaza. In the current prototype, this is only a dummy function and the prototype does not recognize the drawn modifications. However, the drawing does have a function because users will learn to acknowledge the context and constraints of the changes they propose. The design interaction and common understanding of the difficulties involved in the urban planning of a plaza are very clear when using this interface. Future work should be carried out on evolving this interface.

In the second option – picking a new state from a list – the user first selects the element to be changed. After a selection was made, a list of possible states for that element is shown. The user picks a new state and the system updates the evidence for that element in the network. After this, the Bayesian Network is recompiled and the user is given the newly expected situation.

The third option – getting help from the 'Wizard' – is for users who do not want specific changes of elements but who wish to achieve a certain state of an aspect. The Wizard will help these users to create a situation that suits their vision. First, the user indicates the state of the aspect that the user wants to achieve (for example: the aspect 'Status' should be experienced as 'rich') the system gives a list of elements which influence this aspect directly. These elements are in fact the parent-nodes of the aspect in the Bayesian Network. The system responds with a number of suggested states of elements, one suggestion for every element that affects the aspect. The suggested state of an element is that state that contributes most to achieving the desired state of the aspect. The user can choose to apply a suggested change, so that the desired state of the aspect will become effective. The result of each change is shown as the percentage of the population that will actually experience this new state of the aspect.

8.2 Implementation Techniques

The Bayesian Network was implemented using the Netica software-library (www.norsys.com). Netica provides an API for using a BN in a Java environment. Java is also a suitable environment for web application development. The user interface was created using Java Server Pages (JSP). The use of JSP allows the complete separation of interface and technical functionality. Schematically the system interaction looked like this:

JSP Pages ←→ JAVA classes ← API → Netica Bayesian Network.

To ensure uniformity of the web application, it was made XHTML 1.0 compliant, uses an XML configuration file to configure the starting situation of each plaza, and has a separate language file for altering it easily for use in other languages. The dynamic composition of the graphs for the representation of predicted effects (see next section) is done with the help of the open source Java class library JfreeChart (www.jfree.org/jfreechart).

8.3 Presentation of Predicted Effects

The presentation of the predicted effects and other interactions with the user are the most difficult issues of this project. Charts are potentially a very effective and flexible way of presenting the predictions. A potential disadvantage of using charts is the fact that they are not always easy to read and understand. It cannot be assumed that the intended user-group, i.e. neighbourhood inhabitants, is accustomed to reading charts. Therefore, much caution is necessary with the visual ergonomics of the presentation using charts.

Two different types of charts are used in the prototype. One is called ‘Level 3’, using lines, and the other ‘Level 2’, using bars. These are the most useful views of the tool. In these two charts, the presently experienced liveability is shown in black. The experienced liveability that is expected after the situation is changed, is shown in red. A third view is called ‘Level 1’, showing a simplified presentation using stars.

In Level 3, for each aspect a chart like the ones in *Figure 5* is shown. Every chart represents one aspect (indicator of liveability). The possible states of the aspect are on the horizontal axis. The percentage of people that experiences the aspect in each of these states is shown on the vertical axis.

The green, vertical line represents the state that is preferred most by the section of the population that is currently selected by the user. In other words, it shows the preferred state for this aspect for the selected wish-profile. In the example of *Figure 5*, the teenagers want the ‘Space’ as ‘stringent’ as possible. The task that is assigned to the user, is to achieve a situation in which the chart shows a summit at this vertical line. However, in other wish-profiles the preference may be different and the green line will be in another place. Users of the system have to try to deal with all the desires from the different sections of the population. In the example of *Figure 5*, the teenagers will experience the proposed changes positively (left chart). But the elderly people will experience a negative effect for the aspect 'Space', since their optimal experience of 'Space' is 'natural' as shown on the right in *Figure 5*.

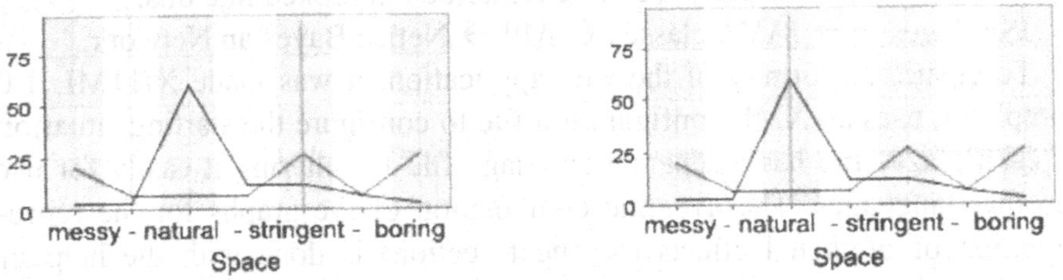

Figure 5. The aspect 'Space' in Level 3 with a green, vertical line indicating the state that is desired most in the selected wish-profile: teenagers on the left; elderly on the right.

The charts in Level 3 are always given for all the aspects at once. This way the users can see the mutual effects between the aspects. Users can select different wish-profiles to inspect the effects in relation with the desires of the different sections of the population. That way the green lines in the graphs will change according to the optimal state for the selected wish-profile.

Level 2 is a somewhat easier-to-read presentation of the effects. In this view, each chart also presents the effects for one aspect, but shows these effects in bars and for all wish-profiles at once.

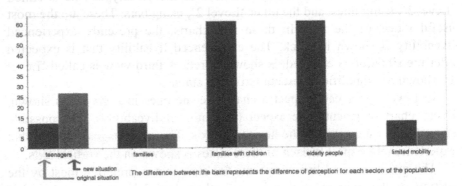

Figure 6. The aspect 'Space' in Level 2 with the effects for all wish-profiles.

The chart shows two bars for each wish-profile, the black bar indicating the original situation, the red bar showing the expected effects. This chart presents how the majority in each section of the population will appreciate the effects. Referring to the line-charts in Level 3, these bar-charts show the percentages found at the intersection of the lines with the green vertical line indicating the preference of each population group. This way the user can see directly whether the proposed changes are experienced positively or negatively in the different wish-profiles. In the example in *Figure 6*, teenagers will think that the changes are positive for the given aspect 'Space', but elderly and families with children will think negatively of the changes.

8.4 Navigation

The navigation of the prototype should be self-explanatory and easy to use for every user. Once a plaza is selected, the user is presented the current situation in photographs and in charts. This allows the user to interpret the liveability as currently experienced at this plaza. Through the navigation buttons, the user can start to make changes to elements of the plaza. After each change, the effects will be shown immediately in the charts. The user can select which level of complexity the charts should show (lines, bars, or stars). *Figure 7* shows two screenshots of the web application prototype.

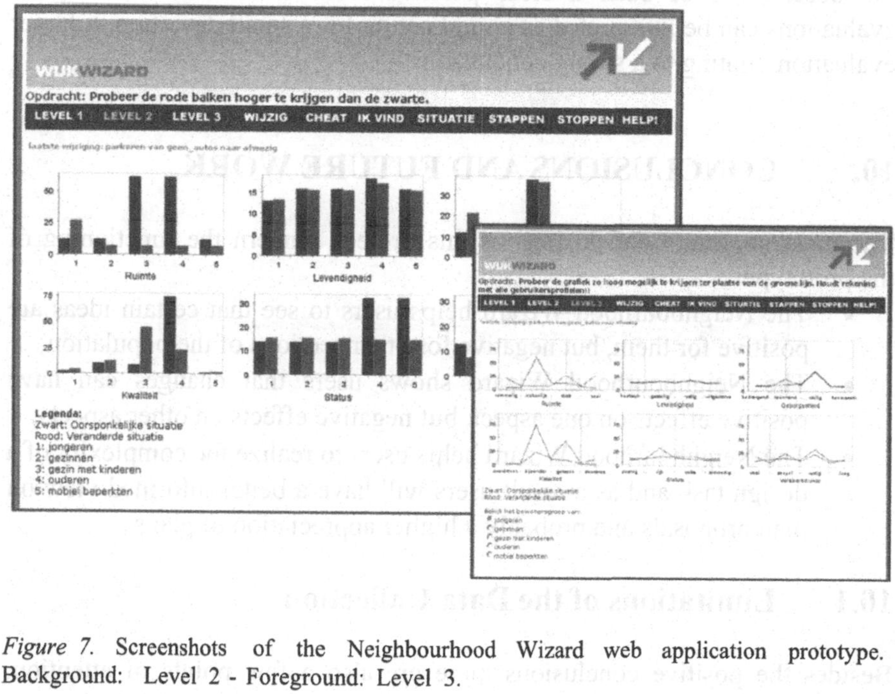

Figure 7. Screenshots of the Neighbourhood Wizard web application prototype. Background: Level 2 - Foreground: Level 3.

9. EVALUATION OF THE PROTOTYPE

The prototype was tested and evaluated by inhabitants of the city of 's-Hertogenbosch. After the test they were asked to complete an online evaluation form. The prototype offered sessions for seven different plazas in the centre of 's-Hertogenbosch. Over one hundred subjects participated in the test. A small number of them also filled in the evaluation form.

The evaluation form contained nine different propositions. The subjects were asked to indicate how much they agreed with these propositions, on a scale from 1 to 10, where 1 represented a total disagreement with the proposition and 10 a total agreement.

A score of 7.4 was given to the proposition "Thanks to the Neighbourhood Wizard, I now see that certain ideas are positive for me, but negative for other members of our community." A score of 7.0 was given to "The Neighbourhood Wizard shows me that changes can have positive effects on one aspect, but negative effects on other aspects."

The evaluation of the test confirmed the educational function of the prototype, but the number of returned evaluation forms is too small, at the time of writing, this paper to make a definite statement about this. However, the observation of such a clear positive result in a small number of evaluations can be interpreted as an indication for a small deviation in larger evaluation, hinting to a strong conclusion.

10. CONCLUSIONS AND FUTURE WORK

The most important conclusions of this project concern the functioning of the prototype:

- The Neighbourhood Wizard helps users to see that certain ideas are positive for them, but negative for other sections of the population;
- The Neighbourhood Wizard shows users that changes can have positive effects on one aspect, but negative effects on other aspects;
- The Neighbourhood Wizard helps users to realize the complexity of a design task and as a result users will have a better informed view on plan proposals and probably a higher appreciation of plans.

10.1 Limitations of the Data Collection

Besides the positive conclusions there are also a few points of attention, mainly regarding the aspect of data collection.

The given data collection is retrieved from an enquiry in one particular city and may not be representative for other cities. The BAI has chosen rather abstract terms to collect the data, which make an unambiguous interpretation difficult. Some people may have misunderstood the used terms and therefore given a derogatory opinion on one ore more characteristics.

Additionally, the inclusion of more concrete elements, such as the number of parking lots or lanterns, can help take away long-living irritations that inhabitants may have. When users cannot express these small irritations, the tool itself will provoke a new irritation. Another shortcoming is that the data collection does not discriminate between the various sections of the population nor were all sections represented.

The data collection is restricted to physical characteristics. However, the liveability of the built environment is also influenced by non-physical characteristics, such as sources of deterioration and social characteristics of the community.

10.2 System Improvements

During the development of the prototype much effort was needed for the design of the user interface. In an early stage of this project the importance of the interface was recognized. Yet, the evaluation of the prototype still pointed out some issues.

The use of charts for the presentation of the predictions was expected to give the most problems. The evaluation revealed that this was not the case. The main issue appeared to be the navigation structure. Users had to click too many times before they came to make a functional action.

A new interface was developed on a short notice. This new interface had one main screen in which all actions could be performed, requiring fewer clicks. Users were also given a specific task to perform. For example "try to raise the red bars above the black bars". This new interface is shown in *Figure 6*. It is not yet evaluated on a large scale.

10.3 Future Work

The causal relations between characteristics are constructed from the analysis of a data collection. It is possible that some relations are found that do not exist in reality. Future work is needed to investigate the relations between characteristics in depth.

Another future research task would be the search for a different technique for the prediction of effects.

Although the predictions achieved through the Bayesian Network are valid and plausible, this technique does not offer an explanation of the expected effects. In many cases the reason for these effects are obvious and users will understand the prediction, but in some cases the predictions are not so obvious and require further explanation. For example: The creation of a quiet plaza has negative effects on the safety of the plaza. This is not a logical, though correct, prediction because the quietness of a plaza will attract criminal behaviour. Future work should be conducted that either adds knowledge to the system that can be used in constructing explanations, or focuses on finding a different approach to model the causal relations that includes explanations.

11. ACKNOWLEDGEMENTS

The project members are indebted to the Bosch Architecture Initiative, in particular its president Aart Wijnen, for their cooperation and making the data collection available. The municipality of 's-Hertogenbosch, in particular Constant Botter, has contributed through discussions and giving insight in their policy for participatory design and planning.

12. REFERENCES

Anderiesen, G and Reijndorp, A, 1990, *Van volksbuurt tot stadswijk, de vernieuwing van het Oude Westen* (From neighbourhood to district, the renovation of the Oude Westen district of Rotterdam), Project Group Oude Westen, Rotterdam.

De Leeuw-Hartog, A, 1988, *Woonmilieu en woonkeuze, een leidraad voor nieuwe lokaties* (Habitat and the choice of dwelling, a guide for new locations), Department of Public Housing, Rotterdam.

Etzioni, A, 2003, Presentation at the conference *Europe, a Beautiful Idea*, September 7, 2004, Den Haag.

Frissen, V, 2003, ICTs, civil society and global/local trends in civic participation, *in Workshop ICTs and Social Capital in the Knowledge Society*, EC IPTS/DG Employment, Seville, 2003.

Jirón, P and Fadda, G, 2000, Gender in the discussion of quality of life vs. quality of place, *Open House International* **25**(4): 76-83.

Keers, G, Hogenes, A, Pouw, N and Giebers, I, 2004, *Het wie, wat en waar van de woonomgeving, hulpmiddel bij integrale planontwikkeling* (The who, what, and where of the habitat, a tool for integral plan development), RIGO report nr. 83760.

Leidelmeijer, K and Marsman, G, 1999, *Beleving van de leefkwaliteit – nadere analyses nulmeting Stad & milieu* (Experience of liveability), RIGO research and consultancy, Amsterdam, report-nr. 73560/99.

Michalos, AC, 1997, Combining social, economic and environmental indicators to measure sustainable human well-being, *Social Indicators Research* **40**: 221-258.

Stoppelenburg, PA, 1982, *Woonmilieu en woongedrag, een evaluatieonderzoek onder bewoners van een aantal naoorlogse woonwijken in Amsterdam* (Habitat and dwellingbehaviour, an evaluation among inhabitants of a number of post-war neighbourhoods in Amsterdam).

Design Interactivity and Design Automation

A Proposal for Morphological Operators to Assist Architectural Design

Jean-Paul Wetzel[1], Salim Belblidia[2], and Jean-Claude Bignon[2]
MAP-CRAI CNRS Research Unit, Nancy, France
[1]School of Architecture of Strasbourg, France
[2]School of Architecture of Nancy, France

Keywords: Morphological operators, Modifiers, Modelling

Abstract: In this paper, we make the assumption that a shape modelling process can rely on the application of a set of morphological operators to initial shapes. We refer to several researches which have attempted to identify such operators. We also attempt to validate this design approach through the analysis of some buildings. A design system based on the combination of these operators could enable the designer to quickly explore a great number of spatial solutions.

1. INTRODUCTION

Since the early 80's CAD tools have provided us with a new means of graphic representation in architecture. Contrary to the usual manner in which spatial forms are represented in a 2D "glance" [in sections, elevations and plans], using this 3D drawing software one is provided directly with a virtual 3D model "constructed" by the computer.

Admitting that graphic figuration plays a key role in architecture (Lebahar, 1983) (Boudon et al., 1994) these recent developments constitute an important milestone in the practice of conceptualisation. In fact today's computer techniques provide architecture with a unique opportunity for "re-tooling" and "re-thinking" its methodologies just as happened with the arrival of perspectives and stereometric projections. 3D model software [Catia, Maya, etc.] are particularly applicable for conceiving enclosed spaces with complex morphologies. Such representations are well beyond the capacities of the conventional tools of the trade. The Guggenheim museum of F.O. Gehry is perhaps the most emblematic example of this novel approach.

From Ph. Boudon's postulation that "The process of conception is a diachronical one that implies a progressive transformation of what a project

Jos P. van Leeuwen and Harry J.P. Timmermans (eds.), Innovations in Design & Decision Support Systems in Architecture and Urban Planning, 409-418.

is" (Boudon et al., 1994), we can hypothesise that the procedures that lead to the production of forms issue from the application of semantic operators on "seed" elements that give rise to "goal" forms. In this article we will at first treat different works of research aimed at identifying these operators. In addition, in the analysis of early drafts of a range of "non-standard" architectural works we will show how such operators can help in the conception of projects and open the way to innovative practices in the creation of forms.

Finally we will present a prototype of a form generator.

2. MORPHOLOGICAL OPERATORS IN ARCHITECTURE

Several works based on project documents have been conducted to try and analyse the various processes and changing phases of architectural projects.

In a study realized on a collection of drawings made for the design of a bungalow by the architect Neimann, Ellen Do et al. have demonstrated a classification of elements, transformations, localization and also colours of elements (Do et al., 1999). This classification led to the naming of the variations realized between each sketch. The authors were able to identify the transformations but these remain simple ones, of the order of rotation, translation or symmetry.

In his work, Philippe Boudon aimed at getting beyond mere geometric operators and to identify a whole range of transformations of state which appear during the design process (Boudon, 1994). He thus proposes to describe through scales how the architect applies measures to an edifice through operations of conceptions. Although called operators, the scales say more about the pertinence of the design operations than indeed about the morphological operations themselves.

Dominique Raynaud in a study about architectural design (Raynaud, 1998) has shown that when changes of the state result in the transformation of the actual structure of a morphological model, the description through scales belonging to the architecturology is barely obvious. Going back to one of the first hypotheses of Herbert A. Simon on the solving of problems (Simon, 1966) he suggests that the description of these transformations calls for another class of operators. These could be schemes if one means the class of prototypical actions which can be expressed through basic verbs like to open, shut, cut, link, etc.

From an empirical analysis of a sample of 162 architectural projects representative of the history of architecture from antiquity to the present times, Raynaud describes the transition from a symbolic representation –

which can be a text as well as an image or an idea – to the formal translation into a project. He thus defines 20 schemes which are: to contain, turn, go up, radiate, cover, go through, open, begin, cut, descend, wave, unite, enclose, go out, link, repeat, end, surround, diminish, cross, battle, grow up.

In his work on the development of a parametric model which permits to generate complex forms, Carlos Barrios (Barrios, 2005) uses similar concepts. He enhances the existence of operators of forms such as "torsion" and "intersection". Working from the analysis of a column of the side-nave of the Sagrada Familia by the architect Antonio Gaudi, he also demonstrates the combination of those operators. This allows him to define a parametric model and to modelize this architectonic part of the cathedral.

Finally we will refer to the research work by John Frazer about evolvable generative architecture which indicates the use of a genetic code to describe the themes of the project (Frazer, 2002). According to an approach close to the one used with genetic algorithms, the project evolves along internal rules confronted to demands imposed by the architect. In this model of morphogenesis multiple mutations will arise, one or several stable phenotypes of the project. This principle of a genetic code in which the future mutations or transformations would be present draws nearer to our own hypothesis about the existence of a collection of shape operators in the architectural design.

We will keep in mind from this quick presentation of works about modelling the architectural form that the existence of morphological operators is at the core of numerous theories which support our own hypothesis. In the following part of this article, we will attempt to verify this principle through the analysis of a few so called "non-standard" architectural projects.

3. MORPHOLOGICAL OPERATORS SEEN THROUGH A FEW EXAMPLES

3.1 Peter Eisenman

Peter Eisenman is a particularly interesting example for our analysis. Indeed, in several "de-constructivist" projects such as the Guardiola House or "Memory of Mak" he conceived a formal system the expression of which is based on materialized recording of conducted operations and on the track of the transformation process itself. It is then easy to detect step by step the morphological transformations in which a formal element replaces one or several others. For example, a volume is divided into faces, the parallel faces become a grid, the grid is turned over and so on. In later projects like "Max

Reinhart House" or "Alteka Office Building", the notion of transformation gained an even more explicit character and allowed us to identify what could be related to schemes.

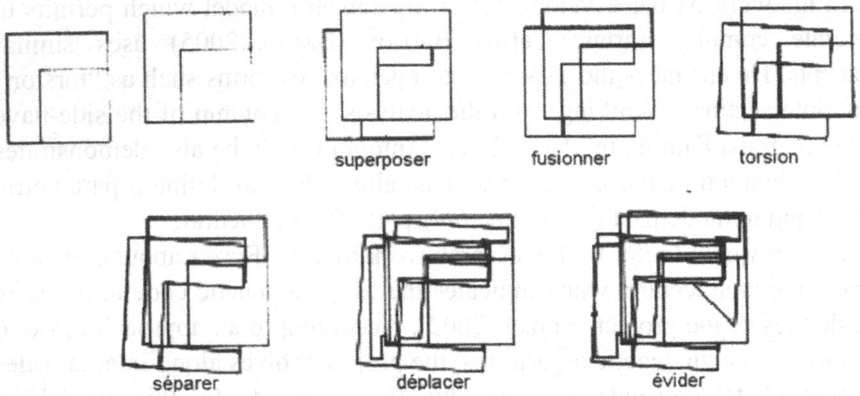

Figure 1. Example "Guardiola House".

It appears then that operators such as "superimpose", "blend" or "twist" are as many tools which assist the project process in its formal genesis.

As far as the meaning of forms is concerned, it is to be related to themes associated with the place of realization or the type of project. Thus, in the bungalow of a visual exposition in Holland, the zigzag of forms is supposed to derive from the paths that electronic beams follow in the display of a tube.

3.2 Marcos Novak

Marcos Novak's work belongs to the field of visual architecture and cyberspace. By creating forms in virtual worlds, the designer gets rid of fetters of reality such as gravity, climate or constructibility. Novak develops the concept of liquid architecture, a fluid and imaginary environment where forms are described in 4 dimensions. They evolve in time and transform themselves through the influence of external factors such as sound-waves or also actions by the user. The forms turn, incurve, stretch, twist themselves … in a way they mutate.

It is this last aspect of Marcos Novak's work which is of great interest to us. We could imagine using similar transformations on architectural seed forms, not in reaction to a continuous phenomenon but as morphological operators being used consciously by the user.

3.3 Frank Gehry

Recent works by Frank Gehry [Los Angeles Walt Disney Concert Hall, Bilbao Guggenheim Museum, Seattle Experience Music project] show on the one hand the possibility to use new forms in architecture, based on curved surfaces which up to then were rather the privilege of fields such as furniture design, car and aeronautics. It is obvious that such forms would be hard to conceive without the help of a computer.

The process of formal research with Gehry is based on an iterative approach on various supports. After the first sketches, the materialization of form is tackled on preliminary models, then refined with the help of the informatics tool (Catia, Dassault Systèmes) before being tested in models of verification on a bigger scale.

Thus the process of formalization is conducted by Gehry mainly through the model. It starts with a first spacialization based on primitive geometrical forms which take into account the functional needs. It then goes through a whole range of morphological operations: to blend, press, stretch, smooth, etc.

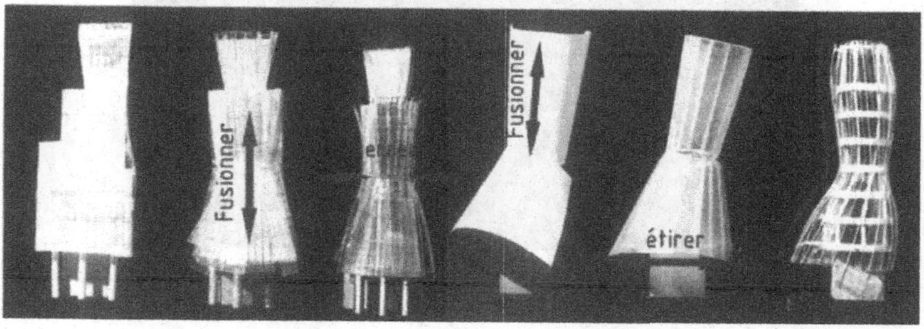

Figure 2. Nationale Nederlanden Building, Praha.

The use of a 3D modeller in a second phase allows a more rigorous and exact definition of the volumes, and the preservation of the various creative stages, as well as the taping of geometrical parameters for a transfer to a CAD-CAM process.

Such an approach shows us that it would be particularly interesting to model the process of morphological transformation so that it could be implemented with the help of the computer and no longer with real models.

3.4 Greg Lynn

Greg Lynn has largely used the ability of 3D modellers to create innovative forms in the design of furniture, interior architecture or buildings. Through his realizations and in his numerous projects for contests, Greg Lynn bases his creative activity on the use of the properties of surfaces and particularly the topological "events" they can generate. Those patterns named folds, knots, buckles, flowers, etc. become elements of a vocabulary which translates itself in architectural terms to become spaces, limits, openings and other spatial elements.

The second characteristic of Greg Lynn's work is his recourse to computer programmes in order to give life to the initial forms. The volumes are modified under the appliance of forces which transform their geometry. Animation is used here as a means to reveal the volume even better and to suggest variants which may be were not even obvious to the designer.

The example of the embryonic house (figure 3) shows the various stages of a morphogenesis and underlines the use of morphological modifiers in a conceptual approach.

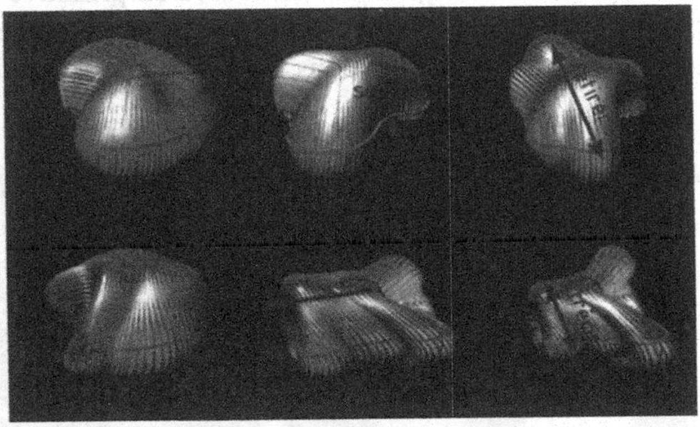

Figure 3. Embryonic house.

4. INTERPRETATION AND INSTALLATION OF A MODEL

4.1 Model

The use of morphological operators or "structural modifiers" (Porada, 2005) appears omnipresent in the process of formal design.

These schemes/operators can differ from the geometrical functions of a modeller because of the architectural meaning they carry. Thus, each transformation Is motivated by the research for a spatial effect on a volume or a space, the user being the outside observer of a spatial composition, or travelling in a 3 dimensional space. A morphological operator can be the combination of several geometrical transformations.

In order to instrument those operators, we have first modelled the identification of operator classes and action parameters.

4.2 Classes

The work by Francis DK Ching (Ching, 1996) on architectural shapes allows us to identify two main strategies of form productions and, potentially, two main operator classes.

The first strategy is metaphorically represented by a lobster and consists in creating forms through adjustment and combination. The seed form is made of unitary forms that we are going, for example, to add, juxtapose, superimpose. We will refer to this as "transformation by composition".

In the second strategy identified by a slug, the seed form will undergo morphological but not topological modifications with operators such as twist, stretch or pinch. We will refer to this as "transformation through metamorphosis". Of note is that some buildings are more concerned by one or other of these strategies but that the 2 production methods are not antagonistic and can be used simultaneously.

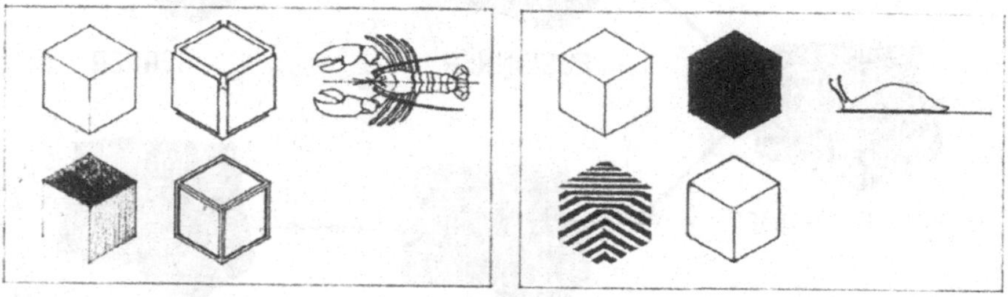

Figure 4. The lobster and the slug.

4.3 Parameters

The use of a morphological operator demands a definition of parameters by the user. We have identified a first family of parameters, linked not to a specific operator but to the way it applies itself to the geometry: the field of application, the intensity, the propagation type, etc. On the other hand, we have detected other complementary parameters which are specific to some operators: vector of direction, angle of inclination, angle of torsion, etc.

4.4 Approach

From this first work, we can propose an approach for a possible formal creation structured in 3 main phases:

To define a first scene from the seed forms chosen among a range of geometric primitives. We have seen from the above examples that the seeds could be geometric primitives as well as organic forms.

To apply an iterative process of transformation by using n-nary and/or unary operators. This process can be mastered step by step by the designer or inscribed in algorithms.

To build the final scene as a result of the process of transformation in which the seed forms, the operators, but also the very order of application of these operators have their importance.

The conjunction of these 3 factors provides many possibilities which justify the role of the computer as "an accelerator for the exploration of solutions" (Frazer, 2002).

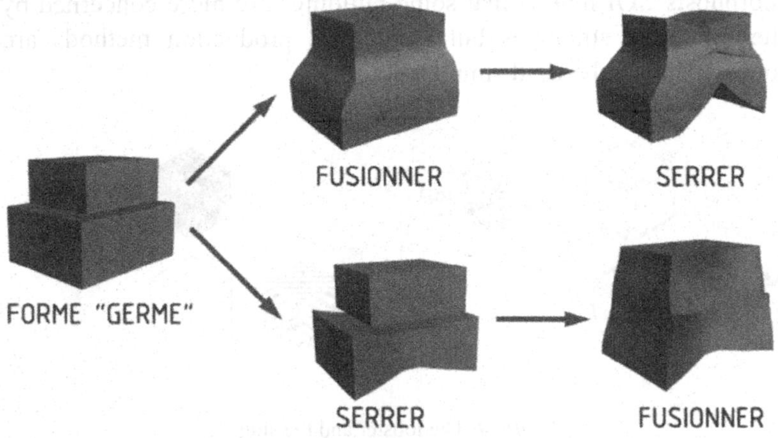

Figure 5. Successive application of the operators "squeeze" and "blend" applied in a different order to the same geometric components.

On the basis of these two main strategies we have identified two main operator classes:

- binary or n-nary operators, already mentioned as composition operators which are related to Boolean operations, and among which can be included actions such as merge, intersect, link, juxtapose, etc.
- unary operations which operate on one single form at a time. Among them are all the common geometric transformations [shifting, rotation, homothety, symmetry] but also the operators we aim at, such as stretch,

compress, dilate, squeeze, bulge, hollow, smooth, roll, split, open, shut, etc. These operators should be – as in Dominique Raynaud's schemes – the transposition of symbolical concepts.

5. CONCLUSION

In this paper, we have attempted to demonstrate the pertinence of the concept of morphological operators and its role in the process of formal research. A first model has been sketched. The installation of a precise experimental process should enable us to define more accurately these operators and to model a defined range.

This step will allow us to reach the definition of an intuitive 3D modelling environment which will enable the architect/designer to explore a number of formal solutions by using combinations of morphological operators.

6. REFERENCES

Aoki, Y. and M. Inage, 2000, "Linguistic Operation System for Design of Architectural Form", in: Timmermans, Harry (Ed.), *Proceedings of the 5th Conference on Design and Decision Support Systems in Architecture and Urban Planning*, Nijkerk, the Netherlands.

Asimow, W., 1962, *Introduction to Design*, Prentice-Hall, Englewood Cliffs, New Jersey.

Barrios, C., 2005, "Transformations on Parametric Design Models", in: *Proceedings of CAAD Futures '05*, Vienna, Austria, 20–22 June 2005, pp. 393-400.

Boudon, Ph., P. Deshayes, F. Pousin, and F. Schatz, 1994, *Enseigner la conception architecturale*, Editions la Villette, Paris.

Ceccarini, P., 2003, *Catastrophisme architectural : l'architecture comme sémio-physique de l'espace social*, Éditions L'Harmattan, Paris.

Ching, F.D., 1996, *Architecture: Form, Space & Order*, Van Nostrand Reinholds, New-York.

de Vries, B., A.J. Jessurun and M. Engeli, 2000, "Development of intuitive 3D Sketching Tool", in: *Proceedings of the 5th Conference on Design and Decision Support Systems in Architecture and Urban Planning,* Nijkerk, The Netherlands.

Do, E.Y., 1996, "Drawing as an interface to knowledge based design aids", in: *Proceedings of ACADIA '96*, University of Arizona, Tucson, pp. 191-199.

Do, E.Y., M.D. Gross, and B. Nieman, 1999, "Sketches and their functions in early design – A retrospective analyses of a Pavillion House", Sundance Lab, University of Colorado, Boulder, pp. 255-266.

Estevez D., 2001, *Dessin d'architecture et infographie. L'évolution contemporaine des pratiques graphiques*, Éditions CNRS, Paris.

Frazer, J., X. Liu, M. Tang, and P. Janssen, 2002a, "Generative and Evolutionary Techniques for Building Envelope Design", in: *Proceedings of the International Conference on Generative Art*, Milan.

Frazer, J., X. Liu, M. Tang, and P. Janssen, 2002b, "A generative design system based on evolutionary and mathematical functions", in: *Proceedings of the International Conference on Generative Art*, Milan.

Jacobs, J., 1961, *The Death and Life of Great American Cities*, Random House, New York.
Jencks, C., 1979, *Le langage de la architecture post-moderne*, Academy editions, Londres.
Laseau, P., 2001, *Graphic thinking for architects and designers*, Wiley, New York.
Lebahar, J.C., 1983, *Le dessin d'architecte - simulation graphique et réduction d'incertitude*, Collection architecture outils, Éditions Parenthèses, Paris.
Porada, S., 2005, "L'instrumentation de la création architecturale," in : *Proceedings of SCAN '05*, Paris.
Pranovich, S., 2004, "Structural sketcher: a tool for supporting architects in early design", PhD Thesis, Technische Universiteit Eindhoven, Eindhoven.
Raynaud, D., 1998, *Architectures comparées: Essai sur la dynamique des formes*, Éditions Parenthèses, Marseille.
Rossi, A., 1981, *L'architecture de la ville*, Éditions L'equerre, Paris.
Xenakis, I., 1971, *Musique Architecture*, Éditions Casterman, Paris.

Generative Design in an Evolutionary Procedure

An approach of genetic programming

Hung-Ming Cheng
China University of Technology

Keywords: Artificial intelligent, Genetic algorithm, Generative design tools, Procedural design studio, Design exploration

Abstract: This study describes a procedural design studio using Genetic Programming as the evolutionary mechanism and formal generation. This procedural design is integrated with a visualisation interface, which allows designers to interact and select from instances for design evolution. Evolutionary design facilitates designers in three areas: 1) diversify instances of design options; 2) inspect specific goals; 3) and enhance the possibility of discovering various potential solutions.

1. INTRODUCTION

In essence, design and designing involve different disciplines that are influenced by participants, knowledge, and information from various domains. For such design frameworks, design problems require a procedure to reconcile multiple viewpoints that are distinguished by particular interests and emphases. For example, a design team consists of architects, engineers, and constructors, with each of them concerning issues/aspects of design from different angles. An architect would be interested in aesthetic and figural aspects of design, a structural engineer concerned about structural members and the underlying reasonable, while a constructor focusing on building costs and other construction issues. They derive their views based on different disciplines and professions that require an integrated method for searching potential solutions. The design task with problem frameworks provides an exploratory nature of design procedure, which also offers

Jos P. van Leeuwen and Harry J.P. Timmermans (eds.), Innovations in Design & Decision Support Systems in Architecture and Urban Planning, 419-431.

profound approaches and solutions. This is why we employ such mechanism to explore the design space from various domains of design knowledge.

Genetic programming provides a way to genetically breed a computer program to solve a wide variety of problems. The recently developed genetic programming search the space of possible computer programs for a highly fit individual computer program (Koza, 1992). Search is a framework for problem solving where alternative solutions are evaluated in a trial-and-error iterative circulation. The meaning of design search is to find out alternatives that lead to good solutions. Searching in enumerative space of design solutions is unlikely, the ideal heuristics such as genetic programming use rule-based method as a guidance for searching alternative solutions. Genetic programming is derived from the hint of Natural Selection (Darwin, 1859) to decide the ordering of alternative solutions, which has been applied to a wide range of problems in combinatorial optimization, automatic programming and model induction.

The evolutionary procedure applies genetic programming as algorithmic method that evaluates and refines the design during conceptual formation. The design procedure integrates synthesized solutions of design teams. The evolutionary procedure reflects different disciplines of designers that collect the fitness from each designer's selection. To explore the design space, we need appropriate representations and procedures to generate the representation of the design task. The generative design tool uses an iterative approach that refines design by evaluating candidates in the process of genetic programming. An algorithmic method is implemented as a structure with regard to the evolution of genetic characters. All told, we implement genetic programming as a platform to search for consensus of design.

2. OVERVIEW

Genetic programming is a heuristic search technique; it is also a common application of artificial intelligence, where search is the framework for problem solving with computers. A design solution based on such concept potentially explores the enumerative space of generative design. Efficient search depends on the evolutionary procedure on the dynamic procedure of designers and algorithmic implementation. The following section will clarify the relationship and interaction between genetic programming and generative design.

2.1 Genetic Programming

Genetic programming inspires problem solving, but this also implies the limitation of its applicability. The strengthening in computing power, enhancement in applications, and the ability to collect tremendous amount of scientific data all require a computational framework, such as genetic programming. This is why in this paper we intend to apply computer to derive massive generative solutions.

There are two key issues in the genetic programming. 1) selection of a population for alternative solutions; 2) how to generate and evaluate individuals of fitness.

A stochastic selection method chooses better solutions from the population that fetch stochastic variations to produce new alternatives. In this way, the design procedure is responsible for guiding a parallel search throughout the design space. However, the evaluation of each population member becomes increasingly expensive and ineffective. The whole processes of evaluation depend on the direct encoding of criteria in the computational environment. Beside, the amount of data produced by a population based search over an enormous design space makes the prediction and analysis difficult.

With the ability to generate and evaluate a possible solution, a search strategy must be defined. Search methods repeatedly generate solutions, evaluate them and generate more by computation mechanism. Genetic programming uses representation of previously generated solutions when a solution meets a particular criterion, which depends on the design quality threshold or designers' consensus. In addition, alternative solution is evaluated by means of a fitness function that allows it to be compared with previously evaluated solutions. The programming procedures with evolutionary steps are the design objective should achieve that is also the goal of evolutionary design.

2.2 Generative Design

Generative systems offer a methodology that produces design space via dynamics and their outcomes. As such, generative systems offer an information processing theory to problem-solving design. Based on the information processing theory, some scholars define design process as a cyclical process from specification, generation and evaluation. (Mitchell, 1992) For designers, generative design would involve reconsidering the permutation of potential types. Conceptualisation that shifts from the primacy of objects to interacting components, and from systems to processes, generates new artefacts with special characteristics.

Encapsulated in a navigating structure of paths and landmarks, design space offers an exposition for actions and intentions associated with design (Chien and Flemming, 1996). To explore the design space we need appropriate representations and procedures to fulfil design task. Algorithmic method is thus implemented as the structure for design generation. A formal representation presents design computing in term of type, symbol, colour and status. In all, generative design provides a possibility that walks through the generation of solutions.

The generative methodology offers an unconventional way for conceptualising and operating in design process. Research in generative design is closely related to the general concept of synthesis, mostly presented in the form of nature and/or discreet. Natural selection develops a specific mechanism for generalised synthesis, using the physical apparatus of DNA, protein synthesis, and biochemistry. The discreet of generative design diversifies into numbers and scalable shape, which demonstrates the capability of generative design to overcome design problems, and to construct diverse forms from relatively simple units.

2.3 Designing

Designing is a reflective conversation that involves the recursive processes of seeing, moving and seeing (Schön and Wiggins, 1992). Choices, alternatives and versions emerge from the interaction between designing (acting) and discovering (reflecting). Exploration encompasses the formulation of requirements and the generation of solutions based on these requirements. Exploration in the design space is an integral part of the process of solution reformulation, and solution reformulation requires efficient searching and selection.

Exploration rationale (Smithers, 2002) and design selections are critical supporters for exploration. Designers must have the capability to exploit from numerous options through navigating and recombining the paths of exploration. Design representation must provide a unified model for representing potential solutions, which designers take into consideration and reformulate. Designers should make decision to select alternative problems and solutions. In addition, representation captures the characteristics of problems (solutions) as well as selections that designers made during exploration. The above issues require designers' reconsideration in order to address representations in design process.

- Characteristics of problems – Problems must correspond to designers' view to problem formulation. As such, the characteristics of problems must correspond to the initial, intermediate and final stages of designs. In essence, an evolutionary procedure must capture the frame of problems and connect the problems with potential solutions.

- Selections – During the exploration process, problems and requirements of design create a large design space that requires a criterion to decide whether solutions fit or not. During a designing process, a problem formulation may have no solutions, a finite number of solutions, or enumerative solutions. The intentional selections that designers made in the reformulation of problems and commitments to solutions during exploration must be recorded in an evolutionary design procedure.

3. METHODOLOGY

Genetic programming is an evolutionary algorithm that applies either a procedural or functional representation. This section describes design representation and the specific algorithm components used in the canonical version of algorithm. The fundamental of genetic programming are initially presented, followed by a discussion of algorithm and description of two evolutionary procedures. Issues with regard to design research and metaphors of genetic programming applications will also be discussed.

3.1 Evolutionary Algorithms

Darwinian evolution applies the principles of competition, inheritance, and variation within a population. These concepts are often used to define iterative improvement in computer programming. These methods, evolutionary algorithms, use a population of solutions and genetic operators to carry out selection and evaluation. The evolutionary algorithm employs the following items: (Gustafson, 2004)

- A population of candidate solutions called individuals,
- A fitness function that evaluates and assigns each individual a score, or fitness value,
- Transformation operators that produce offspring individuals from parent individuals, implementing the concept of inheritance through stochastic variation, and
- A stochastic selection method for selecting individuals with better fitness to produce offspring.

With evolutionary procedure, we adopt a similar search strategy as a genetic algorithm, uses a program representation and special operators. The representation of evolutionary design process makes genetic programming unique. The basic algorithm is refined by design process and shows as follows:

- Initialise a population of solutions
- Assign fitness value to each population member
- While the convergence is not met
- Produce new individuals using operators and the existing population
- Place new individuals into the population
- Assign new fitness value to each population member, and test for the convergence satisfied
- Return the best fitness found

3.2 Design Model with Genetic Programming

Genetic programming has become a popular search technique since early 1990s thanks to the work by Koza (Koza, 1992). Nowadays, genetic programming is applied mostly related to adaptive system and optimization, where representation of programs is used in conjunction with hybrid crossover to evolve a multiplication function. This research employed partial weighting assignment as functional activity and convergence of selections. In addition, the design with genetic programming is not traditionally considered in canonical genetic programming.

Theories of evolutionary algorithms use abstract representations of the solution space, called schemata, to describe various components and behaviours of algorithm. Holland's (Holland, 1975) notion of schema for genetic algorithms was extended by Koza (Koza, 1992) to include syntax trees. Syntax trees are the most popular representation in genetic programming, and these schemas were intended to represent trees that refine data structure as computer programming language. In this case, the code of computer programming language allows the geometry representation to become schema in order to represent the design figures and characters.

The schema of design model is developed into two evolutionary processes of design operations which include natural selections and the evolutionary mechanism. Natural selections provide the tournament for the distribution of designers' weighting that calculates fitness of each population. The parallel process is evolutionary mechanism that interact fitness (selection) and individuals to evolve the population. These processes present actual execution of the algorithmic method with all its characteristics and degrees of freedom *(figure 1)*. Thus, the process will end once it meets the design convergence. This evolutionary process of design is also coherent with the rational design model that is an iterative cycle of design analysis, synthesis, and evaluation (Asimow, 1962).

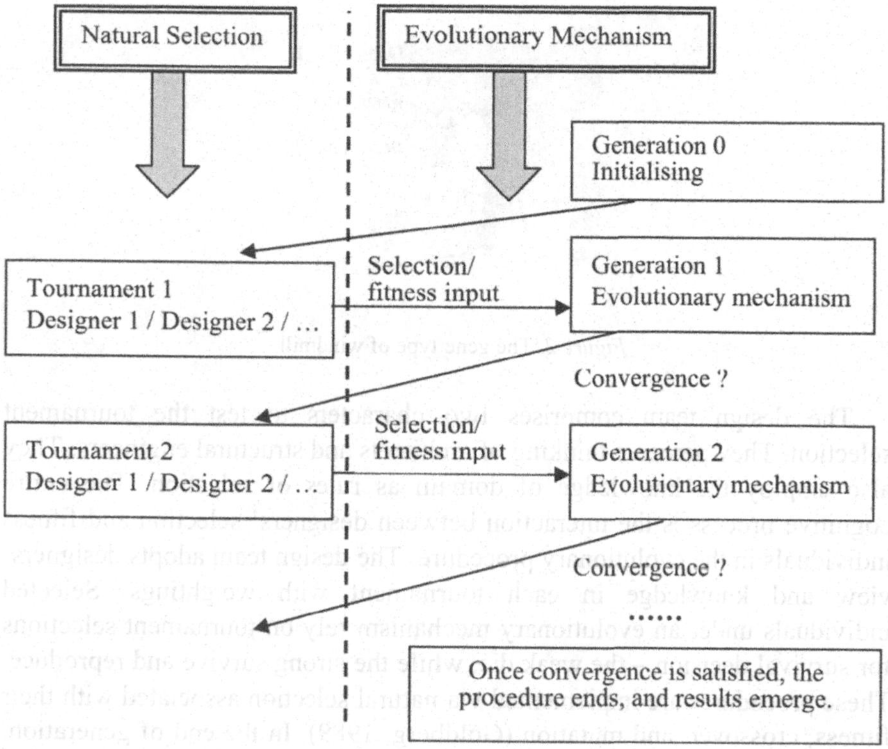

Figure 1. Generative design model.

4. EXPERIMENTAL DESIGN

4.1 Experiment Installation

We start our experiment with a studio assignment "windmill design" to seek for formal solutions. The windmill evolves its possible forms in an evolutionary design process. We implement genetic programming and derive 15 generations for observation. The gene types of the windmill are defined as follows *(see figure 2)*:

- Gene type 1: The legs of windmill could range from 2 to 8.
- Gene type 2: The leaf shapes of windmill could be either square, rectangle, circle or triangle.

- Gene type 3: The relative of each leg could be connected by a circle.
- Gene type 4: The foundation of windmill may change the width of the windmill.
- Colour: The colours in all segments of the windmill are changeable.

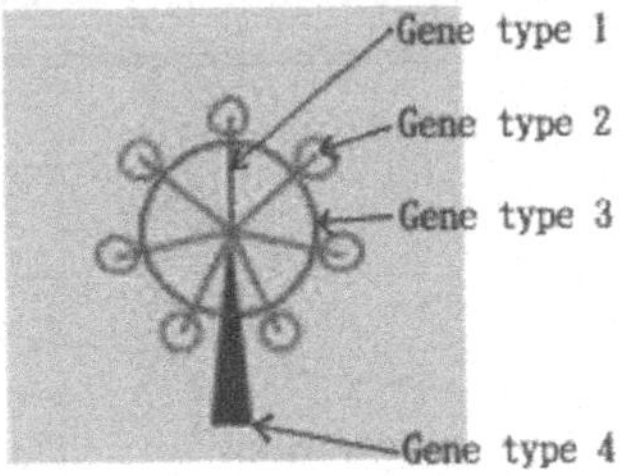

Figure 2. The gene type of windmill.

The design team comprises two characters to test the tournament selection. They perceive thinking of architects and structural engineers. They also employ the knowledge of domain as rules of selection. The entire cognitive process is the interaction between designers' selection and fitness individuals in the evolutionary procedure. The design team adopts designers' view and knowledge in each tournament with weightings. Selected individuals under an evolutionary mechanism rely on tournament selections for survival decision – the weak die, while the strong survive and reproduce. These procedures are implemented via natural selection associated with their fitness, crossover, and mutation (Goldberg, 1989). In the end of generation, potential solutions arrive that correspond to the design team and genetic programming.

4.2 Experimental Procedures

The experimental procedures of design are employed in order to examine the efficiency of design selections and that of searching between designers and computer supporting system. We implemented the schema of genetic programming based on previous study on genetic programming. With computational operators and structure, genetic programming includes mutation, reproduction, selection/fitness, and other representations in evolutionary procedure (Holland, 1975). We express in *Figure 3* (Procedure of Genetic Programming) for implementing programming (Michalewicz, 1992) in evolutionary mechanism (*Figure 4).*

```
Procedure of Genetic Programming
Begin
  T = 0
    Initialize p(t)
    Evaluate p(t)   //p(t) = w1*p1(t)+w2*pa(t)+...
    While (not termination-condition) do
    Begin
        T = t+1
        Select-parents from p(t-1)
        Form p(t): reproduce the parents   //+mutation
        Evaluate p(t)   // p(t) = w1*p1(t) + w2*pa(t)+...
    End
End
```

Figure 3. Procedure of genetic programming.

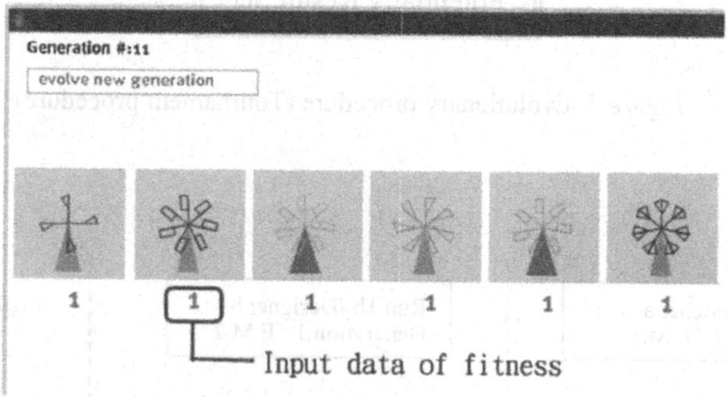

Figure 4. Implementation of the evolutionary mechanism.

To understand the experimental procedure, we developed two procedural settings – tournament procedure (*Figure 5)* and independent procedure (*Figure 6)*. Both intermediary activities and final queries were recorded in these two settings in order to analyse the evolutionary procedure. Observation and discussion of the two evolutionary designs are presented in the following section.

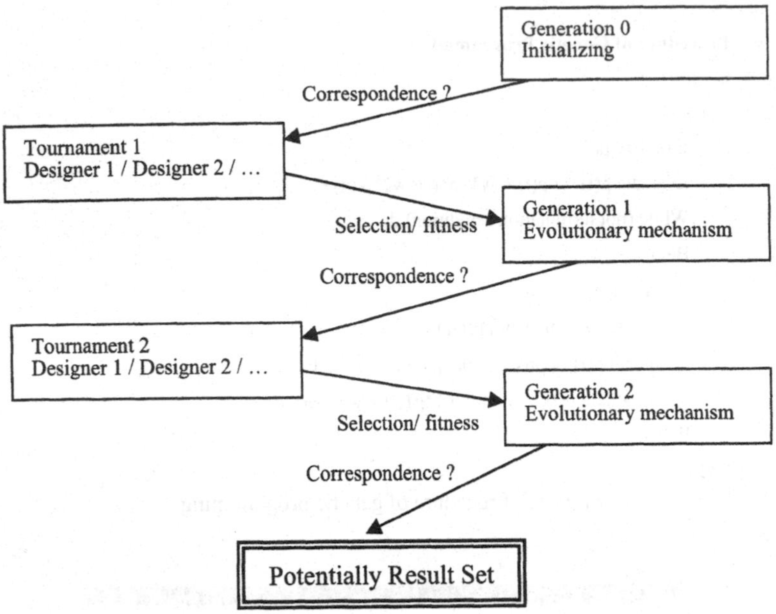

Figure 5. Evolutionary procedure (Tournament procedure).

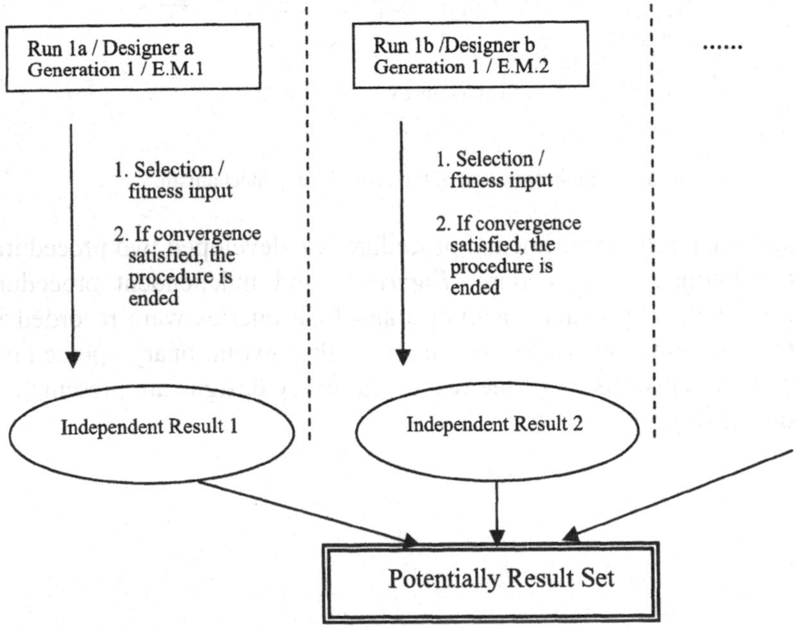

Figure 6. Evolutionary procedure (Independent procedure).

4.3 Results and Discussions

The two experiments above employ the evolutionary procedure to seek for possible outcomes, which however demonstrate different characters and individuals. The tournament procedure truly reflects the fitness/selection of designers as well as correspondence of the evolutionary mechanism. On the other hand, the independent procedure intensifies potential results whereas falls short of integration in the same process of evolution.

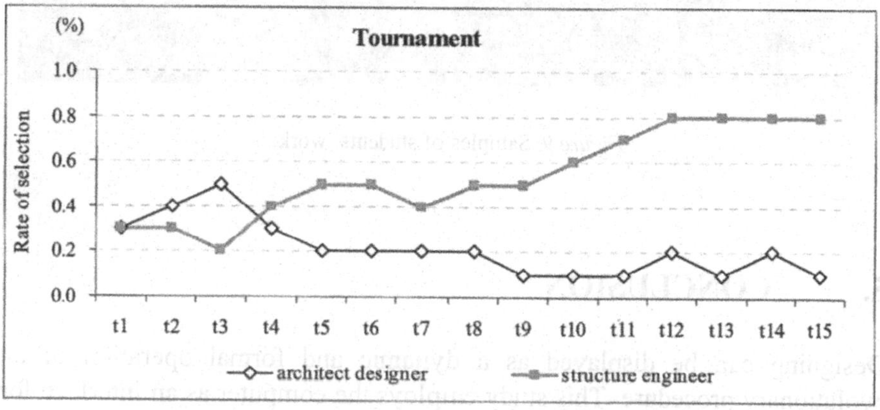

Figure 7. Final result of tournament procedure.

Fitness and weighting of selection decide the survivability and continuity of population. Designers thus are required to exchange their intuitions and/or concepts – to some extent this looks like a cooperative design process. For example (*Figure 7*), the initial selection suggests architects and engineers adopt fairly different strategies – architects intuitively select five or more legs and circle-like wing. On the contrary, engineers chose a three-leg windmill while dislike the one with more than five legs. The counteraction from the above tournament impacts individuals and gene pool. However, counteracting selections cannot produce high rates of survival, although the selection is normally processed. On the other hand, an independent procedure would lead selected individuals to become more homogenous (*Figure 8*).

Figure 8. The final result of tournament procedure result.

The above design processes generated numerous conceptual options, but resulted in distinguishing outcomes at the subsequent design stage. Although inspired by the evolutionary mechanism as exhibited in *Figure 9*, students eventually produced totally different artefacts.

Figure 9. Samples of students' work.

5. CONCLUSION

Designing can be displayed as a dynamic and formal operation of an evolutionary procedure. This study employs the computer as an interface for genetic programming to generate a canonical population for selection. As for generative design with genetic programming, concepts of evolutionary selection are developed that explain different knowledge behaviours. These behaviours are categorised into two evolutionary procedures. First, the evolution of populations towards a stable state corresponding to the designers' consensus. Second, once such a stable state is reached, the fitness solutions emerge and terminate the programming procedure.

An ideal design process is to reflect designers' consensus while evolving with principles and concepts of design. Still, this could leave the design space incorrectly defined. Generative design or hierarchic organisation may help solve this dilemma.

The genetic programming as an evolutionary design process can be transformed into the analysis of modified genetic algorithms. As such, genetic programming in generative design that reconsiders certain changes to the selection operator may produce the fittest population in the evolutionary cycle. Mutation-control selection schemes, including the selection with a divergent election operator, ensure that at least the first breadth individual of a population will become a member of the next generation's population. On the other hand, strait-forward selection schemes reveal not enough breadth samples to become the fitness selections. This also explains that some populations have no chance to be transitory populations. In this case, we need evolutionary strategies to dynamically adjust the mutation rate in order to reach asymptotic stability in every single evolutionary procedure.

We propose in this study that a design research in the dynamics of tournament selection will develop solutions for evolutionary procedure design. With these instruments, we observe interesting design processes that could lead to numerous potential works.

6. ACKNOWLEDGMENTS

We are grateful to insightful discussions with Prof. Sheng-Fen Chien and the research group CODE, Department of Architecture, National Taiwan University of Science and Technology.

7. REFERENCES

Asimow, M., 1962, *Introduction to Design, Englewood Cliffs*, Prentice-Hall, New Jersey.

Chien, S. and Flemming, U., 1997, "Information navigation in generative design systems," *CAADRIA 97*, Vol. 2, Hsinchu Taiwan, pp. 355-366.

Goldberg, D., 1989, *Genetic Algorithms in Search, Optimization and Machine Learning*, Addison-Wesley, Canada.

Gustafson, S., 2004, *An analysis of diversity in genetic programming*, PhD. thesis, University of Nottingham, Nottingham.

Holland, J., 1975, *Adaptation in Natural and Artificial Systems*, The University of Michigan Press, London.

Koza, J., 1992, *Genetic Programming: On the Programming of Computers by Means of Natural Selection.* MIT Press, Cambridge.

Michalewicz, Z., 1992, *Genetic algorithms + data structures = evolution programs*, Springer-Verlag, London.

Mitchell, W.J., 1977, *Computer-Aided Architectural Design*, NY: Van Nostrand Reinhold. New York.

Schön, D.A and Wiggins, G. 1992, "Kinds of seeing and their functions in designing," *Design Studies*, **13**(2): 135–156.

Smithers, T., 2002, "Synthesis in designing," in: JS Gero (ed.) *Artificial Intelligence in Design '02*, Kluwer Academic Publishers, pp. 3–24.

Interactive Rule-Based Design

An experimental interface for conceptual design

Eric Landreneau, Ozan O. Ozener, Burak Pak, Ergun Akleman, and John Keyser

Texas A&M University

Keywords: Fractal geometry, Conceptual design, Generative systems

Abstract: In this paper, we present a method that allows designers to interactively create partially self-similar manifold surfaces without relying on shape grammars or fractal methods. The modellers that are based on traditional fractal methods or shape grammars usually create disconnected surfaces and restrict the creative freedom of users. In most cases, the shapes through conventional fractal or shape grammar methods are defined by hard coded schemes that allow limited interactivity for the design process. We present a new approach for modelling such shapes. With this approach, we have developed a simple generative tool with given adjustable parameters to achieve variety of conceptual forms. Using this tool, designers can interactively create a variety of partially self-similar manifold surfaces.

1. INTRODUCTION

In the framework of conceptual modelling there is always a motivation for achieving unconventional shapes through a generative design process. There exists a strong interest in contemporary architecture to extend the limits of conceptual design by utilizing rule-based generative systems. Several design studios in some architecture schools experiment with rule-based approaches and try to include computational and algorithmic conceptual design content. Designers in these studios use a wide variety of rule-based techniques such as L-systems or cellular automata. For designers, it is important to easily develop new generative procedures to have variety of alternatives. The design of rules and iteration depths may return interesting results, however it is hard to identify the rules and iteration depths. Our key question motivating

Jos P. van Leeuwen and Harry J.P. Timmermans (eds.), Innovations in Design & Decision Support Systems in Architecture and Urban Planning, 433-445.

this paper was "Is it possible to blend rule based techniques such as fractals and L-systems with increased interactivity for designers?" Although, our motivation comes from fractal and L-system, we wanted to give designers interactive control of modifiers to achieve conceptual shapes.

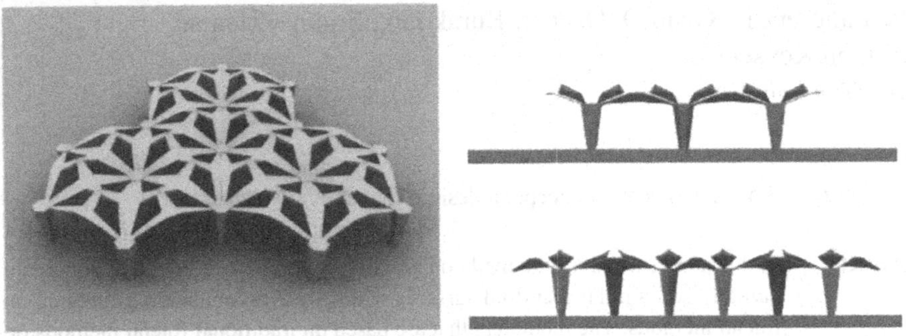

Figure 1. Perspective and two sectional views of a conceptual structure formed from repeated modules that are created with our system.

In this paper we present a simple approach that allows interactive extensions to fractal and L-system methods. Using this approach, we have developed a system that enables designers to control each step of the shape generation process, to increase the interactivity level during generation and to utilize a flexible set of modifiers to achieve shape variety. Our approach is composed of 3 steps: colour based face grouping, modifier applications and remeshing.

- Basically, colour based face grouping is the breakdown process of the given 3D model into certain groups that act together in each iteration. We start with a geometric shape which will be the basis for the final product. The desired face group in this object is labelled by a colour in the material assignment. This allows the designer to create a layout for object development.

- The second step is under control of the designer: in our own interface we give designer the opportunity to apply a set of classical extrusions or extended extrusion methods, face deletion, or crust generation tools.

- Remeshing schemes are for smoothing the surface structure to have different look and surface topology.

Our system is implemented in C++ and FLTK. The system can run on UNIX, Linux and Windows platforms. For face grouping operation and further development of the 3D model we use MAYA software. The software is capable to generate Alias OBJ files to establish a connection between popular 3D software like MAYA and 3D Studio MAX/VIZ. Examples in this paper were created interactively using this prototype system.

One of the main advantages of our system is to create models which are ready to be prototyped trough a 3D Printer. We are using manifold surfaces as a core concept and they gave us a connected surface structure ready for 3D printing. This important feature allows us to create and evaluate models in our system and extend them for fabrication.

The usability of the system was tested in a graduate level shape-modelling course in which a majority of the students had an architecture undergraduate background.

2. MOTIVATION

Our goal in this paper is to blend rule based techniques such as fractals or L-systems with increased interactivity for designers. Although our motivation comes from fractals and L-systems, we want to give designers interactive control of modifiers to achieve conceptual shapes. We also want the resulting shapes to be physically constructible using 3D printers.

In this paper we present a simple approach that allows interactive and 3D extensions to Fractal and L-system methods. Using this approach, we have developed a system that enables designers to control each step of the shape generation process with a high level of interactivity.

With the new approach, novice users can easily create a large set of connected "self-similar" manifold surfaces. Disconnected surfaces are acceptable for "virtual" computer graphics applications in which the objects are used for only display purposes. However, in Architecture we usually want to physically construct the resulting shapes. To be able to construct the shapes, the shapes need to be connected and manifold surfaces. Figure 1 shows one example of how users can add finer details with our method.

Disconnected manifold surfaces (if individual surfaces are manifold) can be printed but it is not possible to guarantee that the resulting physical object will stay together. If individual surfaces are not manifold they will not even be suitable for 3D printing. For instance, two methods based on Iterated Function Systems (IFS) (Barnsley, 1988) create a set of disconnected points or shapes, which can never be printed. In contrast, our method allows us to construct connected manifold surfaces which can be realized a 3D printer.

Our approach is based on face replacements, which are a generalization of the line replacements of 2D fractal geometry. Face replacements are created by using local mesh operators. These operators can be applied to one face of the mesh without affecting the rest. They replace the face with multiple faces. A local operator can be defined by a set of insert edge and create vertex operations (Akleman and Chen, 2005) (Akleman, Chen et al., 2003). The local mesh operations such as extrusions guarantee that the resulting shape continues to be connected and manifold.

Landreneau et al. recently introduced Platonic extrusions as local mesh operators (Landreneau, Akleman, et al., 2005). These extrusions, except tetrahedral extrusion, are generalized pipes in which bottom and top polygons have the same number of sides (Landreneau, Akleman et al., 2005).

For this paper, we have extended Platonic extrusions to certain Archimedean extrusions as shown in Figure 2. Having a large variety of extrusions provides novice users a simple way to make face replacements. In addition to using these general extrusions, we introduce four new concepts for interactive modeling of connected and self-similar manifold surfaces: (1) Face Grouping using colors; (see Figure 3) (2) Group extrusions; (3) Automatic Face Regrouping and (4) Remeshing Schemes. Our method based on these concepts is guaranteed to create connected manifold surfaces.

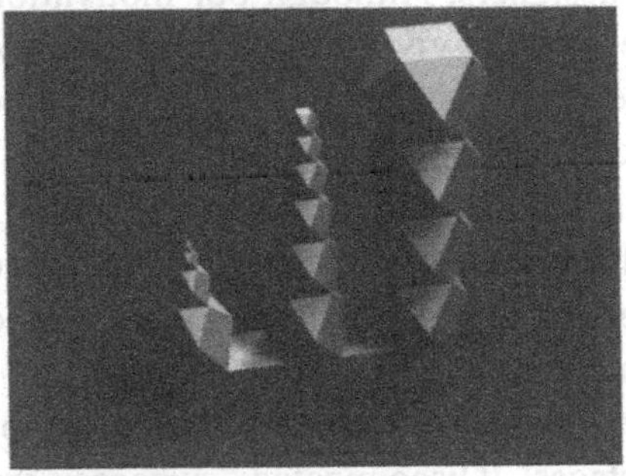

Cuboctahedral extrusion

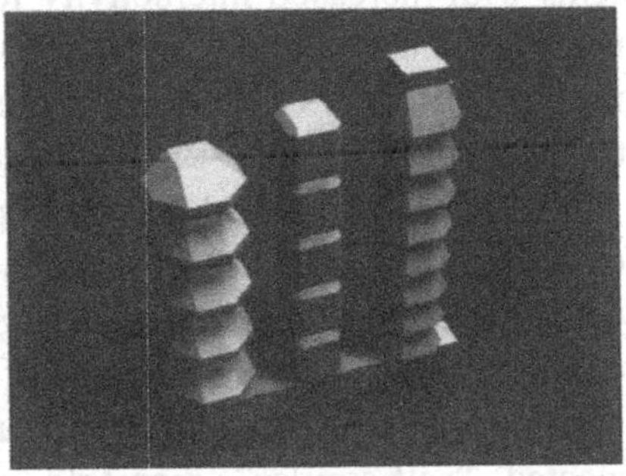

Rhombicuboctahedral extrusion

Figure 2. Examples of local mesh operators.

This modelling approach moves towards a more hands-on approach to grammar based surface modelling. A user can assert much more control over the surface beyond the traditional plug-in-a-formula-and-wait method of generating grammar based models. Our approach will particularly be useful in Architectural concept modeling, in which users can quickly determine the effects of various approaches by simply recoloring the faces. With this approach, users can rapidly learn how to create a wide variety of polygonal meshes that resemble grammar based shapes. The approach is not only limited to fractal looking shapes, however, it can also be used for creating a wide variety of shapes.

3. PREVIOUS WORK

From an architectural viewpoint, generative systems and fractals provide an extreme experimental domain for architectural design. The link between mathematical generation methods and architectural design is significant in contemporary architecture (Batty, 2005). For instance only three years after Benoit Mandelbrot has coined the term "Fractal", Peter Eisenman has designed House 11a, a composition of L shapes combined in rule based rotational and vertical symmetries (Eisenman, 1999). On the other hand, revolutionary designers like Greg Lynn, Karl Chu fully utilize the power of computational shape generation methods in their design studios (Achten, 2001). Chu uses a combinatorial system that branches recursively according to a primitive set of rules that include replication, combination and nesting to generative architectural forms (Chu, 1998). Generative systems in 2D pattern creation and 3D massing often experimented by design researchers (Datta et al., 1998) (Carlson and Woodbury, 1994). Implementations of different methods contribute to design/form variety, return unconventional results with aesthetical complexity. We observe that all the methods that allow creation of fine details are based on replacements. We will discuss previous work in three categories: Iterated Function Systems, Fractal Algorithms that depend on initial shape, and L-Systems.

3.1 Iterated Function Systems (IFS)

Fractal shapes most commonly are constructed with Iterated Function Systems, a simple procedure that exploits their self-similarity property, and introduced by Barnsley (Barnsley, 1988). The Iterated Function Systems approach is based on the concept that self-similar shapes can be considered as a union of transformed (e.g. scaled, rotated, translated and mirrored) copies of itself. For instance, if a self-similar Fractal shape can be given as

$$S = \bigcup_{k=0}^{K} A_k S$$

where A_k is a 4×4 transformation matrix in homogenous coordinates. Barnsley introduced deterministic and probabilistic algorithms to construct such self-similar shapes. In other words if a shape can be expressed as the union of its transformations (e.g. affine transformations like scale, rotate, translate and mirror) it is possible to create recursive algorithms for creation of fractal shapes.

Because of their simplicity, IFS are widely used to create fractal shapes. Another important property of these algorithms is that they are dimension independent. The same conceptual algorithm can be used both for 2D and 3D shape construction. For a successful application of IFS in 3D, see the XenoDream software (Thornton and Sterling, 2005). There are two major problems in this approach prevents designers to use them effectively. First these algorithms do not allow construction of different target shapes from different initial shapes. The second problem with IFS based algorithms is that the constructed shapes are usually not connected, and thus cannot be prototyped in 3D.

3.2 Fractal Algorithms that Depend on Initial Shape

Fortunately, IFS is not the only method for constructing fractal shapes. There exist alternative approaches in which the resulting shape depends on initial shape. However, these alternative approaches are usually not dimension independent and are hard to implement in 3D; they have not been widely used in 3D applications.

3.3 Line Segments Replacements

The line segments replacement method introduced by Mandelbrot is one such algorithm. Using line replacements, a wide variety of 2D fractals can be constructed (Mandelbrot, 1980), however, line replacements are useless in 3D.

For 3D, instead of line replacements we have to use face replacements. However, faces are not as simple entities as line segments. Line segments are always the same. They are straight and have a starting point and an ending point. After the replacement, one line segment is transformed into a set of line segments, which is exactly the same entity. Therefore, we can apply the algorithm iteratively.

It is not the same for faces. A face can have any number of corners and it may not be planar. Therefore, a face replacement method must be able to work on any type of face without affecting the rest of the mesh. Local mesh operations such as extrusions can exactly satisfy this criterion. They can be applied to any face and they do not affect the rest of the mesh. For face replacements it is also important to create a similar version of the original face. Most extrusions also provide this property since they usually are generalized pipes in which the bottom and the top polygons have the same number of sides (Akleman and Chen, 2005). Although they cannot be applied locally, some subdivision algorithms such as the Doo-Sabin and

Loop subdivision schemes (Doo and Sabin, 1978) (Loop, 1987) (Zorin, 2000) (Sabin, 2000) can also be used as face replacements since they can provide smaller versions of faces of an initial mesh. Checkerboard subdivision, a scheme we have also introduced, provides the same property.

3.4 L-Systems

L-systems are a much more powerful version of Fractal geometry's line segment replacement algorithms. In L-systems each line segment can have a label; based on the labels, each line can be replaced by different sets of lines. Because of their grammar based nature, L-systems can include context sensitivity and even parameters. These properties of L-systems make them very useful for designing the shapes of plants and trees. Despite their power over simple line-replacements, L-systems suffer similar problems to those of line-replacements. 3D shapes from L-systems are created by replacing the original lines with surfaces (usually cylinders). Since its is hard to connect these surfaces, the shapes described by L-systems usually consist of disconnected pieces.

4. METHODOLOGY

Our goal in this paper is to achieve the grammar based power of L-systems for constructing connected and manifold surfaces by combining face replacements with face grouping. In this section, we discuss four concepts introduced together in this paper: (1) Face Grouping using colors; (2) Group extrusions; (3) Automatic Face Regrouping and (4) Remeshing Schemes. Our method based on these concepts is guaranteed to create connected manifold surfaces.

4.1 Face Grouping Using Colors

Face grouping using colors allows users to easily group the faces in any modeling software. In face grouping, users classify faces by assigning a color to each. Faces are classified by colors, with identically colored faces belonging to a common group. Note that this stage is completely under the user's control. The faces do not have to be geometrically or topologically similar, so there are no restrictions for assigning a color to a face.

4.2 Group extrusions

Group extrusions simplify multiple extrusions. The users can apply the same extrusion operation to all identically colored faces by selecting only one face (see Figure 3).

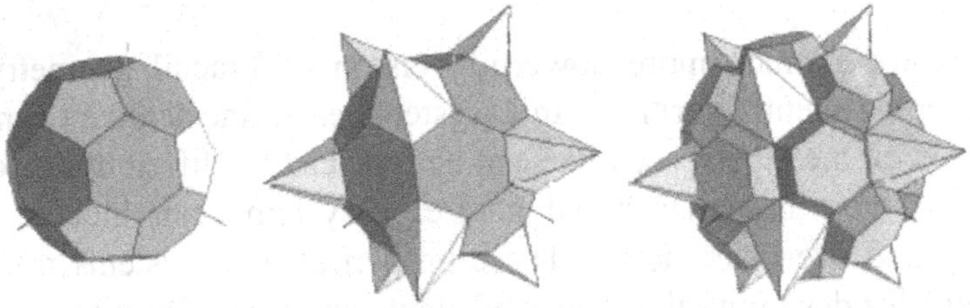

Figure 3. Face coloring and group extrusions.

Extrusions (except tetrahedral) produce a "top" face similar to the parent face, which is connected to the parent edges by "side" faces. The top face can inherent the group of the parent face. However, after a few iterations of remeshing or grouped extrusions, the number of side faces increases exponentially. We have provided automatic face regrouping to simplify the users' job regroup newly created side faces (see Figure 4).

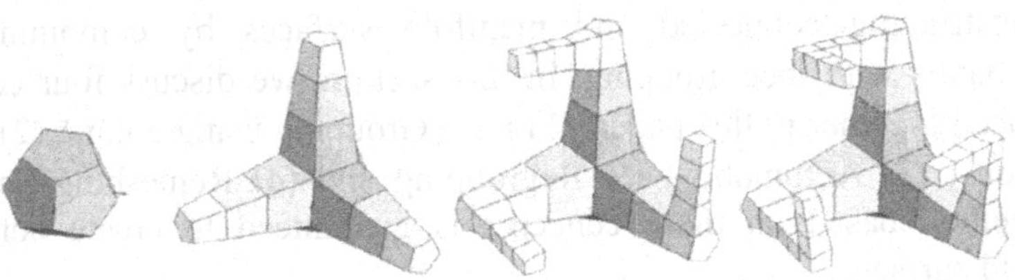

Figure 4. Automatic coloring newly created faces and group extrusions based on automatic coloring.

4.3 Automatic Face Regrouping with Modulus Colouring

To regroup side faces, we introduce the modulus colouring concept. The side faces are regrouped according to a modulus scheme. Starting from a randomly chosen side face, new groups are generated using a user supplied modulus. With a modulus of 1, every side face would share the same group. A modulus of 2 would generate two alternating groups. A modulus of 3 will make every third side face the same group, and so on. Using a modulus equal to the number of side faces of the extrusion, equal to the number of edges in the parent face, will assign a unique group identity for each side face. The modulus ensures that side faces will exhibit radial symmetry, due to side faces sharing colours. The modulus operation is not unique since the regrouping can be different based on the choices of initial side faces. Because of this, regrouping can introduce slight irregularities all allow to break overall symmetry of the object as seen in Figure 4.

4.4 Remeshing Schemes with Group Extrusions

Based on the recent research on subdivision surfaces, there now exists a wide selection of remeshing schemes that can be used in interactive applications. It is possible to view these remeshing operations as face replacements that are applied in parallel. By combining these schemes with group extrusions we provide additional flexibility. In fact, Loop style remeshing is particularly common in fractal algorithms (Fournier, Fussel et al., 1982). Using Loop style remeshing, it is possible to create generalizations of Koch islands (see Figure 5). Loop style remeshing with random vertex displacements is widely used for terrain generation.

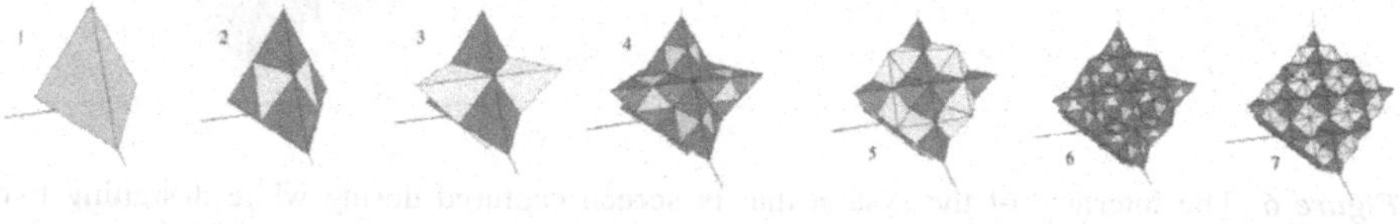

Figure 5. Loop remeshing scheme with tetrahedral extrusion. Using this procedure one can create Generalized 2-Manifold Koch Islands.

Existing subdivision remeshing algorithms can be categorized into two main classes (Akleman, Srinivasan et al., 2004). These are conversion schemes and preservation schemes. Each of these can further be divided into primary and dual schemes. The most apparent reason behind the popularity of Loop style remeshing among Fractal algorithms is that Loop preserves initial faces in every iteration. This property is particularly useful for face replacements since some of the newly created faces inherit the properties of initial faces. However, Loop is not the only one that can provide this property. All dual conversion schemes such as Corner Cutting, Simplest or Honeycomb and all preservation schemes (Loop belongs to this group) preserve initial faces in every iteration. They all can easily be used for interactive face replacement applications.

This classification is also helpful to identify missing remeshing schemes that can be useful for rule-based application. One such scheme we have identified and implemented is the so-called "checkerboard scheme".

Checkerboard is a quadrilateral preservation scheme and it preserves initial faces. Checkerboard is not really a new scheme; it is used in Fractal geometry to create generalized Menger Sponges.

5. IMPLEMENTATION AND RESULTS

The concepts that are discussed in methodology section are implemented and included in our existing 2-manifold mesh modelling system (Akleman, Chen et al., 2003). Our system is implemented in C++ and OpenGL. All the examples in this paper were created using this system. Figure 6 shows the interface of the system with two different examples.

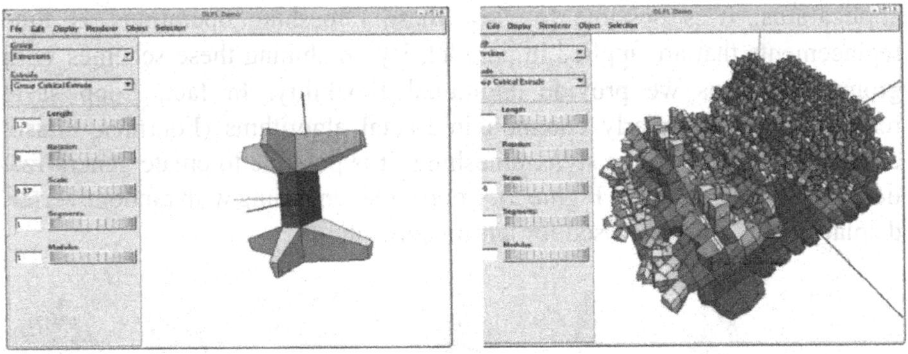

Figure 6. The interface of the system that is screen captured during while designing two different objects.

Our system is used in a short term design experiment by graduate design students. They are given a small sketch problem to create self similar spatial structures. Starting from simple geometric objects, students were able to design structural modules. Results from the experiment showed that our system is very robust to create modular objects. Figures 7, 8 and 9 show design examples created with the system. The main goal of this design experiment was to show that the power of our system to create rule-based architectonic spatial structures. Figure (6) is an example for coloured modular object under development.

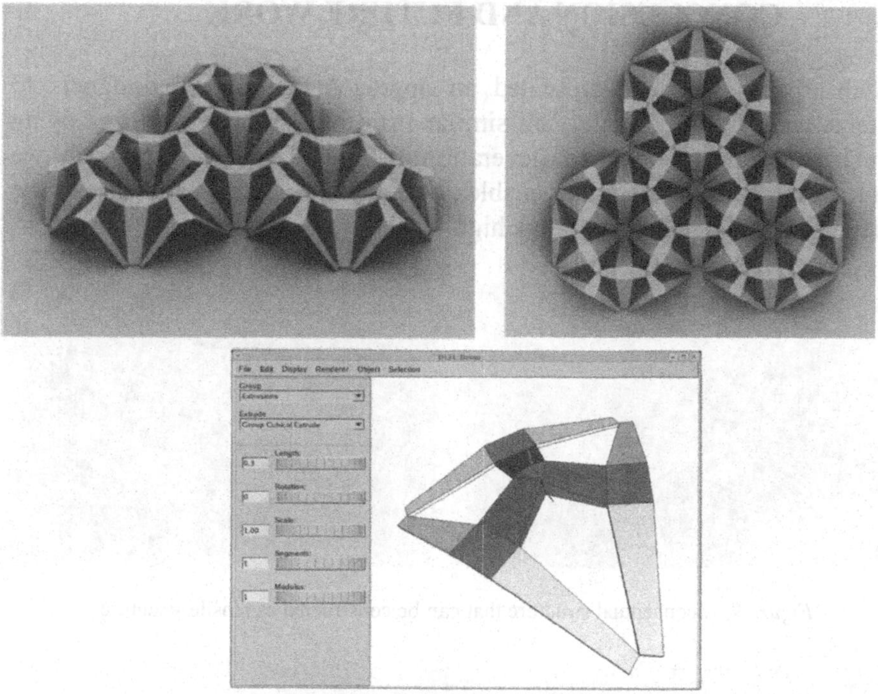

Figure 7. A conceptual Spatial Structure formed from repeated modules that are created using our system. In this case, shell itself is considered as the main architectural form. Interface snapshot is from the mass creation process with our software

Figure 8. A roof canopy created by modules that are constructed with our system.

6. CONCLUSION AND FUTURE WORK

In this paper we have presented an approach that allows designers to interactively create partially self-similar manifold surfaces without relying on shape grammars or fractal generation methods. Using this approach, we have developed a system that enables designers to control each step of the shape generation processes with a high level of interactivity.

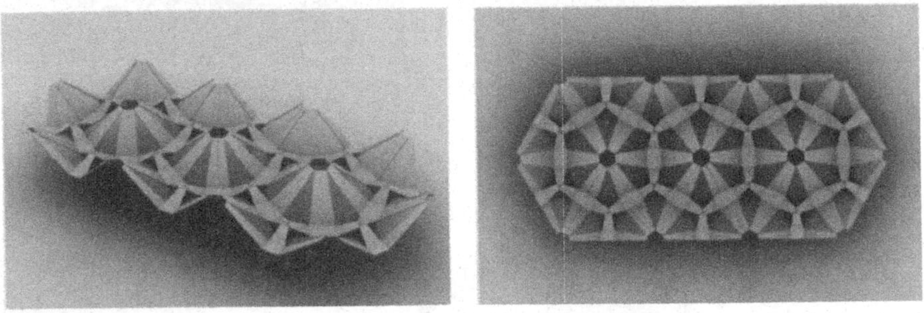

Figure 9. A conceptual structure that can be constructed as tensile structure.

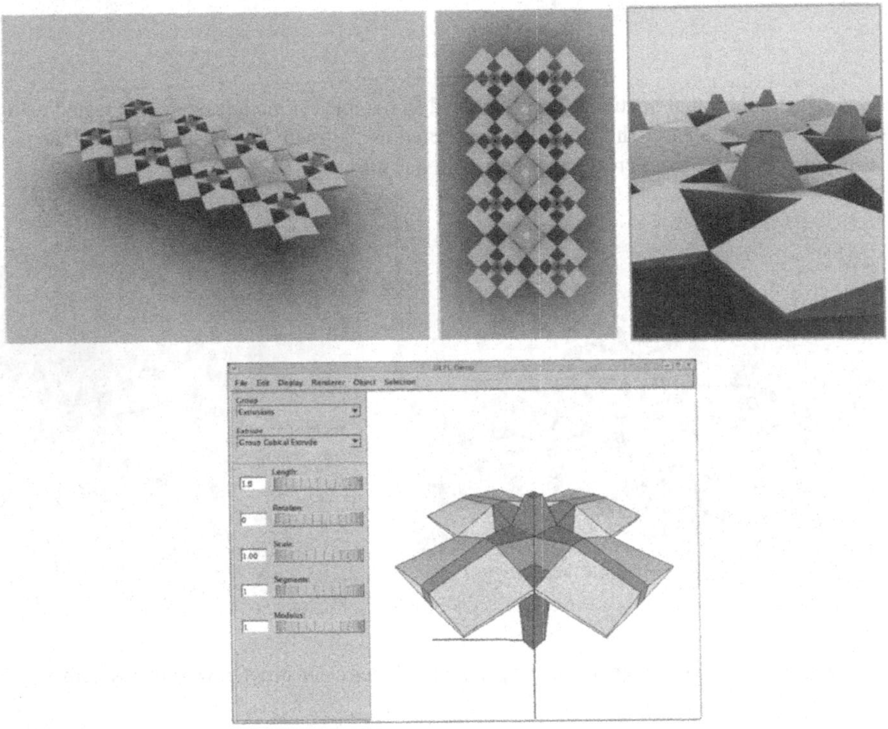

Figure 10. A roof canopy created by modules that are constructed with our system. Interface snapshot is from the mass creation process with our software

With the new approach designers can easily create a large set of connected "self-similar" manifold surfaces. Future work includes implementing more modifiers for form variety and connecting our approach with small scale manufacturing methods. Our method allows us to construct connected manifold surfaces which can be realized by rapid prototyping processes.

7. REFERENCES

Achten, H., 2001, "Normative positions in architectural design – deriving and applying design methods", *Proceedings of 19th eCAADe Conference: Architectural Information Management*, pp. 263–268.

Akleman, E., J. Chen, 1999, "Guaranteeing the 2-manifold property for meshes with doubly linked face list", *International Journal of Shape Modeling*, **5**(2):149–177.

Akleman, E., J. Chen, 2005, "Regular meshes", *Proceedings of Solid Modeling and Applications*, pp. 213–219.

Akleman, E., J. Chen, V. Srinivasan., 2003, "A minimal and complete set of operators for the development of robust manifold mesh modelers", *Graphical Models Journal, Special issue on International Conference on Shape Modeling and Applications* 2002, **65**(2): 286–304.

Akleman, E., V. Srinivasan, Z. Melek, P. Edmundson., 2004, "Semi-regular pentagonal subdivision", *Proceedings of the International Conference on Shape Modeling and Applications*, pp. 110–118.

Barnsley, M., 1988, *Fractals Everywhere*, Academic Press, Inc. San Diego Ca.

Batty, M., 2005.*Cities and Complexity: Understanding Cities with Cellular Automata, Agent-Based Models, and Fractals*. The MIT Press.

Carlson C., R.F Woodbury, 1994 "Hands-on exploration of recursive patterns" *Languages of Design* **2**: 121-142.

Chu, K., 1998, "Genetic-space: Hourglass in demiurge", *AD Magazine: Architects in Cyberspace I*, **68**(11).

Datta, S., R.F Woodbury, 1998 "Reducing Semantic Distance in Generative Systems: A Massing Example" Digital Design Studios: *Do Computers Make a Difference? ACADIA Conference Québec City (Canada)* October 22-25, 1998, pp. 164-171.

Doo, D., M. Sabin, 1978, "Behavior of recursive subdivision surfaces near extraordinary points", *Computer Aided Design*, **10**: 356–360.

Eisenman, P., 1999, *Peter Eisenman : Diagram Diaries*. Universe Publications.

Fournier A., D. Fussel, L. Carpenter, 1982, "Computer rendering of stochastic models", *Proceedings of Computer Graphics, Siggraph*, pp. 97–110.

Landreneau E., E. Akleman, V. Srinivasan, 2005, "Local mesh operations", *Proceedings of the International Conference on Shape Modeling and Applications*, pp. 351 – 356.

Loop, C., 1987, Smooth subdivision surfaces based on triangles. Master's thesis, University of Utah.

Mandelbrot, B., 1980, *The Fractal Geometry of Nature*. W. H. Freeman and Co., New York.

Sabin, M., 2000, Subdivision: Tutorial notes. Shape Modeling International 2001, Tutorial.

Srinivasan,V., E., Akleman, 2004,"Connected and manifold sierpinski polyhedra", In *Proceedings of Solid Modeling and Applications*, pp. 261–266

Thornton, G., V. Sterling, 2005, *Xenodream software*, http://www.xenodream.com.

Zorin, D., P. Schröder, 2000, "Subdivision for modeling and animation," *ACM SIGGRAPH 2000 Course #23 Notes.*

Automatic Semantic Comparison of STEP Product Models

Application to IFC product models

G. Arthaud and J.C. Lombardo[1]
Ecole Nationale des Ponts et Chaussées
[1] Centre Scientifique et Technique du Bâtiment

Keywords: Industry Foundation Classes (IFC), STEP models, EXPRESS language, Semantic comparison, Design process

Abstract: This paper introduces an original method to compare IFC models and more generally any STEP models. Unlike common "diff-like" tools which compare textual files by proceeding line against line, our approach compares actual graphs created from STEP-files. Therefore added, removed, and changed objects can be tracked between two versions of the model. Besides, this standalone tool does not need any heavy database to work so it is fully adapted to design methods of construction projects, where actors are free to modify a local version of their project without any dependence on the database. Moreover it is reusable for other industrial fields thanks to its compatibility with any STEP model. This tool is a part from a more global project which tends to improve accessibility and sustainability of IFC therefore it can be used as a support for VR based design tools.

1. INTRODUCTION

Our team carries out research on Industry Foundation Classes visualization and real-time simulations in virtual environments, in order to improve accessibility and sustainability of IFC. We investigate a major issue in this paper which delays the use of IFC in the AEC community: comparison and merging of IFC models. After a short reminder of IFC and design process in a construction project, we will address the main problem of our research: *semantic comparison* of product models.

Jos P. van Leeuwen and Harry J.P. Timmermans (eds.), Innovations in Design & Decision Support Systems in Architecture and Urban Planning, 447-463.

1.1 Interoperable Project

In normal practice, information is exchanged between engineers, architects and clients, in the form of verbal and hardcopy programs, sketches, diagrams and drawings. This often requires interpreting, re-documenting and re-entering information into software and systems to make it useable for the project team. Great loss and corruption of data may occur during this translation.

Thanks to the Industry Foundation Classes data model specified by the International Alliance for Interoperability, software can exchange standardized product models during a civil engineering project. This standard avoids multiple specialized interfaces between stakeholders specific tools as shown in the figure 1.

Central IFC databases support such models (Cruz, Nicolle et al., 2002) (Vanlande, Cruz et al., 2003). They bring the project team a significant help for document management. But services provided by these databases are not fully adapted to usual design methods of a construction project. Therefore many AEC stakeholders are still reluctant to exploit IFC in their own project. Next subsection is dedicated to a study of the typical workflow during a construction project.

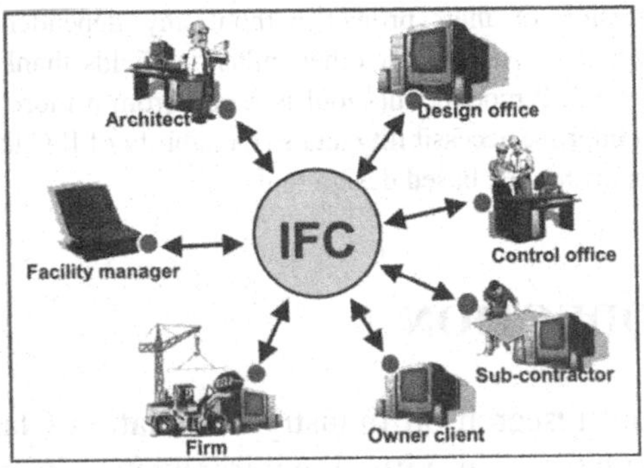

Figure 1. Model of a shared project with IFC (Lebegue, 2004).

1.2 Design Process

Common design processes usually exploit the repeated succession of two important phases (Hanser, 2003):

- Co-design
- Distributed design.

Co-design phase can be a meeting where objectives are defined and work is split. Communication between actors is synchronous here. Afterwards actors can work in parallel and communication between them is asynchronous: fax, e-mails... This is the distributed design phase. Following meetings could define new objectives to restart a new design cycle.

Central IFC databases of an IFC-based shared project only exploits distributed design phase. It makes the process too restrictive, according to many AEC project teams.

That is why our research team suggests a new system. Instead of working around a unique object-oriented model, design team develops the first model (an architectural model for example) and every group works on a local version of the original model. Actors asynchronously communicate these to each other. These product models are freely modifiable without any coherence checking. Then models are merged at the next meeting. This method gives the project team the required flexibility. Figure 2 illustrates that process.

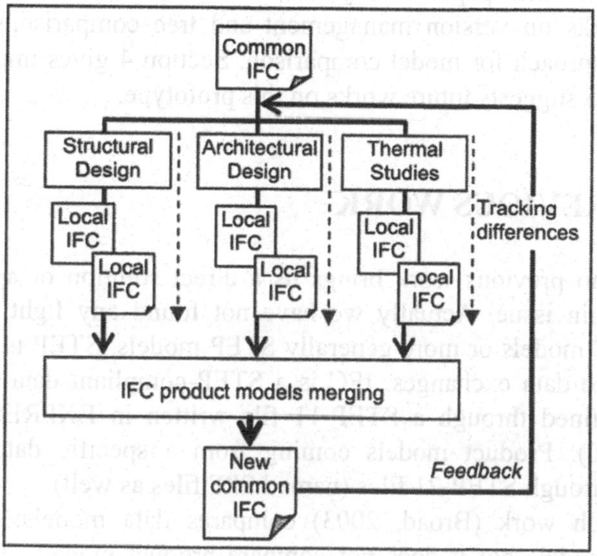

Figure 2. Suggested workflow system. It is fully adapted to the design process.

1.3 Semantic Comparison

Concerning this suggested system, two issues should be considered:

1. Track the local models evolution in the distributed design phase.
2. Merge local models in the co-design phase.

Merging local models is not a trivial problem especially for coherence checking between architectural models and structural models (Chen, 2005). But if the first problem is brought a good solution, the second point will be far much easier to be addressed. In this paper we focus on the first point.

The goal is to track added, changed, removed, and moved objects between the original version of a construction object and the modified one as made with *diff* for ASCII files. Unlike *diff*, the comparison should be semantic here because the serialization of an IFC product model is not unique: Numbering of entities is not unique. Besides the size of IFC files can reach over 100 megabytes so their handling is a heavy task. Therefore semantic comparison should be optimized if the product model has little changed. Lastly design process could be definitely improved through an ergonomic graphical visualization of these differences.

Moreover we focus on light tools which do not need any heavy database to work. Input data are single files coming from standalone software in order to provide the project actors the required flexibility.

The rest of this paper is organized as follows. Section 2 describes previous works on version management and tree comparisons. Section 3 shows our approach for model comparison. Section 4 gives implementation tips. Section 5 suggests future works on this prototype.

2. PREVIOUS WORK

First of all no previous work brings us a direct solution or application to solve the main issue. Actually we have not found any light tool able to compare IFC models or more generally STEP models. STEP is a formalism to standardize data exchanges. IFC is a STEP-compliant data model. Data model is defined through a STEP-11 file written in EXPRESS language (Pierra, 2000). Product models coming from a specific data model are exchanged through STEP-21 files (named SPF files as well).

A research work (Broad, 2003) compares data models: It compares EXPRESS models but it does not compare product models. This research deals with other problems like ontology and taxonomy. A typical application of this work could track evolution of IFC data model.

Therefore we should look for low-level comparisons: ASCII files comparison, trees comparison, and version management.

2.1 The Status-Based Comparison with *Diff*

Concerning ASCII files, classical diff-like tools allow you to track differences between two textual files. Similar tools are intensively used by cooperative software development in order to build versions of source code files: Concurrent Versions System.

Diff assigns a status to each line of both original and modified file. Four statuses are available: IDENTICAL, CHANGED, REMOVED and ADDED.

However ASCII files are basic structures. Connected semantic data could be described by a graph whose nodes numbering and position in file is arbitrary, according to STEP specifications.

We would like to compare elements from these product models and assign status as *diff* does. Therefore we have to compare two data graphs.

2.2 Trees Comparison

Comparing trees is the origin of a lot of famous problems and many papers deal with them. The most known one is "Tree Edit Distance" (Bille, 2003). It compares labelled trees based on simple local operations of deleting, inserting and re-labelling nodes. One common condition is required: nodes must be ordered. Otherwise the problem is in general NP-hard. Usually, an ordered labelled tree can be compared in $O(n^4)$ in time.

Our problem is quite different because the graph (or the tree after a pruning phase) is not a free graph but a constrained one because of its model structure, for example IFC model. Data model defines the skeleton of the product model (STEP-21 file), even if the graph of the product model could contain aggregation nodes: A building contains a list of floors whose size is not fixed by the data model. Nevertheless global structure of a product model is fixed and comparison could be much more effective than a generic tree comparison.

2.3 IFC Servers and Databases

Because of their size IFC files are often stored in robust databases and several researchers tried to improve data content (Ting, Yang et al., 2003) (Tanyer, Aouad, 2005), interfaces (SABLE project), and accessibility to data (Vanlande, Cruz et al., 2003), (Cruz, Vanlande et al., 2004).

More generally many research projects were carried out on objects comparison and versions. Complex Entity Versioning (Urtado, 1998) suggests a flexible approach to manage data evolution, compatible with object oriented data models like IFC. Every stored entity creates a version of entities with various dependences. These dependences imply a propagation mechanism of operations described by propagation rules and strategies. However this approach tends to create heavy version tools which cannot be used as a single diff tool between two models. Few applications separate the comparison mechanism from the version process. We need a light standalone tool in order to obtain a flexible and fast system.

3. APPROACH

As was introduced in the previous section, IFC model comes from a STEP-11 model. This model is described in EXPRESS language. Figure 3 shows an extract from such an EXPRESS file. Product models are exchanged through STEP-21 files.

```
ENTITY IfcOrganization;
 Name          : STRING;
 Addresses     : LIST [0:?] OF IfcAddress;
 Roles         : LIST [0:?] OF IfcActorRole;
 Description   : OPTIONAL STRING;
END_ENTITY;

ENTITY IfcPerson;
 FamilyName     : OPTIONAL STRING;
 GivenName      : OPTIONAL STRING;
 MiddleNames    : OPTIONAL STRING;
 PrefixTitles   : OPTIONAL STRING;
 SuffixTitles   : OPTIONAL STRING;
 Addresses      : LIST [0:?] OF IfcAddress;
 Roles          : LIST [0:?] OF IfcActorRole;
  WHERE
    WR1: EXISTS(FamilyName) OR EXISTS(GivenName);
  END_ENTITY;
```

Figure 3. Extract from the STEP-11 file of IFC 2.0.

Many data models are defined using STEP: IFC, IFC-Bridge, AP203, AP210 (for electronics) and Step-TAS (for satellites) ... Besides a specific data model often evolves: current version of IFC is 2x2 Addentum1. Anyway developing a comparator for a fixed version of IFC is not relevant whereas a more general formalism was created. That is why we have chosen to design and implement a generic comparator of STEP product models. However our analysis has been influenced by IFC structure and specific properties. We validate implemented algorithms with construction projects.

3.1 Analysis of Data Structure

A product model is an oriented graph. Figure 4 shows the transformation from an extract of a STEP-21 file into a graph. A node contains its state and its elementary attributes. An oriented connection creates a link from an entity to another entity. In the figure 4, the graph has two more links (dotted lines). They illustrate specific attributes: the inverse attributes. They are defined in the STEP-11 file and provide a bidirectional link between entities. But only explicit attributes are written the STEP-21 file. For instance, node 17 is an inverse attribute of node 15.

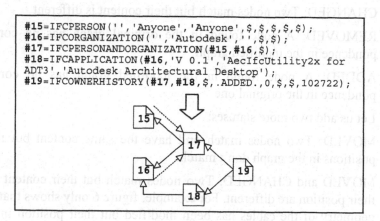

```
#15=IFCPERSON('','Anyone','Anyone',$,$,$,$,$);
#16=IFCORGANIZATION('','Autodesk','',$,$);
#17=IFCPERSONANDORGANIZATION(#15,#16,$);
#18=IFCAPPLICATION(#16,'V 0.1','AecIfcUtility2x for
ADT3','Autodesk Architectural Desktop');
#19=IFCOWNERHISTORY(#17,#18,$,.ADDED.,0,$,$,102722);
```

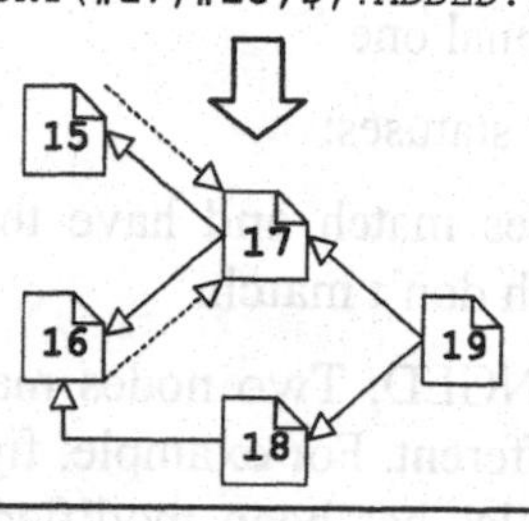

Figure 4. From a STEP-21 file to the associated data graph.

We may need these bidirectional links to traverse the graph, especially with IFC product models: Comparing two graphs starts with the root node comparison. Root node is not set by STEP specifications but the most relevant is IfcProject. Besides connections between important objects use relationship entities: IfcRelContains. It needs an IfcProject as a container and an IfcSite as a contained object. Figure 5 shows the corresponding diagram.

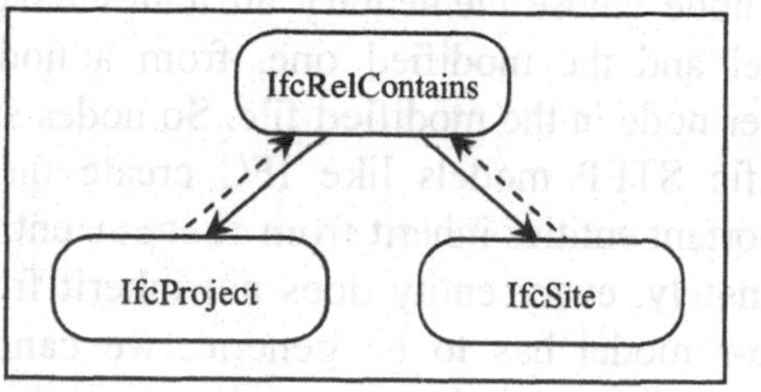

Figure 5. An IfcProject and an IfcSite connected together by an IfcRelContains.

Therefore a graph traversal from the `IfcProject` needs inverse links because direct links come from `IfcRelContains` and not from `IfcProject`. Let us introduce how we will handle the graph nodes.

3.2 Status

We assign a status to each node of the graph. There are four statuses:

- IDENTICAL: Two nodes match and have the same content.
- CHANGED: Two nodes match but their content is different
- REMOVED: A node found in the original graph has no correspondence in the modified graph.
- ADDED: A node found in the modified graph has no correspondence in the original one
- Let us add two more statuses:
- MOVED: Two nodes match and have the same content but their positions in the graph don't match.
- MOVED and CHANGED: Two nodes match but their content and their position are different. For example, figure 6 only shows that the geometry of the cables has been modified but their position in the abstract graph has changed as well because of the removal of the first cable.

Without these "Moved" status and "Moved and Changed" statuses, we describe only the node state without any consideration about connections in the graph. However in this model we do not assign status to connections because we only focus on objects. As was mentioned before, graph structure is constrained by model structure so we do not need to track differences between connections. Figure 6 shows a trivial example with a bridge.

Actually the real challenge is to find a way of matching nodes. The aim is to differentiate a node whose elementary attributes have changed between the original model and the modified one, from a node which has been replaced by another node in the modified file. So nodes should be *identified.* Fortunately specific STEP models like IFC create `UniqueIDs` for many entities: most important entities inherit from `IfcRoot` entity. It has a globalId attribute. Unfortunately, every entity does not inherit from `IfcRoot` entity. As our comparator model has to be generic, we cannot use `UniqueIDs` directly. In section 4, we explain how to work around this problem thanks to *helpers*. Anyway we suppose that several objects are identifiable and others are not. Next section describes how to manage non-identifiable nodes.

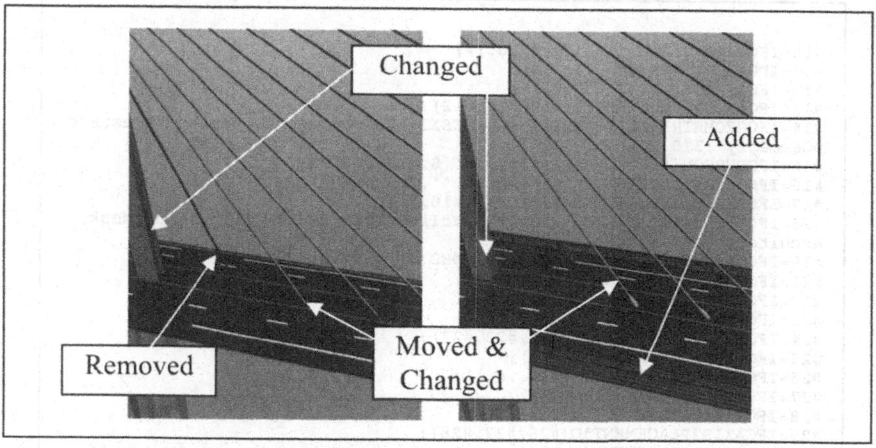

Figure 6. Assigning status to objects on an architectural bridge model, the original model is on the left and the revised one is on the right.

3.3 Graph Pruning

As described in first section, an IFC file can be larger than 100 Mb. The associated graph can have more than 1,000,000 nodes with cyclic links. So another challenge is the simplification of the graph. There are two categories of simplification:

- To take off elementary nodes
- To cut redundant connections

Actually elementary nodes are not assigned a status because they are absorbed by the parent entity. Therefore we differentiate the composition of a node from its state: The content of an elementary node is inserted in the state of its parent entity. If there is a change in this node, the parent becomes changed. If a node has several parents, every parent gets a copy of its child node. But what is an elementary node? How can we differentiate a composition from a state? The sharpest way is to assume elementary nodes are non-identifiable nodes. Therefore we solve the previous problem of assigning a status to non-identifiable node. Figure 7 is another extract of an IFC file and figure 8 shows the associated graph with inverse links.

```
#10=IFCCARTESIANPOINT((0.,0.,0.));
#11=IFCDIRECTION((0.,0.,1.));
#12=IFCDIRECTION((1.,0.,0.));
#13=IFCAXIS2PLACEMENT3D(#10,#11,#12);
#14=IFCGEOMETRICREPRESENTATIONCONTEXT('TestGeometricContext','TestBre
pGeometry',3,0.,#13,$);
#15=IFCPERSON('','Anyone','Anyone',$,$,$,$,$);
#16=IFCORGANIZATION('','Autodesk','',$,$);
#17=IFCPERSONANDORGANIZATION(#15,#16,$);
#18=IFCAPPLICATION(#16,'V 0.1','AecIfcUtility2x for ADT3','Autodesk
Architectural Desktop');
#19=IFCOWNERHISTORY(#17,#18,$,.ADDED.,0,$,$,1032772322);
#21=IFCCARTESIANPOINT((0.,0.,0.));
#22=IFCDIRECTION((0.,0.,1.));
#23=IFCDIRECTION((1.,0.,0.));
#24=IFCAXIS2PLACEMENT3D(#21,#22,#23);
#25=IFCLOCALPLACEMENT($,#24);
#26=IFCCARTESIANPOINT((0.,0.,0.));
#27=IFCDIRECTION((0.,0.,1.));
#28=IFCDIRECTION((1.,0.,0.));
#29=IFCAXIS2PLACEMENT3D(#26,#27,#28);
#30=IFCLOCALPLACEMENT($,#29);
#31=IFCBUILDING('08xiz9C0z02wxhslb2YMDz',#19,'build','Building',$,#25
,$,'',.ELEMENT.,0.,0.,$);
#32=IFCSITE('11Nk2_XwfFseAUbJ$kz9w8',#19,'site','Site',$,#30,$,'',.EL
EMENT.,$,$,0.,'',$);
#33=IFCRELAGGREGATES('1Z7EsmAUPAPvkmuhSls_Se',#19,'Testrelation','Def
ault Site',#20,(#32));
#34=IFCRELAGGREGATES('3QWRAOndj7KA2U$TGXE09S',#19,'Testrelation','Def
ault Building',#32,(#31));
#20=IFCPROJECT('3j1ffkM2rDQPn37k7tj2W6',#19,'','Testing','','Testfile
','design',(#14),#9);
```

Figure 7. Extract from an IFC file.

In this graph, there are only four identifiable nodes: 20, 31, 32, 33 and 34, identifications are sequences of characters as first parameter of these nodes (cf. figure 8). Consequently, we obtain the following graph in figure 9.

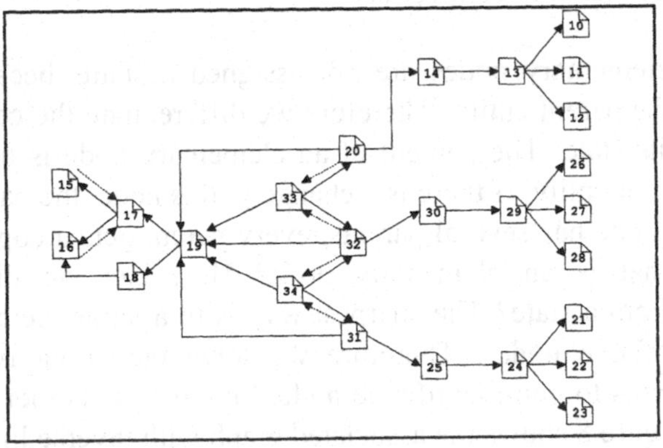

Figure 8. Graph resulted from the parsing of the figure 7 file.

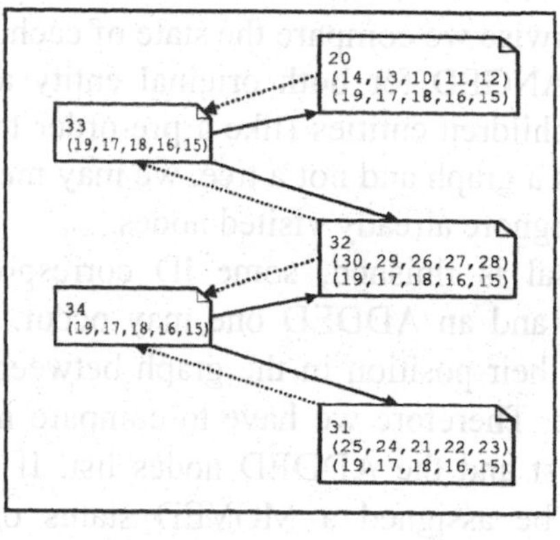

Figure 9. Graph after a pruning. Old nodes are absorbed by the parent entity and become new elements for the state of parent entity.

The element 19 was referenced by 5 parents so this element is copied 5 times in these parents.

Lastly the semantics of a model are not impacted by this pruning since absorbed nodes from a product model are instances of unidentifiable objects, so they never can be compared between them.

3.4 Graph Traversal

Previous paragraphs gave some details concerning the traversal, especially the inverse links role. This section suggests an optimized method to traverse this kind of semantic graph.

Our approach is definitely different from classical trees comparisons and traversals because:

- The global structure of the graph is fixed by a data model.
- We focus on nodes traversal, not on links.

A first approach would compare nodes without caring for connections between them. A quadratic-time computation would compare lists of nodes from the original model to lists of nodes from the modified model. We reject this extreme approach because MOVED and MOVED AND CHANGED status cannot be assigned in this case.

This leads us to suggest a true graph traversal: we drive a simultaneous traversal in the original graph and the modified graph. This can be done if and only if root nodes from original and modified graphs are same type. During this process only four statuses are assignable: IDENTICAL, CHANGED, ADDED, and REMOVED. During a node comparison, we compare at first identifications. If there is no matching of IDs, then original entity is assigned a REMOVED status and the modified one is assigned an

ADDED status. Otherwise we compare the state of each node, and we assign IDENTICAL or CHANGED for both original entity and modified entity. Lastly, we compare children entities (like a pre-order tree traversal) and so on. As the structure is a graph and not a tree, we may meet an already visited node. We use tags to ignore already visited nodes.

When the traversal is finished, some ID correspondences between a REMOVED element and an ADDED one may occur. It means that some nodes have changed their position in the graph between the original model and the modified one. Therefore we have to compare in quadratic-time the REMOVED nodes list and the ADDED nodes list. If two nodes have the same ID, they will be assigned a MOVED status or a MOVED AND CHANGED status depending on the state comparison result.

Therefore complexity is quadratic in the worst case (all nodes are assigned ADDED and REMOVED during the traversal), but the graph traversal visits every node only one time. This is much more efficient if a model is hardly modified between the original version and the modified one.

4. IMPLEMENTATION

This section gives some details concerning our implementation of the STEP-21 Semantic Comparator. First we need to know how C++ IFC classes could be generated from the STEP-11 file.

4.1 Early Binding or Late Binding?

STEP models can be exploited in two ways (Loffredo, 2005):

- The package can work directly on any STEP objects: IFC, AP203, Step-TAS… without knowing the model structure before, this is Late Binding. In this case the STEP-11 model is used at runtime as a dictionary to interpret the STEP-21 file.
- The package code has been generated from a program which reads the model structure (STEP-11 file) and then can only work on STEP objects compatible with this model structure. This is Early Binding. In this case the STEP-11 model is hard coded in the package.

Late Binding is very useful when application is intended to work on multiple EXPRESS schemas, but Early Binding is easier and faster to process. So we would rather work on early binding implementation.

Several applications are able to parse an EXPRESS schema to generate C++ classes. We have chosen Expressik from MINT Group of the University of Manchester. This is a Java package which contains an EXPRESS file parser and a C++ classes Generator (Withers, 2005).

Generated C++ classes contain entities and various get/set methods which give access to attributes. Our comparison model adds a new behaviour to these generated classes. Therefore we use the Visitor design pattern (Gamma, Helm, et al., 1995) whose implementation is prepared during the generation of classes.

4.2 Comparator Generator

Like C++ model classes, we have to choose a way to implement the semantic comparator: Early-binding or Late-binding. As we previously chose Early-binding for C++ model classes, we generate specific comparator classes and hard code the STEP model into these classes.

We used the EXPRESS parser library provided by Expressik package. We have created a Java application which needs three input data: the EXPRESS schema (STEP-11 file), C++ template files and helpers.

Figure 10 illustrates the global mechanism.

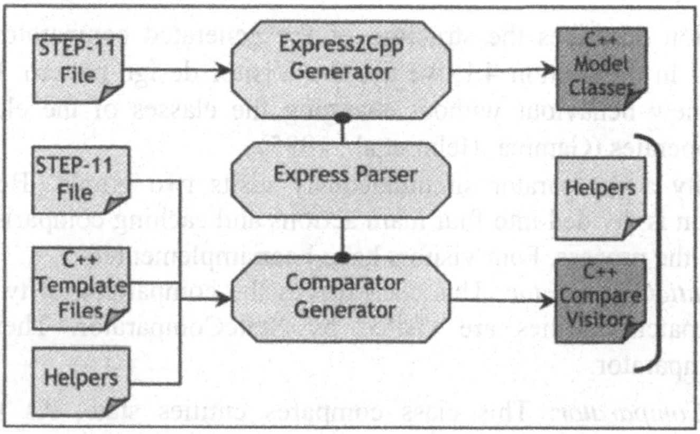

Figure 10. Generating C++ classes and the semantic comparator.

C++ template files are encapsulated into a XML file where specific tags are handled by the comparator generator. C++ template files set a specific syntax compatible with the Express2Cpp Generator. Therefore our comparator generator is reusable for any C++ model classes if specific C++ template files are created.

Lastly helpers are encapsulated into a XML file as well and give some guidelines concerning special properties of specific EXPRESS schema: the class name of the root node, the class name of identifiers, a function to compare identifiers... So we have to create a helper for every EXPRESS schema. Concerning IFC, we choose `IfcProject` as root node, and `IfcGloballyUniqueId` as identification. The comparing identification function is a string comparison. Figure 11 shows a simple helper for IFC. Helpers could be improved to provide subtle behaviours for the comparator.

```
<helper>
<Id name="IfcGloballyUniqueId">
bool haveSameId(
IfcGloballyUniqueId* obj1,
IfcGloballyUniqueId* obj2)
{
   return*(obj1)==*(obj2);
}
</Id>
<Root name="IfcProject"/>
</helper>
```

Figure 11. A simple helper for IFC data model.

4.3 Generated Classes

This section describes the structure of the generated comparator. As was mentioned in the section 4.1, we use the Visitor design pattern. It lets you define a new behaviour without changing the classes of the elements on which it operates (Gamma, Helm et al., 1995).

Actually a comparator simultaneously visits two graphs. Besides, the comparison is divided into four main actions and caching comparison results optimizes the process. Four visitors have been implemented:

- *SemanticComparator*: This class drives the comparison of two entities. If IDs match, entities are visited by StateComparator. Then it calls ChildComparator.
- *StateComparator*: This class compares entities state. As soon as a difference is detected, the method returns a CHANGED tag. If no difference was found, it calls then NidComparator.
- *NidComparator*: This specific visitor prevents the system from copying multiple times non identifiable nodes. It visits these nodes by comparing elementary attributes and complex attributes recursively. It returns a CHANGED tag as soon as a difference is detected.
- *ChildComparator*: This comparator calls Semantic Comparator to visit the children nodes sequentially.

Figure 12 shows a simplified UML diagram of the comparator system. "`m_old`" is the original graph. There are as many "`visit`" methods as the number of model entities.

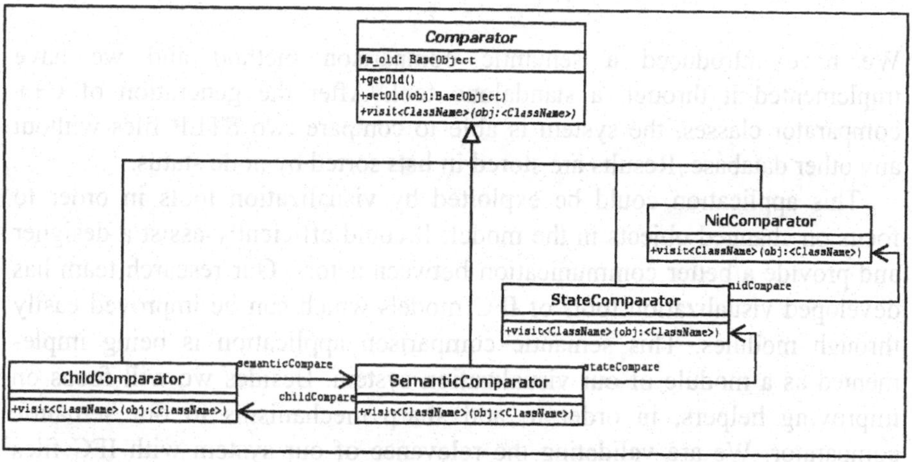

Figure 12. UML diagram of the semantic comparator system.

5. RESULTS

The current prototype is running on a PC (Pentium D 820 3.0 Ghz, 2 Gb RAM, Windows XP Pro). We evaluated the performance of two IFC files comparison. We have chosen to measure performance in the worst case: n^2 comparisons, where n is the number of identifiable nodes, no ID correspondence between nodes has been found. Results are reported in Table 1.

Table 1. Measure of IFC files comparisons.

Model Name	BLIS Helsinki	Residential House	Orchard Project	Level 2 Step 14
File Size (Mb)	1.94	8.66	20.5	31.6
Entities	38 785	156 831	399 221	539 553
Identifiable nodes	3 928	5 350	10 226	15 545
Loading time	1.16 sec.	5.11 sec.	12.3 sec.	19.9 sec.
Comparison time	0.328 sec.	0.841 sec.	4.52 sec.	12.54 sec.

This evaluation simulates a comparison of very different models. Therefore performance should be usually better. These results show that the comparison is a fast operation comparing to file loading. It allows us to improve helpers in order to create a smarter semantic comparator.

6. CONCLUSION AND FUTURE WORKS

We have introduced a semantic comparison method and we have implemented it through a standalone tool. After the generation of C++ comparator classes, the system is able to compare two STEP files without any other database. Results are stored in lists sorted by node status.

This application could be exploited by visualization tools in order to focus on changed objects in the model. It could efficiently assist a designer and provide a better communication between actors. Our research team has developed visualization tools of IFC models which can be improved easily through modules. This semantic comparison application is being implemented as a module of our visualization system. Besides we will focus on improving helpers, in order to add sharp mechanisms to our semantic comparator. We are validating the relevance of our system with IFC files generated by commercial software and helpers are the fundamental keys for the achievement of the semantic comparator.

Other tests will validate the semantic comparator with other STEP models, like IFC-Bridge (IFC extension for engineering structure) and Step-TAS (STEP model for satellites).

Lastly this application tends to improve design process. We should take into consideration the other phase of the design process: co-design. Complementary tools and algorithms will be developed to merge local version of IFC files. During this merge cohesion problems could occur so we need a semi-automatic way to check cohesion between construction objects. Once again design process could be improved and AEC actors may be more motivated to exploit IFC in their own project in the near future.

7. REFERENCES

Bille, P., 2003, "Tree edit distance, alignment distance and inclusion", IT University of Copenhagen, *Technical Report Series*.

Broad, A.P., 2003, "Comparative Code Understanding of Information Models". PhD. thesis, University of Manchester, Manchester, UK.

Chen, P.H., L. Cui and C. Wan, 2005, "Implementation of IFC-based web server for collaborative building design between architects and structural engineers", *Automation in Construction*, **14**(1): 115–128.

Cruz, C., C. Nicolle, and M. Neveu, 2002, "The active3d-build : A webbased civil engineering platform", *IEEE MultiMedia*, **9**(4) :87–90.

Cruz, C., R. Vanlande and C. Nicolle, 2004, "Active3d : Semantic and 3d databases for civil engineering projects", *In Hamid R. Arabnia, editor, IKE*, pages 56–61. CSREA Press.

Gamma, E., R. Helm, R. Johnson, J. Vlissides, 1995, *Design Patterns, Elements of Reusable Object-Oriented Software*, Addison-Wesley.

Halfawy, M.R. and T. Froese, 2002, "Modeling and implementation of smart AEC objects: An IFC perspective", *CIB W78 conference.*

Hanser, D., 2003, *Proposition d'un modèle 'd auto coordination en situation de conception, application au domaine du bâtiment.* PhD thesis, Institut National Polytechnique de Lorraine, Nancy, France.

Lebegue, E., R. Vankeisbelck, 2004, "Standard d'échange IFC et contrôle de vérification d'accessibilité des locaux", *Technical Report*, CSTB.

Loffredo, D., 2005, *Fundamentals of STEP Implementation,* http://www.steptools.com/library/

Pierra, G., 2000, *Spécification de données, le langage EXPRESS.* ENSMA, http://www.lisi.ensma.fr/ftp/enseignement/A3_Master_Ingenierie_donn%E9es/EXPRESS 1.pdf

Tanyer, A.M. and G. Aouad, 2005, "Moving beyond the fourth dimension with an IFC-based single project database", *Automation in Construction,* **14**(1): 15-32.

Ting, S.K., Q.Z. Yang, P.H. Chen, 2003, "Ifc-based information modelling and model server technologies for product lifecycle information support", *Technical Report 064*, PDD Group, SIMTech.

Urtado, C., 1998, "Versions d'entités complexes. Approches microscopiques et macroscopiques", PhD. thesis, Université Montpellier II, Montpellier, France.

Vanlande, R., C. Cruz, and C. Nicolle, 2003, "Managing ifc for civil engineering projects". *In CIKM,* ACM, pages 179–181.

Withers, D., 2006, *OSS-Express, C++ Code Generator*, http://mint.cs.man.ac.uk/ossexpress/cpp/index.html

Virtual Environments and Augmented Reality

Design Tools for Pervasive Computing in Urban Environments

A. Fatah gen. Schieck, A. Penn, V. Kostakos[1], E. O'Neill[1], T. Kindberg[2], D. Stanton Fraser[3], and T. Jones[3]

The Bartlett Graduate School, University College London, UK

[1] Department of Computer Science, University of Bath, UK

[2] HP labs, Bristol, UK

[3] Department of Psychology, University of Bath, UK

Keywords: Urban space, Pervasive systems, Urban computing, Space Syntax, Interaction space

Abstract: In this paper we report on ongoing research in which the implications of urban scale pervasive computing (always and everywhere present) are investigated for urban life and urban design in the heritage environment of the city of Bath. We explore a theoretical framework for understanding and designing pervasive systems as an integral part of the urban landscape. We develop a framework based on Hillier's Space Syntax theories and Kostakos' PSP framework which encompasses the analysis of space and spatial patterns, alongside the consideration of personal, social and public interaction spaces to capture the complex relationship between pervasive systems, urban space in general and the impact of the deployment of pervasive systems on people's relationships to heritage and to each other. We describe these methodological issues in detail before giving examples from early studies of the types of result we are beginning to find.

1. INTRODUCTION

Developments in computing and architectural environments have evolved together to a surprising degree since the 1990s: Virtual Reality 'VR' represents one of these developments, however, this has remained of greater academic than practical interest for a number of reasons such as cost and the

Jos P. van Leeuwen and Harry J.P. Timmermans (eds.), Innovations in Design & Decision Support Systems in Architecture and Urban Planning, 467-486.

unfeasibility of working in a completely immersive environment (Penn et al., 2004). Virtual reality was soon followed by a growing interest in combining real with virtual environments, allowing people to interact with digital information within real physical space, and bringing live video imagery into virtual environments. It was argued that this type of Mixed Reality interaction fits more naturally with the way people act and interact with everyday objects offering them a greater sense of 'embodiment', a state of being in the world, compared to interacting with more abstract virtual representations (Dourish, 2001). At the turn of the millennium the focus has shifted again and attention has been drawn to the increasing importance of interactivity beyond the scale of the task at hand. Recently we have witnessed a shift in focus of interest from 'cyber space' to 'ubiquitous computing'; digital technology is built into our environments, embedded in our devices, everywhere. Increasingly these technologies are networked.

Ubiquitous computing (also known as 'ambient, physical, embedded, environmental or pervasive computing') was first introduced by technology visionary Mark Weiser. He envisioned a world of fully connected devices with cheap wireless networks where information is accessible everywhere. A world in which computers and information technologies become invisible, and indistinguishable from everyday life: 'any time, any where and always on' (Weiser, 1991). Weiser proposed that computing would follow the evolutionary path of the electric motor. Early motors were very large and serviced the needs of many users. Today, a family is surrounded by hundreds of 'invisible' motors in the appliances around them. Similarly, in a pervasive computing environment, computers and information processing become ordinary, and penetrate into every object in our daily lives. Until recently the word 'ubiquity' was seldom heard, but now, like the word 'cyberspace' before it, ubiquity has quickly come to mean just about anything having to do with universal connectivity. As a result, information technology contexts are no longer valued so much for the immersiveness they offer as for how peripheral they appear to be, and in this way reducing information overload (Weiser and Brown, 1996). McCullough, 2004, noted that architecture has acquired a digital layer, which involves the design of organisations, services and communications and it seems that both architecture and interaction design together can help compose the necessary framework for a better integration. In this respect we believe that building pervasive systems into our urban environment requires a new way of thinking about the design and use of digital flow and how it interweaves with the built environment.

In this paper we report on Cityware, a current research project within the VR Centre, Space Syntax Laboratory and Centre for Sustainable Heritage at the Bartlett, UCL, in collaboration with the University of Bath Departments of Computer Science and Psychology, Imperial College London and HP

Labs Bristol, Nokia, Vodafone, IBM and Node Ltd. This project aims to develop a better understanding of the urban landscape augmented with the digital landscape of a city, by providing tools, methods and a theoretical framework for designing pervasive systems as an integral part of the urban landscape. To achieve this we draw on two sets of relevant work, Space Syntax (Hillier and Hanson, 1984; Hillier, 1996) and the Public Social Private 'PSP' framework (Kostakos, 2005). In our research we are interested in designing, not just the architectural space in which people move and interact, but also the interaction spaces (Kostakos and O'Neill, 2004; O'Neill et al., 2004) for information which they discover and use and which support their movements, behaviors and interactions within architectural space.

To design these new integrated systems, we need to extend and adapt our understanding and practice of urban design by observing the existing situations and practices, experimenting with wireless, mobile and located technologies, as well as constructing installations to experiment with new forms of human interaction. Bath is interesting for this kind of experiment because first, it is (and always has been) a tourist city. Second, it is a heritage site – the only city in the UK currently on the UNESCO list of World Heritage Sites – this sets a hard context for the introduction of new technologies in that they must respect the fabric and sensitivity of their context. Third, the city has a rich tradition in literature figuring in the works of Jane Austen, and that literature provides one of the main attractions for some tourists. Finally, Bath is manageable in size and this makes it possible for us to experiment at the urban scale, but within a well constrained area. Here we are developing an approach based around the use of Space Syntax methods for the analysis of the spatial morphology of the city of Bath. In the next section we describe recent research projects that have addressed some aspects of pervasive systems in urban contexts.

1.1 Related Work

Recent research has addressed some aspects of pervasive systems in urban contexts but has not considered the design of pervasive systems as an integral facet of urban design. Research to date has been mainly conducted through 'experiences' or 'performances' that cover a small area and in almost all cases are held over short timescales. When architecture has been considered in relation to pervasive technologies, it has typically been focused on the relatively small-scale architecture of individual buildings (Rodden and Benford, 2003) or even rooms (Krogh, 2000; Streitz, 1998) and has at times been used vaguely as a term simply reflecting the notion of built physical space (Wisneski, 1998). Our work includes architecture but focuses strongly on urban design at the scale of cities (Hillier, 2000). Previous studies have explored people's social behavior and relationships with urban space and pervasive technologies on which we can build (for instance: eGraffiti, Guide, Equator IRC, Mobile Bristol, Urban Tapestries and Intel's Urban Atmospheres). However, these studies have not solved the engineering challenges of implementing city-scale pervasive systems.

In our research we seek to extend our understanding and practice of urban design by looking at the urban environment as an integrated system mediating both the built environment and pervasive systems. It differs from other approaches by addressing three key research issues:

- It addresses pervasive systems design as a facet of urban design. Urban design is a key perspective for Cityware. We need to understand how pervasive technologies interact and interweave with the built environment to create the spaces that frame and influence people's behaviour.
- It addresses the methodological challenges of longitudinal cohort studies.
- It addresses technical and engineering challenges of implementing city-scale pervasive systems.

In the next section we describe the first and third of these concepts in detail. First, we explain the PSP framework, used to conceptualise, design and evaluate pervasive systems, followed by a brief description of Space Syntax tools and methods as currently applied to understand the effect of the built environment on what happens in cities. It is this which we hope may throw further light on the intersection between the digital and spatial layers of behavior as part of the urban landscape. We then review early findings from the first studies including a radio survey and qualitative approach in our early observations of the digital landscape in the city. Finally, we draw conclusions on certain generic characteristics of both facets and highlight related issues that need further research.

2. SPACE AND PERVASIVE SYSTEMS

A systematic approach to designing the urban environment as an integrated system demands a coming together of Architecture and Computer Science. Key to this interdisciplinary integration is the concept of space, by which we mean not only physical location but also the social protocols, conventions and values attached to a particular physical space (Cole and Stanton, 2003; Harrison and Dourish, 1996; Kostakos and O'Neill, 2004; O'Neill et al., 2004). In this section we provide an overview of the two sets of ideas we draw on: the PSP framework and Space Syntax.

2.1 Understanding Pervasive Systems

The Public Social Private framework is used for understanding and designing pervasive systems (Figure 1). This framework is extensively reported in (Kostakos, 2005), and its application has been reported in (O'Neill et al., 2004). In the following we provide a brief overview of its main concepts.

To illustrate the PSP framework in use imagine the following scenario: consider John who lives in London and wishes to visit the city of Bath. John arrives at Paddington train station (public architectural space) in London, and observes the departure times on the big screens at the station (public interaction spaces). At the same time John receives a text message on his mobile phone (private interaction space) about the weather in Bath. Once John is on the train (public architectural space), he is informed of the train's schedule via the intercom system (public interaction space).

John arrives in Bath, and observes a group of tourists discussing and pointing on a map they are sharing (social interaction space). John's phone receives a message asking him to send an SMS to a specific number if he wishes to receive further notifications throughout the city. John decides to sign up. He makes his way to the main Cathedral, where he receives a short message on his phone informing him that he has reached the main Cathedral. After spending a few minutes outside the Cathedral, he is identified as a 'persistent' user, and thus his phone is sent information about the Cathedral's history. Because John's phone has been identified as having a small screen, the information contains low-resolution photographs. Users owning a PDA with larger screens are sent photographs of higher resolution.

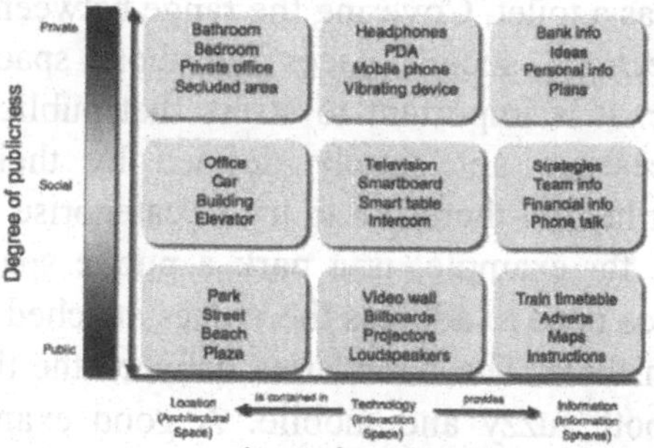

Figure 1. The PSP framework. Y axis represents the degree of publicness. X axis represents three main features of pervasive systems and the relationship between them.

The PSP framework provides the designer with two main insights. First, that a pervasive system can be conceptualised by considering three main aspects: architectural spaces, interaction spaces and information spheres. Secondly, each of these aspects can be classified as Public, Social or Private. Designing a pervasive system involves understanding which types of architectural spaces, interaction spaces and information spheres are being designed for. The PSP framework enables the comparison of alternative design decisions, and draws the designer's attention to potential problems. In the following we focus on architectural spaces and interaction spaces.

2.2 Architectural Spaces

Architectural spaces have values attached to them and as such tend to convey cultural meaning and frame our behaviour. In addition, the presence of others within an architectural space has an effect on how we behave and perceive the space. One might say that architectural space has embedded understandings and protocols of what is regarded as appropriate behaviour. Despite the variety of characteristics found in individual spaces, we can still abstract over them in attempting to support systematic design.[1]

Green, 2002, in her sociological work discusses how architectural spaces can be thought of as public or social. Building on that the PSP framework provides a top-down approach that categorises all possible architectural spaces in three main groups: public, social, and private architectural spaces. These notions carry with them a great number of characteristics and understandings that are peculiar to each society or social group. Public spaces are open to everyone, mainly because they usually belong to the community itself, e.g. a town square is a public space. On the other hand, private spaces are spaces controlled by an individual, which can be used in whatever way the owner sees fit. Private spaces promote a sense of security and privacy, such as a toilet. Covering the range between the extremes of the public-private spectrum, social spaces are those spaces that are neither private nor public. It is important to stress that public, social and private architectural spaces are not simply defined by their geographical co-ordinates. It is not helpful therefore to try to categorise 'pure' public, social or private spaces; for example, is a park a public space? The criteria for categorising a space need to address the values attached to the space and the things that happen there. The boundaries between the three types of spaces we propose are both fuzzy and mobile. A good example is 'the office', which at different times may be a social and a private space (e.g. a group of colleagues discussing in an office vs. a single person alone in an office).

[1] We do not, of course, use "space" to carry mere 3D volume attributes.

2.3 Interaction Spaces

Interaction spaces describe spaces that are created by designed artefacts. These spaces define the physical boundaries within which the device or artefact is usable (O'Neill et al., 1999). Interaction spaces can also be categorised as public, social or private. For instance a large display, in a train station, showing departure times could create a public interaction space. On the other hand, time information presented on people's mobile phones would create a private interaction space despite being in a public architectural space. An example of a social interaction space is the one created by a map viewed by a group of tourists, since the group of tourists are included in it but not everyone in the area. Interaction spaces may also be auditory or wireless. For instance the wireless interaction spaces generated by GPRS or 3G masts define volumes of space within which certain artefacts (such as phones and PDAs) and services (email, browsing) are usable. We can differentiate between such mast-based wireless interaction spaces and more mobile wireless interaction spaces. The former tend to be static in relation to location within the city. On the other hand, the wireless interaction spaces created by technologies such as Bluetooth[2] are often mobile, and move around with users and devices. For instance, Bluetooth interaction spaces can be used for accessing information and services, as well as for communication between users. They move around with users, and come in contact with various other features of the digital landscape: services beaming out of an interactive poster, Bluetooth phones belonging to friends, colleagues or even strangers, as well as various Bluetooth devices such as headsets, keyboards and mice. An artefact, such as a mobile phone, generates various interaction spaces. The phone's screen generates a visual interaction space, the phone's speaker generates an auditory interaction space, while its wireless capabilities generate wireless interaction spaces.

2.4 Understanding Space with Space Syntax

Space syntax, first developed by Hillier and Hanson (1984), analyses cities as systems of space created by the physical artefacts of architecture and urban design. It takes the position that the key to urban function, at the level of movement of people, is the way in which each space is accessible from every other space in the city, not merely in terms of metric distance, but rather in terms of topological distance, or the number of changes of direction needed to move from one space to another. Space syntax analysis methods

[2] Bluetooth: a radio standard for short-range of up to 100 meters wireless communication of cellular phones, computers, and other electronic devices.

represent the open space of the city as a graph and use graph measures to quantify morphology. Using a range of observation methods Space Syntax gathers data on people's overt behaviours and revealed preferences. Using the quantitative analysis of spatial morphology one can then investigate the degree to which behaviours appear to be related to spatial design, and the degree to which other explanatory factors must be invoked. Some key studies in Space Syntax are reported in (Hillier et al., 1987, Hillier et al., 1993, and Read, 1997). For instance if we ask an individual in Bath about her pattern of movement, she is likely to respond in terms of purposes of journeys. However, cities are very dense and heavily populated, and hence the collective activity gives rise to a pattern of use and movement that is independent of the intentions of individual but reflects morphological factors in the setting itself (Hillier, and Hanson, 1984).

Figure 2. Left: Axial map of Bath. Right: gate locations in the city of Bath.

Figure 2 is a representation of part of the street network of Bath as the 'fewest and longest' lines that cover the system (the axial map). Observations are then made at different times of day of movement flows along each street segment by counting people passing points on a street, 'imaginary gates[3]', and indexing them in flows per hour through that gate. The various spatial values for the lines are then compared to the movement flows by simple and multiple regression. The only other variable required to model movement is knowledge of the special attractors, for instance average building height, development densities and land uses in an area. The reason for the use of Space Syntax as an aid to understanding the urban space, is to a great degree based on correlations found between the measures generated by this analysis, and flows of people counted in real urban space. This is

[3] "A brief lesson in Observations" Space Syntax Lab, 1997.

supported now by many studies, mainly of pedestrian movement, indicating that under normal conditions the spatial configuration of the urban grid is in itself a consistent factor in determining movement flows (Hillier, 2000).

2.5 Design Support with the PSP Framework

The concepts of architectural spaces and interaction spaces – public, social and private – aid us in mapping from locations to the technological artefacts that are available to us and the forms of interaction we wish to support. With the advent of mobile and pervasive technologies, interaction spaces can range from being 'persistent' (such as the displays at the train station, or a desktop computer's WiFi network), to having 'medium-term presence' (such as the mobile phone or laptop of a coffee shop customer), to being 'highly transient' (such as the phones of people walking down a street). The PSP framework, however, does not describe 'operationalised methods' that the designer can use to identify the architectural spaces and interaction spaces in an urban environment. Therefore, a challenge we face as designers is recognising the different types of interaction spaces scattered throughout a city. How can we identify, both at design and run time, the appropriate architectural and interaction spaces through which information may be delivered or accessed? Furthermore, how can we evaluate our design decisions in deployed pervasive systems?

In order to address these questions, the Cityware project is applying a series of different methods. These include:

- **Mapping the physical and digital flows in Bath** (macro level): Using Space Syntax methods for spatial observation and analysis (axial and visibility). Mapping the digital flow, by scanning Bluetooth devices, allows us to identify the presence of potential interaction spaces.
- **Observation of static space use** (micro level): Using methods drawn from ethnography, including people following and observation of static activities, local movement and the pattern of social behaviour and interactions. This method is complemented by mapping the types of mobile device usage behaviours that occur at various locations in order to understand the regular and persistent patterns of space use.
- **Longitudinal cohort studies**: Using 30 participants for engagement throughout the duration of the project (3 years). This method puts the individual's view in the foreground. Data collected through the cohort will be regularly analysed and integrated in the analysis.
- **Experimental installations and interventions:** Installations will be used to engage the general public in our research, encouraging playful

use of technology. Various installations will be introduced using plasma screens in building interiors, and very large projected displays in public exterior space. In these experiments it is hoped that interventions can serve as a methodology for better understanding of social and digital interactions and underlying affordances.

3. METHODOLOGY AND DATA

In this section we report on the first studies that were carried out in Bath. To date, three pilot studies have been conducted with the intention of capturing data on both the physical and the digital flow represented by pedestrian movement and Bluetooth presence. These represent the first steps towards building an integrated spatial and functional database for the study area bringing together observation-based surveys of land-use, space-use and pedestrian flow in addition to the information related to the digital landscape.

3.1 Understanding the Physical Flow in Bath

As part of the first phase of understanding the city, we have carried out a two-day observation study and spatial analysis of the city of Bath using Space Syntax methods. The study was aimed at establishing:

- patterns of movement at different times of the day and assessing the degree to which the spatial layout influences the patterns of pedestrian movement and social behaviour of visitors in comparison to local people.
- patterns of space use and interaction by observing static use of space.
- and observing existing social practices (e.g Graffitti) and its relation to space in respect to the heritage value in the city of Bath.

The structure of the built environment in Bath is by no means a unified, planned whole. The expansion that took place in a relatively short time span in the 18th and 19th c. created a collage of separate, largely unconnected, and often incomplete pieces of speculative development, each shaped apparently by accident (Forsyth, 2004). Data about pedestrian movement were gathered using an observation-based pedestrian survey conducted in the study area. The study used the 'gate method' for 96 different street segments between street intersections. The observations were made throughout 5 time periods between 8:30 am and 4:00 pm – spread over 2 working days – by a group of 12 observers. Pedestrians were classified as locals, and tourists, for both men and women. This study was coupled with a rapid survey of land use for the main retail area, which was supplemented by the retail survey of central Bath undertaken in 2004 by the Bath Council. The city of Bath was analysed embedded into its surroundings as well as an independent system. The local integration was correlated with pedestrian movement data (Figure 2). The configurational analysis of its spatial patterns shows a background level of correlation (R^2=.47) with outliers above and below the regression line. Some spaces attract more people because they may contain or lead to important tourist attractions (Figure 3). Other streets appear under used given their context (Fatah gen. Schieck et al., 2005).

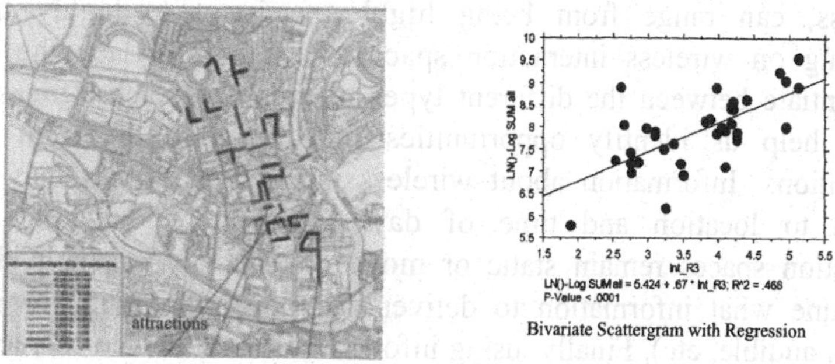

Figure 3. Movement flow and scattergram with correlation of movement and local integration.

Early findings indicated clear differences between the behaviour of tourists in comparison to locals, which for the purposes of the study includes students currently based in the city. This appears to be determined, to some extent, by the spatial configuration of the city. Identifying the social interactions within various social groups and their locations would allow us to recognize opportunities for delivering appropriate types of information and services. For instance, some of the streets were mainly used by locals. Other static locations were dominated by students (Figure 4).

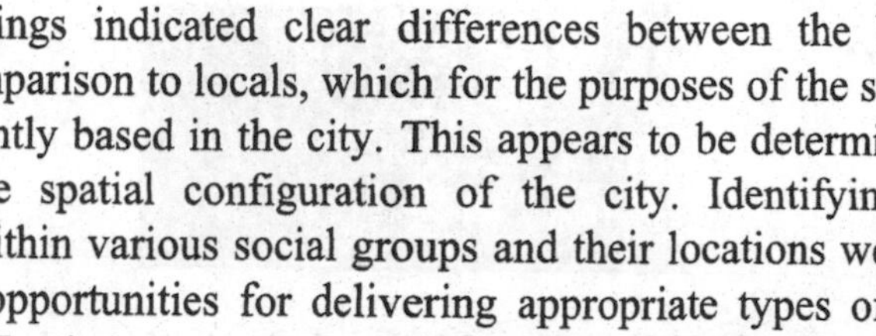

Figure 4. Left: passage mainly used by locals. Right: static space used by students.

It has been argued that new technologies tend to undermine existing social practices, requiring new ones in their place (Feenberg, 1999). In our study an attempt was made to map and understand existing social practices in relation to the space and in respect to the heritage value in Bath. We believe that identifying these practices and their location within the city would help gain a better understanding of the underlying affordances, which can then be addressed and taken advantage of in designing the new technologies and interventions.

3.2 Understanding the Digital Flow in Bath

In section 2 we noted that interaction spaces, whether visible, audible or wireless, can range from being highly persistent to highly transient. Focusing on wireless interaction spaces, we currently lack the tools to differentiate between the different types of presence. Such a measurement would help us identify opportunities for delivering the right type of applications. Information about wireless interactions spaces may also be related to location and time of day. Furthermore, knowing whether interaction spaces remain static or move dramatically can help designers determine what information to deliver and how present it (for instance, visible, audible, etc). Finally, using information about device characteristics, for instance, mobile phones vs. laptop computers might help achieve a greater understanding of the digital landscape.

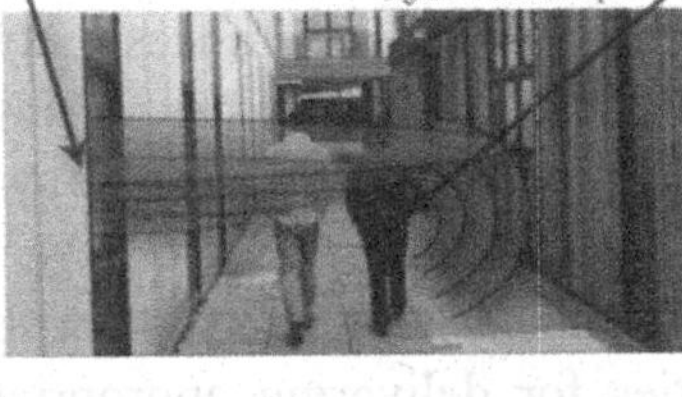

Figure 5. Our gate consisted of a scanner placed on the inside of a window in a long narrow passageway between two buildings.

In the following we discuss our attempts to identify the presence, type and distribution of wireless interaction spaces over space and time. Here we focus on wireless interaction spaces generated by Bluetooth devices. The vast majority of Bluetooth interaction spaces are created by small, personal, mobile devices such as mobile phones. Thus, in contrast to the interaction spaces created by typically static WiFi access points, the wireless interaction spaces created by Bluetooth devices map very closely to the movements of people around the city, which in turn can be measured by Space Syntax methods. We carried out a trial measuring Bluetooth activity on a single gate located on Bath University campus over a 6-week period (Figure 5).

During the six-week trial, our scanner continuously searched for and recorded Bluetooth devices within a 10-metre range. A device was recorded if it had Bluetooth and was set to discoverable mode (Figure 6). For each record we had the following information: serial number, name of device (provided that the device was within the range), date, time, class of device (e.g. phone, laptop), and services offered by the device.

Furthermore, by applying filters on the recorded data, we were able to classify devices as persistent or transient (Figure 7). In this case, persistent activity was due to the presence of laptop and desktop computers in nearby offices, which generated high Bluetooth activity. Another example of a persistent device is the mobile phone of someone who came into the office for a relatively long time.

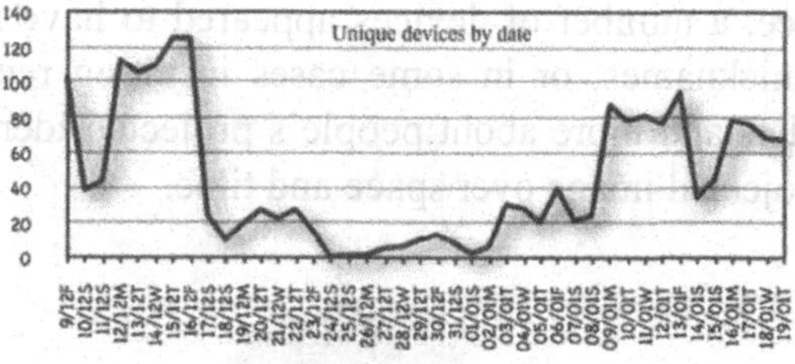

Figure 6. Bluetooth activity during our six-week trial. X axis show the date, y axis shows number of unique Bluetooth devices. Number of devices dips during weekends, with low values for the week leading up to Christmas (19-23 Dec.).

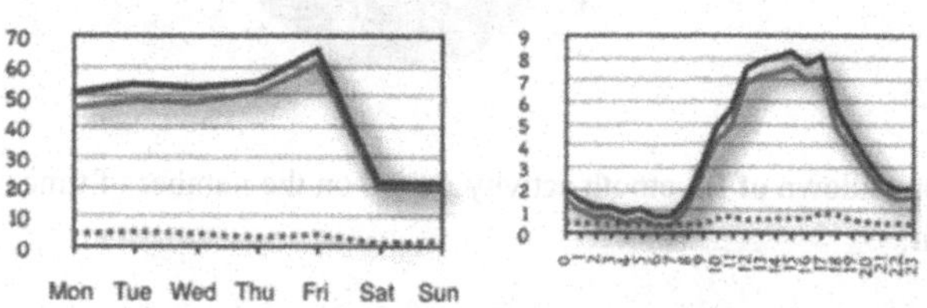

Figure 7. Summary of the data shown on a per day basis (left) and per hour basis (right).The top lines show total activity, the lines below show transient activity, while the dashed lines shows persistent activity.

During our study a number of issues were identified:

Accuracy of Bluetooth scanning: A number of issues relating to the accuracy of Bluetooth scanning became apparent during our trial such as the effectiveness of Bluetooth dongles for scanning purposes. Our scanner consisted of a single Bluetooth dongle, which meant that if many Bluetooth devices passed our gate simultaneously, less information (or in some cases no information) would be recorded. One way of overcoming this limitation is to scan with more than one Bluetooth dongle simultaneously. Another one would be to reduce the amount of information recorded for each device.

Persistent vs. transient interaction spaces: It is easy to differentiate between highly persistent (such as a printer or desktop) and highly transient interaction spaces (a user passing by our gate once and never seen again). However, we still need to explore how to usefully interpret the data that lies between these two extremes (Figure 8).

Device attributes and people attributes: A characteristic of Space Syntax observations that we were not able to achieve in the Bluetooth study was the ability to classify pedestrian flow depending on people's characteristics. For instance, human observers can classify traffic in terms of men, women, locals, tourists, children or adults. However, Bluetooth data reveals information such as custom names that users assign to their devices. In our study, for instance, a number of devices appeared to have names that were catchy phrases, nicknames, or in some cases insulting remarks. This data may help us understand more about people's projected identity, and also to trace people's projected image over space and time.

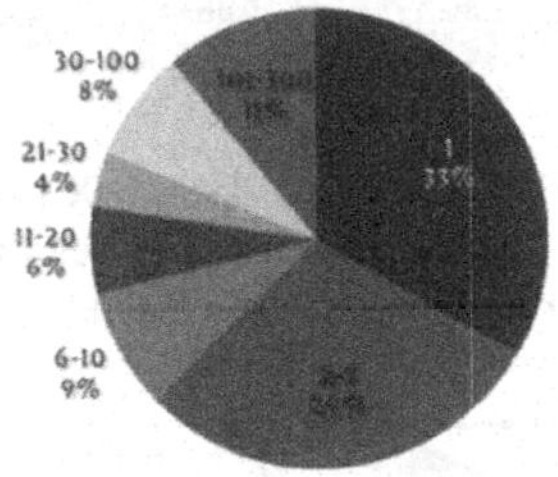

Figure 8. A breakdown of Bluetooth activity, based on the number of times a device was seen within an hour.

Static vs. mobile Bluetooth scan: In planning to deploy numerous Bluetooth gates throughout a city, a secure and convenient setup is not always possible. We are therefore considering how the 'long-term' gates we discussed in this section can be combined with 'short-term' mobile scans of the city. In planning such a mobile-scan study, we need to take into consideration that Bluetooth activity may be very limited in certain parts of the city. More recent, an attempt was made to generate a 'snapshot' of the city's digital landscape. Various locations in the city were selected based on the study described in section 3.1. An observer covered these locations by moving throughout the city with a laptop and recording Bluetooth activity at each one of them. Early findings suggested that ~10% of people in Bath seem to have Bluetooth switched on. However, in some places we have identified different percentages, which could be for a number of reasons such as the nature of the location. For instance, in more open areas no single observation 'gate' catches all the flows of people, and in addition there may be static people hanging about within Bluetooth range. In order to verify the effect of these factors on the results more locations need to be scanned.

Bluetooth digital trails over space and time: By establishing a network of gates throughout a city, a specific device can be traced throughout a city (each Bluetooth device carries a unique MAC address) by looking at the recorded data from multiple gates. This data carries with it date and time signatures, and thus we can extend our following of devices over time as well. Recording people's 'Bluetooth trails' could prove useful in two main ways: First, digital trails could be used to study the effects of 'digital attractors' (such as Bluetooth enhanced posters) placed throughout a city. For instance, a poster broadcasting Bluetooth messages could influence people's movement in favour, or against, a specific location. By looking at people's digital trails we can measure this effect. Secondly, Bluetooth trails could be studied to gain insights about people's use of other technologies. A Bluetooth signal usually originates from a mobile phone or laptop. By studying the Bluetooth trails of mobile phone users, or laptop users, it is possible to identify patterns that can then be attributed to other technologies such as GPRS, 3G or WiFi.

Finally, associating the type of device, its interaction space, and the integration of the specific location, we can make use of the PSP frameworks' design suggestions. For instance, we can assume that an interaction space within a not-so-busy street (as predicted by Space Syntax and observed by 'gate counts') will have a greater degree of privacy as opposed to an interaction space within a highly integrated street (Figure 2).

3.3 Exploring Mobile and Digital Behaviour in Bath

As well as capturing quantitative data we are carrying out complementary qualitative analysis of behaviour as an additional part of our methodology. This process addresses people's behaviour towards technology in social and public spaces. A one-day pilot study was conducted with the intention to gather initial evidence about the use of mobile and digital technology in the urban space. Observations were made throughout 5 time periods between 10:30 am and 5:00 pm by a group of 8 observers. Types of behaviour in reference to use of e.g mobile phones and PDAs were observed and their locations were recorded in 7 selected urban spaces. This was complemented by an observation in a public café in Bath. Different types of behaviour were observed in relation to space properties. For instance, in pedestrianised areas mobile device usage was more obvious close to the locations of heavy movement flow. In contrast, in areas close to vehicle movement less mobile phone usage was recorded, these tended to occur near bus stops and waiting locations (Figure10: left). People at the bus stops used the phone to give updated information on temporal uncertainties: 'no sign of a bus yet, I'll let you know when I am on it'.

In these locations the social behaviours and the interaction spaces appear to take a shape which provides the person with more privacy (Figure 9: left). This seems to be supported by the properties and affordance of the physical environment encouraging a certain type of behaviour.

We also noted differences in the walking pace (e.g. fast, slow) of the mobile phone user which may be related to the age (e.g. young teenager, business woman) and apparent social class of the user. But it might also be related to the area in which the mobile was being used (e.g. fast in the integrated areas, slow in the less integrated area with a tendency to move body orientation to be a 'removal from distraction' or avoiding interrupting others with one's conversation. People on the phone tended to place their back to the main space, often facing windows or other objects on the wall, or gazing into shop windows from the street while on the phone. This also applies to the observations which were conducted in a public café in Bath (Figure 10: right).

Figure 9. Mobile phone use supported by the properties of the physical environment.

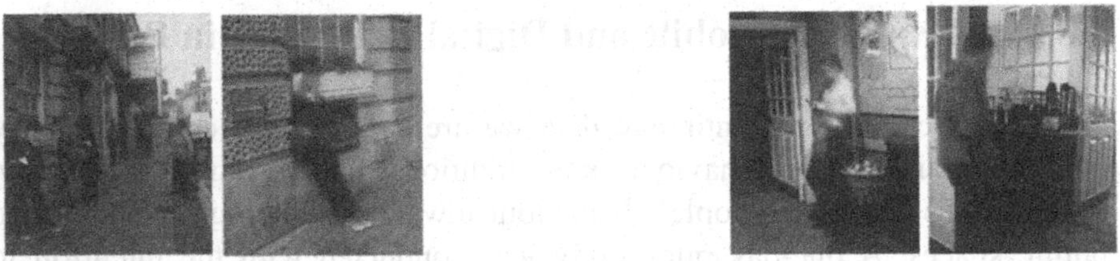

Figure 10. Left: Mobile phone use near a bus stop and close to the ATM, which is situated in a way failing to create a private interaction space for an interaction involving private use, Right: Mobile phone use in the café supported by the properties of the physical layout.

Figure 11. Mobile phone use in the café supported by the physical layout, but also related to the individuals intentions.

In these observations two key factors seem to emerge. First, people tend to use the mobile phone while queuing to place their order. This is likely to happen close to the sale's counter and could be a sign of people giving updated information on temporal uncertainties. 'I am in the queue, where are you?' Second, many people put their mobile phone on the table immediately after arriving and leave it visible. This might be to mark a territory or perhaps to take the place of a 'potential other' person. We also observed differences between singles and groups in this behaviour. In a group we see people toying with their phones when talking to the rest of the group. Other type of mobile behaviour include receiving a phone call or making one or even talking on the phone while approaching their table (Figure 11).

All these reflect a repetitive pattern of generic behaviour, which could take different forms and would reflect different intentions. In this observation based methodology we try to understand these in relation to the properties of the architectural space. However, we are also interested in looking at the social and mobile behaviour from the individual's view. This raises some questions related to the intentions of the individual, for instance the level of certainty and uncertainty in timing of a meeting which would encourage the usage of a mobile phone at a certain point. To address these questions in detail relatively longer term observations are needed to capture the transitions between the states that will allow us to make sensible deductions about these readings of observations such as the 'potential other' status of the phone on the table-the frequency of usage, the differences between singles and groups in this behaviour. Engaging our cohort of volunteers over a three year period will enable more extensive insight into social and digital interaction around the city, alongside more extensive observational studies.

4. CONCLUSION AND FUTURE WORK

We have argued that the urban built environment plays a critical role in the construction and reflection of social behaviours. This can be seen in the way it acts to structure space, which not only reflects and expresses social patterns, but can also play a part in generating these patterns (Hillier and Hanson, 1984). Having said that, the design of pervasive systems may also change the environment for interaction and so stimulate the emergence of new social behaviours. Inevitably then, designing new technologies tends to modify existing social practices, and on occasion stimulate new ones. However, the design of new technologies is often accompanied or guided by speculations about their social effects, and it is here that designers working with radically new and disruptive technologies tend to work on the boundary of existing knowledge. With the advent of mobile and pervasive technologies and the rapid adoption of Bluetooth mobile devices which can be detected in public and busy urban spaces, we need to gain a better understanding of the role of context as an emergent situation – both physical and social – of surrounding aspects that give meaning to our activities. We then need to address its impact on shaping social patterns. The mobile phone, for instance, as a particular socio-cultural object, as Ito has argued, is not so much about a new technology as much about a personal and intimate techno-social device supporting constant, lightweight, and mundane presence in everyday life (Ito et al., 2005). Public spaces such as cafes or public transport hubs are the 'stages' on which people negotiate boundaries of a social and cultural nature, in an attempt to carve out personal territories in public space. In order to understand these facets of socio-technical behaviours we need to deploy various methods covering different perspectives related to the physical and social context.

In this paper we have presented a methodology providing tools, methods and a framework for designing pervasive systems as an integral part of the urban landscape. We described three pilot studies and illustrated the methods deployed for mapping the physical and digital flow in relation to the proposed methodology. Here we draw on two main concepts that address issues of urban space, interaction space and the relation between the two.

However, at this stage we are only able to report our methodological intentions and the earliest of results. We cannot, for example, fully assess the effectiveness of a set of observation gates scattered throughout a city as a methodology to capture 'context'. Part of our ongoing work involves establishing multiple gates throughout the City of Bath and combining data from multiple Bluetooth gates. This should help us form a better picture of the digital landscape in the city. However, we are certain that spatial sampling of this kind will only give a very partial view of the complexity of social and technological interaction. We are therefore also applying 'snapshot' observation techniques in the context of cafés and restaurants where people are likely to use mobile and wireless technologies in order to address questions such as 'how do people physically orient their laptops in public spaces', and 'what wireless activities do they feel safe with'. To answer these questions we are planning to deploy scanning equipment that will record both Bluetooth and WiFi activity, complemented by human observations in the café or restaurant.

Of particular concern are the security and privacy of interaction between people and the private information sphere services in relation to physical space. We believe that the effect such issues have on people's perception of different spaces, and of what informs the definition of boundaries between the personal, private and public domains, requires considerable further research before it can be used to inform the design of new pervasive technologies in the urban realm.

5. ACKNOWLEDGEMENTS

The authors would like to thank the members of Cityware for their contributions. This research project is funded by the EPSRC, UK.

6. REFERENCES

Cole, H. and D. Stanton, 2003, "Designing mobile technologies to support co-present collaboration". *Personal and Ubiquitous Computing*, 7 (6), ACM/Springer.

Dourish, P., 2001, *Where the Action Is?: The foundations of embodied interaction*, MIT Press.

Fatah gen. Schieck, A., I. Lopez de Vallejo and A. Penn, 2005, "Urban Space and Pervasive Systems". *7th Intl Conference on Ubiquitous Computing*, Tokyo. Japan, (poster).

Feenberg, A., 1999, *Questioning Technology*, Routledge.

Forsyth, M., 2004, *Bath*, Yale University Press.

Gaver, W., 1996, "Affordances for interaction: The social is material for design". *Ecological Psycholog,* **8**(2).

Green, L., 2002, *Communication, technology and society*. London: SAGE.

Harrison, H. and P. Dourish, 1996, "Re-place-ing space: the roles of place and space in collaborative systems". *Computer Supported Cooperative Work*, Massachusetts, USA.

Hillier, B., 1996, *Space is the Machine*, Cambridge Press.

Hillier, B., R. Burdett, Peponis, J., and Penn, A., 1987, "Creating life: or, does architecture determine anything? " *Architecture and Behaviour*, **3**(3).

Hillier, B., 2000, "The Common Language of Space: a Way of Looking at Social, Economic and Environmental functioning of Cities on a Common Basis", University College London.

Hillier, B. and J. Hanson, *The Social Logic of Space*, 1984, Cambridge University Press.

Hillier, B., A. Penn, J. Hanson, T. Grajewski and J. Xu, 1993, "Natural movement; or, configuration and attraction in urban space use", *Environment and Planning B*.

Ito, M., D. Okabe and M. Matsuda, eds., 2005, *Personal, Portable, Pedestrian: Mobile Phones in Japanese Life*. Cambridge: MIT Press.

Kostakos, V. (2005). *A design framework for pervasive computing systems*. PhD. Thesis, University of Bath, UK. Technical Report CSBU2005-02, ISSN 1740-9497.

Kostakos, V. and E. O'Neill, 2004, "Pervasive computing in emergency situations", *37th Annual Hawaii Intl Conference on System Sciences*, IEEE Press.

Krogh, P.G., 2000, "Interactive Rooms – augmented reality in an architectural perspective", *DARE 2000*, Elsinore Denmark.

McCullough, M., 2004, *Digital Ground: Architecture, Pervasive Computing, and Environmental knowing*. MIT Press, Cambridge, MA.

O'Neill, E., P. Johnson and H. Johnson, 1999, "Representations and user-developer interaction in cooperative analysis and design". *Human-Computer Interaction*, **14**(1).

O'Neill, E., D. Woodgate and V. Kostakos, 2004, "Easing the wait in the Emergency Room: designing public information systems". *ACM Designing Interactive Systems*, Boston, MA.

Penn A., C. Mottram, A. Fatah gen. Schieck, M. Witkämper, M. Störring, O. Romell, A. Strothmann and F. Aish, "Augmented Reality meeting table: a novel multi-user interface for architectural design", *DDSS* 2004, Kluwer Academic Publishers.

Read, S., 1997, "Space syntax and the Dutch city", *1st Intl Space Syntax Symposium*, UK.

Rodden, T. and S. Benford, 2003, "The evolution of buildings and implications for the design of ubiquitous domestic environments". *CHI 2003, USA*. CHI Letters **5**(1).

Streitz, N.A., J. Geissler and T. Holmer, 1998, "Integrated design of architectural spaces and information spaces". *1st Intl Workshop on Cooperative Buildings*, Darmstadt, Germany.

Weiser, M., 1991, "The Computer for the 21st Century", *Scientific American*, **265**(3).

Weiser, M. and S.J. Brown, 1996, *The Coming Age of Calm Technology*, Xerox PARC.

Wisneski, C., H. Ishii, A. Dahley, M. Gorbet, S. Brave, B. Ullmer and P. Yarin, 1998, "Ambient displays: turning architectural space into an interface between people and digital information". *1st Intl Workshop on Cooperative Buildings*, Darmstadt, Germany.

1:1 Spatially Augmented Reality Design Environment

Chien-Tung Chen and Teng-Wen Chang
Graduate Institute of Architecture, NCTU
Graduate School of Computational Design, NYUST

Keywords: Design & Decision Support Systems, Spatially Augmented Reality, Architecture Education, and Computer Visualization

Abstract: With the development of ubiquitous computing (Weiser, 1991), what will become of the traditional media such as pen and sketches, especially in the design education environment? Or what will they be transformed into? In this research, we focus on the interior design process with a particular type of media—1:1 spatially augmented reality design environment (SARDE). In this research, we tried to implement SARDE and have a scenario experiment to check how designers interact with such design media. Furthermore, through this research, we have come to know more about how designers use design media to represent their design dream.

1. INTRODUCTION

One common problem, especially in interior design domain, encountered by design learning process is the mismatch between what novice designers intend to do and what they have drawn on the paper, as addressed in (Bailey, 2005). Gradually learning expressing spatial concepts in terms of elevations, plans and sections over time, novice designers start to look at the design differently. This is a time consuming process. One recent approach towards this problem is via both interface and the spaces surrounding it.

1.1 Making Design via Direct Manipulation

The common experiences novice designers have problems with are *scale*, *textures* and how they are represented in different situations. One of the

Jos P. van Leeuwen and Harry J.P. Timmermans (eds.), Innovations in Design & Decision Support Systems in Architecture and Urban Planning, 487-499.

reasons is because of their inexperienced expression and insufficient knowledge. For dealing with this problem, design studio often intends to ask for large scale model or drawing to reflect the design outcomes in a suitable form. By direct manipulating large scale physical model and drawing, design is made with simulated representation. How to understand the design with its context implication needs further exploration in addition to the media representation. Amount of time is spent on the site for novice designers to learn the contextual factor such as real scale and texture even after construction. This is a decision gap between learning the design documentation (the drawing) and the contextual information of the site (Figure 1).

Figure 1. Novice designer spends lots of time learning how to use the 2D drawing to *represent site* information and make design decision.

1.2 Deciding Design While You are On Site

For dealing with the decision gap, the contextual site information has been brought for information awareness. However, how the external representations interact with internal representations remains further study (Pearson, Alexander et al., 2001). A general view used in this research is that design process is treated as a process from unmeasurable spatial ideas to measurable spatial construction then to unmeasurable user's perception (Stockli, 1992). Furthermore, the measurable factors in decision making process can then be modelled. With this view, there are three obscures about measurable factors in design:

1. The 1:1 scale: Novice designers cannot apply the metaphor of scales on representing their spatial concepts easily, and often requires the 1:1 scale of representation for the space in their imagination.
2. The texture: Novice designer often has difficulty to associate the textures (represented by a set of images or physical materials) with their representation in the 1:1 scale of model or further the consequence in the design.
3. To associate design outcomes with surrounding environment: Novice designer cannot easily collage the design with physical spatial

environment. On site experience is still the best learning process for novice designers.

Therefore, a design decision making will require a direct manipulation over on-site experiences. Furthermore, a 1:1 on-site design environment is proposed to help novices to learn how to use external representations as decision support system.

1.3 Interactive Design Decision-Making on Site with Direct Manipulation

For the proposed 1:1 on site design environment to help novices learn how to use external representations as decision support system, the direct manipulation of interaction is required. Further, with the hypothesis above, an interacting decision-making on-site will need to fulfil the requirements as: 1) direct manipulating the design elements either in 2D or 3D representation; 2) a 1:1 scale visualization over the simulation is required; 3) the immediately visual feedback and suitable for used in multiple sites; 4) the interaction based on the direct manipulation is sustainable and looping in order to satisfy the design outcome.

Therefore, a review for finding suitable technologies as well as framework in realizing the system above is conducted in next session. To understand how this design system works and how novice interact with this system, the system as well scenario experiment for evaluating the results are implemented and discussed in the later sessions.

2. PREVIOUS WORK

Direct manipulation on site needs two parts work seamlessly: A portable immersive display, and the interactive media for manipulating the design virtually. The reviews are divided following these two groups.

2.1 Portable Immersive Display

For a portable immersive display, Tsai and Chang (2005) has explored and elaborated the requirement for the display system of a design-on-site purpose: a movable immersive unit. However, the result of their visualization (a notebook display) cannot satisfy the 1:1 immersive perception needed for this research. While searching for an immersive display, there are many researches from virtual reality and can be divided into three parts:

1. VR CAVE: and other full scale virtual environments are mainly used for expressing and implementing 1:1 design environment. But as the

1:1 size of the display system, while providing the impressive virtual perception, these virtual environments cannot satisfy the mobility and precise conditions in our research. Further, without the mobility, the virtual reality system will not be able to reflect the contextual information of site. However, the stereo images and several immersive techniques provided by this domain will be important.

2. Augmented Reality (AR): To integrate the virtual and physical objects to take advantage of both sides, the notion of AR is proposed. Many researches such as (Balcisoy, Kallmann et al., 2000; Benford, Schnadelbach et al., 2003) embedded the partial virtual system into physical environment. This creates a possibility to implement a portable immersive display system for our research. However, on the technological side, the Head Mounted Display (HMD), the often used for its mobility, has some important limitation: namely display lag and wearing heavy device. These limitations will reduce users' intuitive behaviour significantly (Voida, Podlaseck et al., 2005) and further reduce the users' spatial perception with its narrow field of view (Low, Welch et al., 2001).
3. Spatially Augmented Reality (SAR): To overcome the weakness described above, the most advanced research about AR is spatially augmented reality. Using projector and programming, virtual elements and physical environment could be easily combined together (Low, Welch et al., 2001) and have a better immersive perception. A portable spatially augmented reality system such as described in (Sukaviriya, Kjeldsen et al., 2004) will provide the necessary needs for used in design on site.

Figure 2. Spatially Augmented Reality (SAR) combined immersive display on physical environment. (Low, Welch et al., 2001).

2.2 The Interactive Media

With the portable immersive display systems like SAR described above, the interaction for direct manipulation will require some attention on intuitive behaviours analysis as well as implementation. This part of researches are various according to the needs of system implemented. With the SAR in mind, our reviews will focus on the interactive media for dealing with SAR directly.

With capture sensors, SAR provides possibility to manipulate the virtual elements directly (Raskar and Low, 2001). Therefore, the interactive media for using with SAR can be divided into gestural and vocal controls (Voida, Podlaseck et al., 2005). Gestural research and analysis have broadened the possibility to extend the interaction with intuitive perception needed in design behaviours. In the main interest of this research, the interaction with portable immersive display on site, the interactive media will be more focused on how to react to virtual objects. In this trend, Lu (2004) further provides the interactive gestural analysis for direct manipulating 3D virtual objects. The analysis is shown in Figure 3. In addition, for measuring purposes and intuitional design behaviour, tangible senses provide a better perception as part of sketch behaviours (Ishii and Ullmer, 1997).

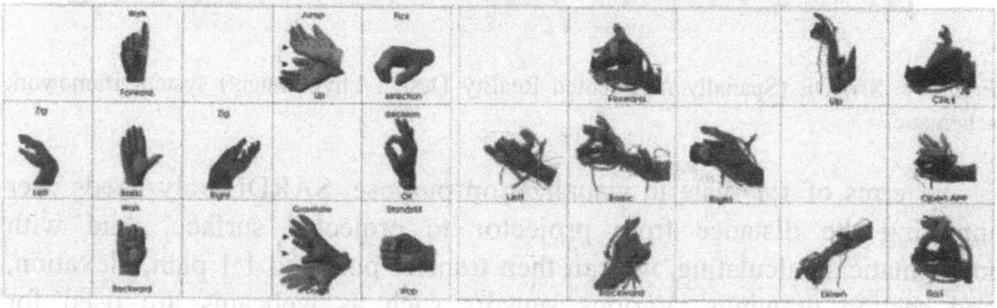

Figure 3. Intuitionally operate virtual objects with gestures (Lu, 2004).

3. THE SYSTEM

The system prototype developed in this research is called *Spatial Augmented Reality Design Environment* (SARDE). SARDE is divided into hardware and software components that will be elaborated in the following sessions. The information flow among these components is also analyzed and implemented for satisfying the requirement unleashed above.

3.1 The hardware structure

By projecting 1:1 virtual images onto physical environment on site, this research implements a spatially augmented reality design environment SARDE with mobility in mind. The hardware structure is shown in Figure 4. The hardware part of SARDE is comprised of three main components including sensors, visualization system, and the computing server. Each component is integrated with other for serving the interaction/reaction needed in our research.

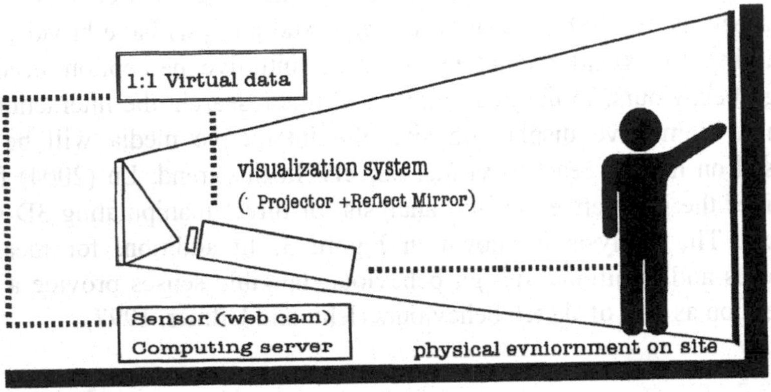

Figure 4. SARDE (Spatially Augmented Reality Design Environment) system framework schematic.

In terms of 1:1 scale in visualization purpose, SARDE only needs user inputting the distance from projector to projected surface. And with mathematical calculating, we can then transfer pixels to 1:1 plan, elevation, and section drawings. Various sensors such as webcams are used for capturing the designers' gestures and other analogue information from site. Reflect mirror is used to enlarge the projected surface, because most interior spaces are quite narrow. In addition, the reflect mirror could be helpful for projecting image on floor surface which is used for plan drawings or ceiling plan. By directly adjusting 1:1 virtual drawing onsite with visual feedback provided by SARDE, this system is not made to replace traditional drawing design media, but to better understand how the drawing design media can help designers in taking design decisions on the site.

3.2 The Software Structure

During the design process, designers usually have a good habit to use their own bodies as a reference, and simulate how users will experience their design. To keep this body reference design behaviour, and intuitional design behaviour, SARDE must have abilities to express designers' willingness clearly for manipulating virtual data through various sources such as gesture, design device, and voice analysis. For a prototype purpose, we used Macromedia Director and Macromedia flash to implement the analysis system. While webcam capturing the different led light on designer's gesture, actionscript in Flash could analyse the gesture location in whole captured pixels. By connecting the led lights and operating code, designer could use simple gesture to adjust the 1:1 projected image.

The design behaviours are collected by interviews and analysis. They are making lines, extending lines, making irregular lines, square, circle, and moving, duplicating, measuring, erasing, plugging in objects. These design behaviours have roughly covered the most interior drafting behaviours. In simulating texture, different behaviours are needed: browse texture, select texture, infill texture area, infill texture, adjust texture direction, input texture library. In addition to the drafting and texturing behaviours, much personal behaviour are also conducted and explored during the experiment process, even they are not frequently used. For demonstrating purpose, we only focused on most fundamentally design behaviours as our interact manipulation in this paper.

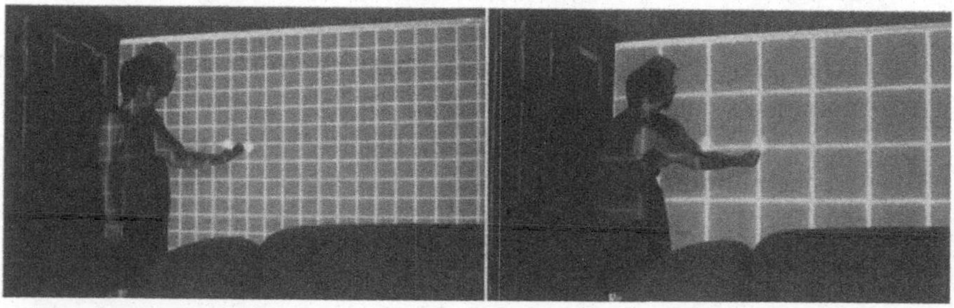

Figure 5. Designer directly manipulates the interactive surface.

3.3 The Information Flow

Design behaviours are quite different in elevation and in plan. In elevation situation, designers raise their hand to test how height is easy for put something, they may repeat stand up and squat to experience the height factor to their body. We organized these important factors considered while designers making elevation design decision: angle of view, height of view, touchable area by hand, touchable area by feet, and sit down situation. In the plan situation, designers prefer to walk around, change posture, and position, such factors are considered: usability of objects such as table, chair, arrange objects. Design behaviours have countless possibilities depending on design project, therefore we only can outline some fundamental principals about elevation and plan design behaviours.

4. A SCENARIO EXPERIMENT

4.1 The Setting and Steps

An experiment has conducted for testing the system proposed. The experiment is aiming on understanding the factors and drawbacks of our approaches, namely the usages of scale representation and media perception. We use video camera to capture whole the detail of their behaviours.

The experiment has five steps to conduct:

1. The users were asked to design with totally drawing represent media such as plan, section, and elevation. At this stage, we only provide 2D data about site information.
2. They were brought to the site, and operate SARDE to readjust their 2D represents. At this stage, novices can see their spatial idea on 1:1 scale, and combined into the site.
3. The experiment will stop as long as designer wants to stop.
4. After the design finishing, we interview with them to get some comments immediately.
5. After the experiment, we check the video to organize the most frequently used design behavior.

4.2 The Participants

We choose five novices architectural designers as our experiment participants. They all learned architecture design within two or three semesters without using computer as design supporting system. This period of students have the experience about making design decision, but still not familiar with traditional design media. For discussing purpose, only one participant is elaborated in this paper.

4.3 The Process

1. The 2D stage: In this stage, participants were asked to design a reception centre of an auditorium. This reception centre has to provide such program: sign in, selling ticket, selling relative product, product display, storage space, and poster area. They only got 2D information such as photos and plan drawings revealed in Figure 6. Participants had 20 to 30 minutes to finish the

2D drawings. One of the participants' outcomes is shown in Figure 7. The participant finished the plan, isometric, and two elevations. The main idea is using glass slab in 30cm width to reframe the design.

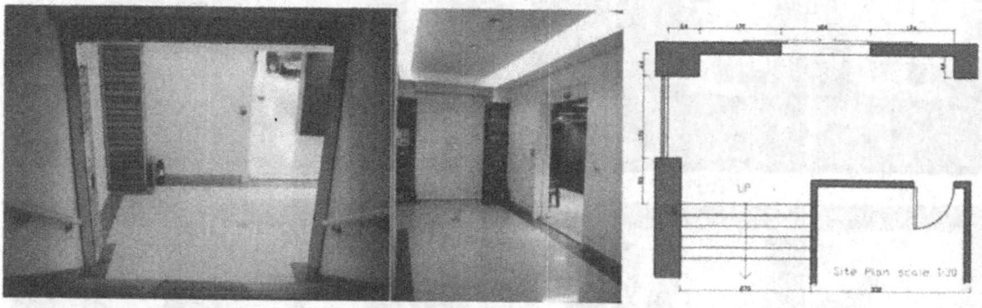

(a) (b) (c)

Figure 6. The first step: 2D site information. (a) (b) photos. (c) plan : scale:1/30.

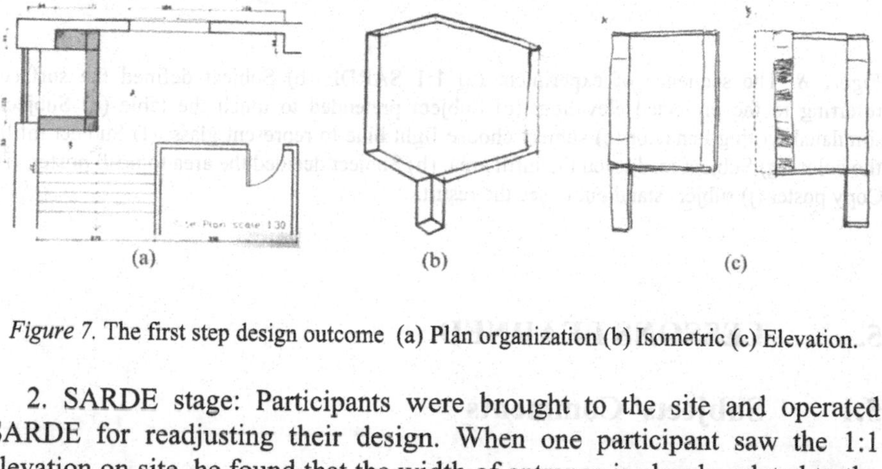

(a) (b) (c)

Figure 7. The first step design outcome (a) Plan organization (b) Isometric (c) Elevation.

2. SARDE stage: Participants were brought to the site and operated SARDE for readjusting their design. When one participant saw the 1:1 elevation on site, he found that the width of entrance is closely related to the width of reception table. He has discovered the easy to simulate the user's view angle is the main strength when he operated SARDE. Then this particular participant readjusted the table height with simulation touching the table. To identify the glass material, he chose light blue colour to represent the glass area. Finally, he put the poster to simulate the speech reception. In the whole process, participant kept walking back and forth to see the result combined with the site as shown in Figure 8.

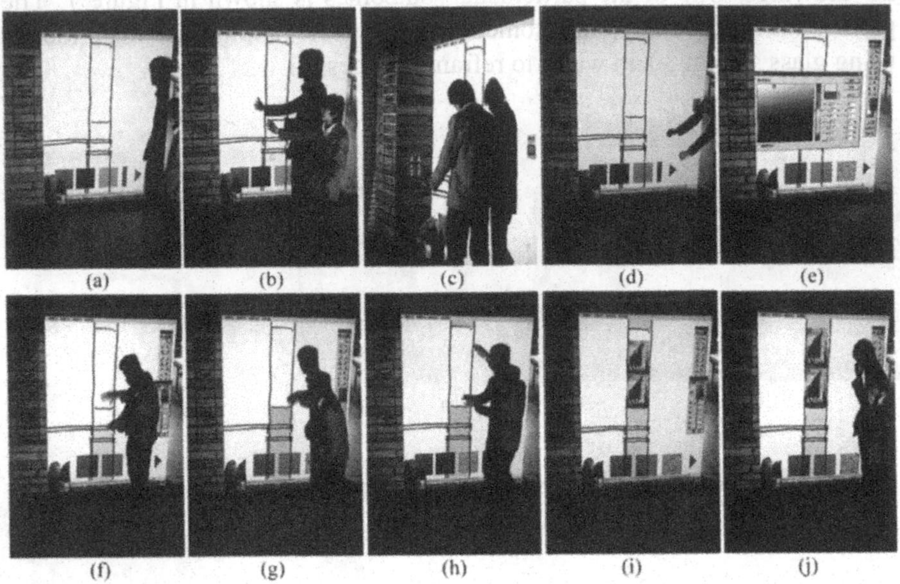

Figure 8. The sequence of experiment (a) 1:1 SARDE (b) Subject defined the surface referring to the projected elevation. (c) Subject pretended to touch the table (d) Subject simulated writing behavior (e) subject choose light blue to represent glass. (f) Subject infill the color. (g) Subject readjusted the infill area. (h) Subject decided the area to infill poster. (i) Copy poster (j) subject stand out to see the result.

5. LESSONS LEARNED

5.1 Subjects' Comments

Some important comments are addressed here. Almost all subjects agreed that 1:1 display let them use their bodies as reference more frequently, and this process made them feel more confident about their decisions. Some feel shock when they first visualize their ideas presented on 1:1 scale, but also motivate to adjust the design they have. One participant pointed out that it is more real when virtual image has physical characters. Some participants thought this design process is more tiresome than sketch, they are afraid that larger scale design will spend far more physical energy than they use to. The most valuable comment is that after experiment they feel more understanding towards how 2D drawings represent spatial ideas. In addition, they have to get right information from 2D drawings and use their imaginary to compose all the space. We consider these comments are very inspiring for us to rethink the combination of virtual and physical elements.

5.2 Behaviour Model

We arrange the design on site behaviour and got such result: a Design on site loop model (as shown in Figure 9). There are three important steps occurred in this cycle.

1. Manipulation, they use gestures and vocal to control the virtual objects.
2. Body reference, they use their body to simulate how user's behaviour in their design situation.
3. See in different distance, they will walk around to check whether the design is match their internal image or not.

In 1:1 drawing situation, design behaviours have evolving for concerning more scale factors. Each subject has different detailed behaviour, but generally they are doing the same loop behaviour as shown in Figure 9.

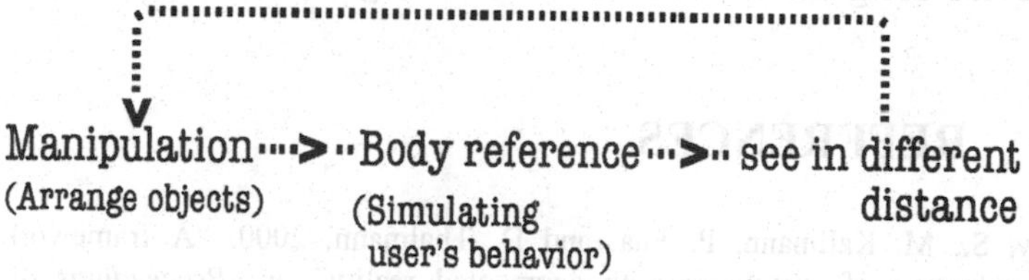

Figure 9. Design on site behavior loop.

6. DISCUSSION AND CONCLUSION

After the experiment, we find that novice can't find the key information even they see the result combined on the site. In participant 1, he noticed that the width of entrance is closely related to the width of reception table, and he feel more easily to simulate the user's view angle when he operated SARDE. But still there are many problems exposed when 1:1 design combined on site, they still can't recognize these contextual problems.

6.1 Significances and Drawbacks

Most adorable rewards are participants' comments after their experience. Every designer seems entering his or her design fantasy, and come back to reality to share with us. We also encounter a contradictory about imagine and physical labour. Most imagery and experienced designer can complete his/her design decisions with only a small sketch. However, no one as a novice can do that.

6.2 Limitation and Future works

In our consideration, SARDE will provide better information for designer to make design decision. But the ability of finding problem still is an important skill that novice lacked. Even experiment time didn't have time limit, but some novice don't know where is the problem. In this point of view, SARDE only can help some novice with problem finding talents, not for every novice.

1:1 SARDE has already provided better environment for designer to make decision, but for novices there are still more active system elements needed to appear in order to help necessary learning and motivation in interior design. We believed that if SARDE can provide more site information (or contextual information) as design hints that will certainly help novice designers to learn decision making better.

7. REFERENCES

Balcisoy, S., M. Kallmann, P. Fua, and D. Thalmann, 2000, "A framework for rapid evaluation of prototypes with augmented reality", in: *Proceedings of the ACM symposium on Virtual reality software and technology, Seoul, Korea,* ACM Press, pp. 61–66.

Bailey, R., 2005, "Digital Tools for Design Learning", in: *Proceedings of the 23rd conference on eCAADe, Lisbon, Portugal,* pp. 131–138.

Benford, S., H. Schnadelbach, B. Koleva, B. Gaver, A. Schmidt, A. Boucher, A. Steed, T. Anastasi, C. Greenhalgh, Y. Rodden, and H. Gellersen, 2003, "Sensible, Sensable and Desirable: a Framework for Designing Physical Interfaces", *Technical Report Equator-03-003, February 2003.*

Ishii, H. and B. Ullmer, 1997, "Tangible bits: towards seamless interfaces between people, bits and atoms", in: *Proceedings of CHI '97*, ACM Press, pp. 234–241.

Low, L., G. Welch, A. Lastra, and H. Fuchs, 2001, " Life-sized projector-based dioramas", in: *Proceedings of the ACM symposium on Virtual reality software and technology , Baniff, Alberta, Canada,* ACM Press, New York, pp. 93–101.

Lu, K-T, 2004, "Navigating 3D Information Space with 6 Degree of Freedom Devices", in: *Proceedings of the 10th International Conference on Computer Aided Architectural Design Research in Asia, CAADRIA '04,* Seoul, Korea, pp. 777–786.

Pearson, D.G., C. Alexander, and R. Webster, 2001, "Working Memory and Expertise Differences in Design", in: J. S. Gero, B. Tversky and T. Purcell (eds.) *Visual and Spatial Reasoning in Design, II – Key Centre of Design Computing and Cognition, University of Sydney, Australia.*

Raskar, R. and L. Low, 2001, "Interacting with spatially augmented reality", in: *Proceedings of the 1st international conference on Computer graphics, virtual reality and visualization, Camps Bay, Cape Town, South Africa,* ACM Press, New York, pp. 101–108.

Stockli, T., 1992, "The Measurable and the unmeasurable or from form to design to existence", in: *Proceedings of the 4th conference on European Full-scale Modeling Association, Lausanne, Switzerland,* pp. 55– 62.

Sukaviriya, N., R. Kjeldsen, C. Pinhanez, L. Tang, A. Levas, G. Pingali, and M. Podlaseck, 2004, " A portable system for anywhere interactions", in: *CHI '04, Vienna, Austria,* ACM Press, New York, pp. 789–790.

Tsai, R-W. and T-W. Chang, 2005, "Land forming while you are on site", in: *Proceedings of the 10th International Conference on Computer Aided Architectural Design Research in Asia, CAADRIA '05, New Delhi, India,* pp. 387–397.

Voida, S., M. Podlaseck, R. Kjedlsen, and C. Pinhanez, 2005, "A study on the manipulation of 2D objects in a projector/camera-based augmented reality environment", in: *Proceedings of the SIGCHI conference on Human factors in computing systems, Portland, Oregon, USA,* ACM Press, New York, pp. 611– 620.

Weiser, M., 1991, "The computer for the 21st Century", *Scientific American,* **265**(3): 94– 104.

AUTHOR INDEX